THE 20 YEARS OF THE SYNCHROTRON SATURNE-2

THE 20 YEARS OF THE SYNCHROTRON
SATURNE-2

Paris, France 4–5 May 1998

Editors

A. Boudard
CEA–SPhN, CEN Saclay

P.-A. Chamouard
CEA — Laboratoire National SATURNE, CEN Saclay

World Scientific
Singapore • New Jersey • London • Hong Kong

Published by

World Scientific Publishing Co. Pte. Ltd.

P O Box 128, Farrer Road, Singapore 912805

USA office: Suite 1B, 1060 Main Street, River Edge, NJ 07661

UK office: 57 Shelton Street, Covent Garden, London WC2H 9HE

British Library Cataloguing-in-Publication Data
A catalogue record for this book is available from the British Library.

THE 20 YEARS OF THE SYNCHROTRON SATURNE-2

ISBN 981-02-3679-4

Printed in Singapore by Uto-Print

CONTENTS

Foreword

SATURNE 2 is well adapted to nuclear physics. It has performances which compare to those of a Van de Graaf although it is a synchrotron of 3 GeV. Almost continuous extracted beam with a narrow energy spread and a good emittance.

Its predecessor SATURNE 1 was an old synchrotron of low focusing technique (perhaps the last one). SATURNE 2 had a completely new ring of the usual strong focusing types. It is very precisely build to provide high reliability.

The history of those accelerators includes the construction of 4 spectrometers as particle detectors. The first, SPES I designed in 1968 is of high resolution. The resolution permits the selection of one nuclear level which is necessary even for elastic scattering. A program (ZGOUBI—***Jean-Claude FAIVRE, Denis GARRETA)*** for precise orbit computing had to be written and was the first one to be fast enough to permit new spectrometer design. The structure of SPES 1 was fitted with an engineer from Berkeley during a few hours conversation, including the use of air pads to rotate the magnet.

But the development for particle localization on the focal plane is as important and I like to emphasize the remarkable achievements of ***Jean SAUDINOS*** for SPES 1 and also the one of ***Jean-Marie DURAND*** for SPES 2 : They remained in use until the end.

The fact that 4 spectrometers were financed is not surprising. It reflects the attention paid to design inexpensive magnets using when possible existing pieces of iron like parts of the old ring at SATURNE 1.

The remarkable polarized beam of Argonne, using a polarized source led us to do the same with SATURNE 2. It had been done at Saclay's cyclotron with ***René BEURTEY*** using ABRAGAM's high frequency transitions but at SATURNE we had to use de old linac injector with its pulsed behavior. We then dreamed of a new storage ring avoiding the linac : MIMAS was to solve all our problems.

Jean-Louis LACLARE was here and became the leader of the design and construction. It is a new type of synchrotron which was an acrobatic project including the problem of finding funds. Every thing was made possible by ***Jules HOROWITZ.***

The final performances of SATURNE 2 with MIMAS are related in this book and are so far the world record for a high energy polarized beam. In those synchrotrons there are a number of depolarization resonances to cross ; one has to treat them in a careful and ingenious procedure. It is at the same time very reliable : this is due to the informatic system which was planed since the beginning of SATURNE 1 and MIMAS ; it is also due to ***Paul-André CHAMOUARD*** constant activity.

Experimental nuclear physics is started at Brookhaven with PANOSKY et al. ; Saclay followed around 1970 with SPES 1 and it was from that on it became a National Laboratory of France which undertook the design and construction of SATURNE 2 and MIMAS. Progressively physicists not only from Europe but also from America and Japan joined us. That developed a new life : A large part of the SATURNE family is assembled here and it is a pleasure to meet them again.

Everybody will certainly read the excellent conclusion of ***Colin WILKIN*** where he reassembled most of the beautiful experiments achieved. I believe there is no need to add anything except our warmest thanks.

The initiative and organization of this meeting is entirely due to ***Alain BOUDARD*** : Life is short but we enjoyed it all.
THANK YOU ALAIN.

Jacques THIRION

INTRODUCTION

The decision to close the "Laboratoire National Saturne" (LNS) has been taken and was effective at the end of 1997.
This was done in spite of the misunderstanding and the sadness of the scientific community, physicists and ingeniors, working with this facility. It was the end of a place where fundamental research were pursued during more than 40 years. Dedicated first to elementary particle physics, the last 20 years after a full reconstruction of the synchrotron (that we called SATURNE – 2) were devoted to nuclear physics At this point a National Laboratory was created by an association between the CEA/IRF and the CNRS/IN2P3, and was actually open to the world community.
The idea of a special meeting to show the scientific results and the place taken by the LNS came in mind of several physicists, especially during the last "Journées d'Etude de Saturne" at Ramatuelle in January 1996. At the end of the same year, this idea was discussed and encouraged by the "Comité de Direction"(Ms. C. Cesarsky CEA-DSM and Mr. C. Detraz CNRS-IN2P3).
Finally "Les 20 ans de Saturne – 2" was organized by the LNS with the financial support of the CEA-DSM and of the CNRS-IN2P3 as 2 days of summary talks given in the Science Museum "Le Palais de la Découverte" in Paris. This book contains the written version of the talks, some pictures of Saturne and a list of the registered experiments with the most recent or significant scientific and technical publication. The aim of this work was to provide a digest of the technical activities and accelerator developments and a comprehensive guide through the scientific literature produced from research carried out at the LNS.
We have taken the opportunity of the Science Museum to organize in there an exhibition for a large audience. The techniques of acceleration, guidance and detection of particles as well as a survey of the main subjects of research were illustrated with real objects, models, pictures and panels. This exhibition could remain accessible to the public of the "Palais" during more than one year. It has been made possible due to the kind hospitality and encouragement of the museum management (Mr. M. Demazure, director) and with the devoted help of Mmes N. Ducatez and M. Verges and of Mr. F.

Bastardi from the museum. For this exhibition we have also received encouragement, financial support and technical helps from the IPN-Orsay, the SPhN and the LNS. Last but not least, many physicists and technicians have contributed, especially Mr. J.C. Duchazeaubeneix, Y. Terrien, J.L. Boyard and P.A. Chamouard who were deeply involved in the organization, the realization and the elaboration of the scientific content.
All this activities were controlled by a Scientific Committee (Ms. M. Soyeur and Mr. S. Harar, G. Fontaine, J. Thirion, P. Radvanyi, J. Arvieux, J. Saudinos, J. Faure, T. Hennino, C. Wilkin, P.A. Chamouard, A. Tkatchenko, and A. Boudard).
I was in charge of the organization with P.A. Chamouard and Françoise Haroutel. We were efficiently helped by Simone Peresse and Bernard Mailleret. We thank them warmely.

A. Boudard

Les 20 ans de Saturne - 2

Monday, May the 4th of 1998

Title	Speaker
Welcome	J. Faure
Introductive speech	C. Cesarsky
Introductive speech	C. Detraz
Genesis and history of SATURNE 2	P. Radvanyi
General theoretical framework	M. Soyeur
The source of polarized particles	P.Y. Beauvais
Depolarizing resonances	J.M. Lagniel
Polarimeters and polarimetry	M. Garçon
Physics with polarized deuterons	M. Morlet
The NN force (including the search for dibaryons)	C. Leluc and F. Lehar
Inelastic channels of NN	W. Kühn
p-Nucleus interaction or the SACLAY-GATCHINA collaboration.	A. Vorobyov

Tuesday, May the 5th. of 1998

Title	Speaker
Pi-Delta dynamics in nuclei	M. Roy-Stephan
eta physics and baryonic resonances	G. Dellacasa
Progress in accelerator theory	A. Tkatchenko
Performances of the control device for accelerators and beam lines	P. Ausset
MIMAS	J.L. Laclare
The DIONE source	J. Faure
Physics with heavy ions	J. Gosset
SATURNE - an intense source of cosmic radiation.	R. Michel
Transmutation (spallation studies)	S. Leray
SATURNE and its external activities	M. Tkatchenko
Concluding remarks	C. Wilkin

Genèse et débuts de Saturne-2

Pierre Radvanyi
Institut de Physique Nucléaire, Université Paris-Sud, 91406 Orsay Cedex

Genesis and beginnings of Saturne-2: The Saturne-2 project resulted from different former reflections, scientific developments and organizational evolutions. The very idea of a national facility had already been put forward in 1963/1964. The plan to replace the old Saturne-1 in Saclay by a modern strong focusing 2.9 GeV synchrotron for nuclear physics was finally jointly adopted at the end of 1973 by the CEA/IRF and the CNRS/IN2P3. A Common Undertaking was then set-up in order to define more precisely the characteristics of the new machine -including polarized and heavy ion sources-, the experimental areas and the experimental equipment (magnetic spectrometers) needed. The Laboratoire National Saturne was officially installed on January 1st, 1978 and the first experiment with the new accelerator started in October of that year. A programme committee started to scrutinize the various experimental proposals submitted by a large number of teams from France and different other countries. A draft of a new injector synchrotron Mimas, was worked out in 1979; this project was approved and funded in 1983.

Ce n'est pas sans appréhension, ni également sans un serrement de coeur, que je remplace René Beurtey au pied levé. Je n'ai pas sa verve et, sur la préhistoire de Saturne-2, il a vécu les développements du côté IRF/CEA, et moi du côté IN2P3/CNRS.

1 Genèse

La genèse de Saturne-2 se trouve en fait au confluent de plusieurs histoires différentes et parallèles :

- Celle de Saturne-1 d'abord, construit pour la physique des particules, lancé en 1954, opérationnel en 1958, qui, après de bons et loyaux services, s'est trouvé, vers 1969, abandonné par les physiciens des particules partis travailler auprès de machines d'énergie plus élevée, au CERN en particulier.

- A partir de 1963/1964 environ, ensuite dans le cadre de la préparation du 6ème plan, sous l'impulsion de A. Blanc-Lapierre, les physiciens des particules discutèrent d'une éventuelle machine nationale en soutien national du nouveau grand accélérateur du CERN. Ils invitèrent les physiciens nucléaires à s'intéresser aussi à une telle machine. Les études, menées par le groupe Gésyn comprenant un certain nombre d'ingénieurs de Saturne et du LAL, étaient financées sur des crédits de la DGRST depuis 1963. Ce fut d'abord un projet de synchrotron à électrons de 15 GeV, qui, sur le point d'aboutir, fut remplacé, in extremis, par un 45 GeV à protons, et enfin par un 23/45 GeV à protons en 1967. Le projet était soutenu par Leprince-Ringuet, Perrin, Grégory, Abragam, Berthelot, Blanc-Lapierre, Teillac et Lévy-Mandel. Saturne et le LAL devaient être arrêtés dans ce cas. Ce fut le projet Jupiter, nourri par une source d'ions qui prit le nom de la chèvre Amalthée. Mais, vers 1969, ce projet ne réussit pas à passer au niveau ministériel. L'avenir de Saturne et de ses ingénieurs devint alors incertain.

- Pendant ce temps la physique nucléaire évoluait. Parallèlement à l'effort consenti pour les sondes électromagnétiques (LAL, puis ALS), les physiciens nucléaires travaillant à moyenne énergie rêvaient de protons d'énergies plus élevées, au delà du seuil des pions. C'était un domaine encore très peu exploré, où des phénomènes nouveaux pourraient apparaître, où l'on pourrait étudier l'interaction des mésons (vecteurs de l'interaction forte) et des résonances avec les noyaux. Les physiciens nucléaires se mirent à prendre la place des physiciens des particules, auprès des accélérateurs d'énergie intermédiaire délaissés ; ce fut le cas de H. Palevsky au cosmotron de Brookhaven avec des protons de 1 GeV; il communiquera ses nouveaux résultats à la conférence internationale de Gatlinburg, en 1966. (Voir lettre de P.-G. de Gennes du 24 novembre 1966).

En décembre 1967 et au début de 1968, toujours dans le cadre du 6ème plan, la préparation du rapport de conjoncture fait état de ce nouveau champ d'activité, évoquant -entre autres possibilités- un Saturne rénové et reconverti.

- A Saclay, J. Thirion et le DPhN/ME ont entrepris de construire un spectromètre à haute résolution à 1 GeV auprès de Saturne (voir figure 2), qui allait devenir le SPES-1. L'étude démarre en 1967/1968; les premières expériences commenceront en mai 1972. Dès les premières mesures, il se confirme que la qualité des faisceaux de Saturne n'est pas idéale pour des expériences de physique nucléaire, qui demandent une bonne émittance, une bonne résolution et un bon cycle utile sans spikes. D'où un projet plus précis de transformation de Saturne, pour en faire une machine de qualité nucléaire . Dès 1971, on commence à étudier un deuxième spectromètre, à grand angle solide et à large bande en impulsions, qui deviendra le SPES-2, opérationnel fin 1975.

- Au CNRS entre temps, sous l'égide successive de Blanc-Lapierre puis de J. Teillac, était né, en 1971, l'IN2P3, qui devenait un partenaire valable pour le CEA, plus précisément pour l'IRF. La transformation de Saturne - en fait la construction d'une machine nouvelle - ne pouvait être réalisée que dans un cadre national, et, c'est pourquoi, à l'automne de 1972, le CEA et l'IN2P3 décident d'ouvrir une concertation à ce sujet. Parallèlement doit être discutée la construction d'un accélérateur d'ions lourds permettant d'étendre les études dans ce domaine jusqu'aux éléments les plus lourds; il deviendra le GANIL. Des groupes de travail sont constitués avec des physiciens appartenant aux deux organismes.

UNIVERSITÉ DE PARIS

FACULTÉ DES SCIENCES

SERVICE DE PHYSIQUE DES SOLIDES

BATIMENT 210
91 - ORSAY

TÉLÉPHONE : 928-53-80 - POSTE ~~3-0~~ 319

Monsieur RADVANYI
Physique Nucléaire
ORSAY

Ce 24 novembre 1966

Cher Monsieur,

Veuillez trouver ci-joint la copie d'une lettre qui vient de m'être adressée par H. PALEVSKY de Brookhaven, concernant les recherches sur la structure nucléaire avec des protons de environ 1 BeV.

Personnellement, je ne suis absolument pas spécialiste de ces questions, mais je pense que les propositions de la page 2 vous concernent assez directement.

Croyez à mes sentiments très cordiaux.

P.-G. de Gennes

Copies : MM. ABRAGAM
MESSIAH
BERTHELOT

P.J. : lettre de H. PALEVSKY

Figure 1: Lettre de P. G. de Gennes.

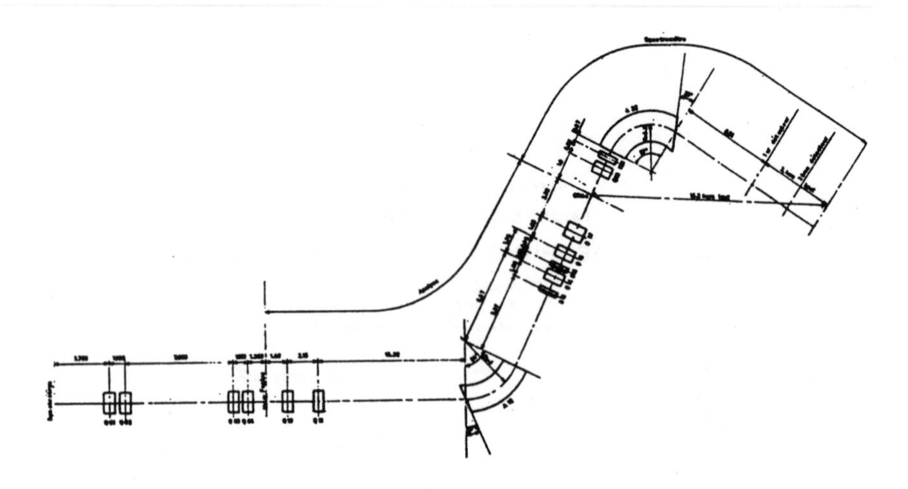

Figure 2: Le spectrometre SPES-1.

- Fin 1973, la décision commune est prise de transformer Saturne en un synchrotron moderne à focalisation forte et une première ébauche de l'Entreprise commune qui doit mener à bien ce projet est définie. Une première convention entre les deux organismes est signée en juin 1974. La constitution d'un laboratoire national autour du nouveau Saturne -Saturne-2- doit être préparée. Une deuxième convention sera signée en 1977. Naturellement, contrairement à ce qui sera le cas plus tard au GANIL, il y a au départ une certaine dissymétrie inévitable : Saturne-2 est construit à Saclay en utilisant les infrastructures existantes pour réduire les dépenses, et le personnel nécessaire existe pour l'essentiel. Du personnel IN2P3 viendra peu à peu se joindre au personnel CEA; ces personnels conservent leurs statuts. Le budget de construction du nouveau Saturne est partagé à raison de 2/3 CEA et 1/3 IN2P3.

2 L'Entreprise commune

Un Comité de projet, présidé alternativement par J. Horowitz et J. Teillac, va superviser les opérations.

Deux structures nouvelles, GERMA (Machine) et GEREX (équipements expérimentaux) viennent se superposer aux structures du Département du synchrotron Saturne et à ses services, le SEFS, le SEDAP et le SOC.

COMITE DE PROJET

J. HOROWITZ (C. E. A.), Président 1974	J. TEILLAC (IN2P3), Président 1975
H. FARAGGI (C. E. A.)	M. JEAN (IN2P3)
M. GERIN-ROZE (C. E. A.)	G. RICCI (IN2P3)
A. MESSIAH (C. E. A.)	J. YOCCOZ (IN2P3)

Le secrétariat du Comité de Projet est assuré par J. PELLERIN (C. E. A.)

CONSEIL DES PHYSICIENS

J. THIRION (D. Ph. N. Saclay) Président
P. RADVANYI (I. P. N. Orsay), Vice-Président
B. D'ALMAGNE (L. A. L. Orsay)
J. ARVIEUX (Grenoble)
E. ASLANIDES (Strasbourg)
R. BERTINI (Strasbourg)
R. BEURTEY (D. Ph. N. Saclay), Secrétaire
J. L. BOYARD (I. P. N. Orsay)
I. BRISSAUD (I. P. N. Orsay)
G. BRUGE (D. Ph. N. Saclay)
J. P. BURCQ (Lyon)
J. C. FAIVRE (D. Ph. N. Saclay)
D. GARRETA (D. Ph. N. Saclay)
L. GOLDZAHL (IN2P3. Saclay)
F. LEFEBVRES (Caen)
F. LEHAR (D. Ph. P. E. Saclay)
R. LOMBARD (I. P. N. Orsay)
J. MILLER (D. Ph. N. Saclay)
J. R. PIZZI (Lyon)
J. SAUDINOS (D. Ph. N. Saclay)
R. SCHAEFFER (D. Ph. T. Saclay)
R. TURLAY (D. Ph. P. E. Saclay)
F. YIOU (C. S. N. S. M. Orsay)

REALISATION DU PROJET

La direction du projet est assurée par le Président et le Vice-Président du Conseil des Physiciens.

GERMA

R. VIENET (Saclay), Chef du groupe GERMA

A. CABRESPINE (Orsay) Chef-adjoint de GERMA

GEREX

B. THEVENET (Saclay), Chef du groupe GEREX

R. BERTINI (Strasbourg) Chef-adjoint de GEREX

Figure 3: Conseil des Physicien 1974.

Un Conseil des physiciens (voir figure 3) est mis en place au début de 1974, pour proposer les grandes lignes du programme expérimental de Saturne-II avec ses répercussions sur les caractéristiques de la machine et sur les équipements communs, ainsi que d'étudier toute question liée à l'utilisation de la machine .

Beaucoup de travail est alors accompli; j'espère ne pas trop en oublier. Pendant ce temps les expériences se poursuivent auprès de Saturne-I, autour notamment du DPhN/ME et de l'ER 54 de L. Goldzahl.

Les caractéristiques de la nouvelle machine sont examinées. Les dipôles seront construits sur appel d'offres par Rade Konçar à Zagreb en Yougoslavie; on se souviendra du géphyrotron (du grec "gephyra" qui veut dire pont, Brücke en allemand, comme Bruck son concepteur) destiné à préparer l'extraction du faisceau. Les performances attendues sont optimisées pour la physique nucléaire. L'énergie maximum de l'accélérateur est définie pratiquement par la dimension du bâtiment existant qui doit le contenir. On n'utilisera pas de cible interne solide; de toute manière l'intensité éventuelle de particules secondaires ne peut pas être importante : l'intensité du faisceau accéléré est limitée par les épaisseurs de béton que l'on peut disposer autour de la machine et des aires. La supériorité internationale de Saturne proviendra de la qualité et de la diversité de ses faisceaux, de l'alliage entre de remarquables performances techniques et un bon programme de physique. Il y aura, à différentes reprises des contacts étroits avec le CERN.

Une nouvelle implantation des aires expérimentales est définie, puis redéfinie. Les postes expérimentaux sont ainsi précisés.

De nouveaux gros équipements sont proposés, puis étudiés : pour compléter la panoplie des spectromètres, il y aura SPES-3 (avec la culasse de l'ancien cyclotron) et SPES-4, dit aussi le 4 GeV/c (avec des éléments de l'ancien Saturne). De vives discussions ont lieu sur la banalisation des gros équipements.

L'informatique est discutée en détail, aussi bien pour la machine et les aires que pour les postes expérimentaux, où les SAR aux performances remarquables feront leur apparition.

La possibilité de produire - au moyen de sources d'ions appropriées (Cryebis, qui sera suivie par Dioné, et Hypérion) - et d'accélérer à grande énergie des ions lourds et des particules légères polarisées (protons et deutons) est discutée et retenue. Les meilleures solutions ne sont pas trouvées du premier coup. On s'enthousiasme pour la cible polarisée du poste nucléon-nucléon.

La possibilité d'une enceinte séparée pour le laboratoire national est étudiée; mais on nous dit : Préférez-vous une nouvelle enceinte ou un nouveau spectromètre? La nécessité d'un nouveau bâtiment pour les physiciens (qui sera l'extension du 144) est soulignée; un projet est préparé puis mené à son terme.

Un effort de communication est réalisé. Des articles sont écrits. Un bulletin annuel d'information Nouvelles de Saturne est édité à partir de février 1975 (voir figure 4). A partir de 1979, paraît aussi périodiquement "En bref de Saturne", préparé par le Bureau des Utilisateurs.

Nous prenons notre bâton de pèlerin pour élargir l'assise nationale du futur laboratoire, expliquer ce que sera le LNS et appeler à des projets d'expériences. Strasbourg, Clermont-Ferrand, Lyon, Grenoble, Bordeaux et l'IPN d'Orsay participent, chacun à sa manière. L'ouverture internationale est rapidement encourageante; des post-docs et des seniors viendront passer un ou deux ans au LNS, avant que des équipes entières (américaine, allemande, scandinave, italienne, russe,...) ne viennent avec leurs détecteurs collaborer (ou proposer et réaliser) à telle ou telle expérience.

nouvelles
de saturne

BULLETIN D'INFORMATIONS FEVRIER 1975

VERS SATURNE II.

Le projet SATURNE II est maintenant en cours de réalisation. Une première convention d'association entre le C.E.A. et l'IN2P3, signée au mois de juin 1974, prévoit :

" - de réaliser la transformation du synchrotron SATURNE, y compris l'aménagement des aires expérimentales,

- d'assurer le fonctionnement du synchrotron après transformation, et l'exécution, dans le cadre d'un laboratoire national, d'un programme d'expériences

nouvelles
de saturne

BULLETIN D'INFORMATION n° 3 JUIN 1978

GRANDES LIGNES DU PLANNING DE SATURNE

Jusqu'au 1er juillet, on continuera le montage des équipements de l'anneau et la mise au point des programmes informatiques pour le réglage et chine, cette dernière restant prioritaire. A cette époque l'informatique sera totalement opérationnelle.

Figure 4: Une première page des Nouvelles de Saturne.

Des Journées d'études Saturne sont organisées dans des lieux sympathiques. Nous assistons aujourdhui aux dernières; les premières ont eu lieu à Aussois en octobre 1977. Elles ont été riches et utiles; c'était aussi un moyen de mieux faire connaissance. Des comptes rendus ont été chaque fois édités.

Il y aura ultérieurement des physiciens assistants de la direction, puis des physiciens détachés ou résidents. Des contacts plus suivis sont établis avec les théoriciens. Par la suite il y aura un juin à Saturne et plusieurs journées théoriciens.

Le Laboratoire national Saturne est créé au 1er janvier 1978. Quatre services et deux groupes (le SM, le SA, le SSG, le SD, un groupe Théorie des accélérateurs et un groupe administratif) viennent remplacer les anciennes structures. Les premières expériences débutent en octobre de la même année.

Un Comité des expériences est mis en place, comportant des utilisateurs et des non utilisateurs, français et étrangers, élus et nommés. Lors de sa première réunion, le 6 avril 1978, il adopte une procédure générale pour examiner les propositions d'expériences et les lettres d'intention. Un Bureau des utilisateurs est créé.

Toute cette organisation débroussaille le terrain pour GANIL, dont la construction à Caen a été décidée en 1975, et qui sera opérationnel en 1983.

Le DPhN/ME réalise une chambre à projection temporelle, le détecteur 4π Diogène.

Un premier projet pour l'anneau injecteur MIMAS est présenté aux deuxièmes journées d'études Saturne, à Roscoff, en mai 1979. Après une étude plus poussée du projet et un certain nombre de démarches et de plaidoyers auprès des instances supérieures, la décision de construire et de financer MIMAS et sa RFQ sera finalement prise en juillet 1983; les travaux commenceront aussitôt.

Nous ne sommes plus alors dans la préhistoire, mais déjà dans l'histoire de Saturne-2.

SATURNE 2 : 20 YEARS FOR PHYSICS

Paul-André CHAMOUARD
CEA/DSM/UGP SATURNE 91 191 Gif sur Yvette CEDEX FRANCE

Abstract: SATURNE NATIONAL LABORATORY (L.N.S) has been created in 1978. The facility was constituted around SATURNE 2 , a 2.95 GeV synchrotron (maximum energy for protons). A large choice of particles was available for experiments : light ions (protons, deuterons, Helium 3 and Helium 4) , polarized ions with the highest polarization and intensities in the world (protons, deuterons and Lithium 6) , heavy ions up to Krypton 84. 11 experimental lines including 4 spectrometers completed the facility. The facility was also characterized by a large range of beam energies, fast and flexible tunings and an excellent reliability. The resonant extraction of the beam used a betatron deceleration to push the beam to the resonant area, by this way the energy of the beam was constant during the spill and the energy spread was reduced ; the duration of the spill was easily adjustable and the regulation of the flux of particles was possible. SATURNE 2 had 2 extraction channels with the possibility of simultaneous extraction with an easy adjustment of intensity sharing . We will describe the performances of SATURNE 2 : beam characteristics , elements and operation results.
Résumé: Le LABORATOIRE NATIONAL SATURNE a été créé en 1978 . Le nouvel accélérateur "SATURNE 2" de 2.95 GeV (énergie maximale en protons) en était l' élément principal. Une grande diversité de projectiles était offerte aux utilisateurs : protons, deutons, Hélium 3 et Hélium 4, des ions polarisés (protons, deutons, Lithium 6) avec les plus hautes valeurs mondiales d'intensité et de polarisation, et enfin , des ions lourds jusqu'au Krypton 84. 11 lignes de faisceau comprenant 4 spectromètres complétaient l'installation. Le fonctionnement de SATURNE 2 était caractérisé également par une large gamme d'énergies, des réglages aisés et rapides, et une excellente fiabilité de fonctionnement . L'éjection résonante du faisceau utilisait une décélération bétatron pour amener le faisceau sur la résonance : cette méthode très originale permettait une énergie constante du faisceau pendant le déversement, une faible dispersion d'énergie et une régulation du flux du faisceau déversé. Les 2 canaux d'éjection pouvaient recevoir le faisceau simultanément avec un réglage facile du partage d'intensité dans chacune des voies. Nous décrirons les caractéristiques des faisceaux , les éléments et les résultats du fonctionnement.

1 INTRODUCTION:

The general display of LNS is represented FIGURE 1.

SATURNE 2 was a strong focusing machine with separate functions totally different of the precedent machine "SATURNE 1" .

2 MAIN CHARACTERISTICS OF LNS FACILITY:

⋆ 2 EJECTION CHANNELS

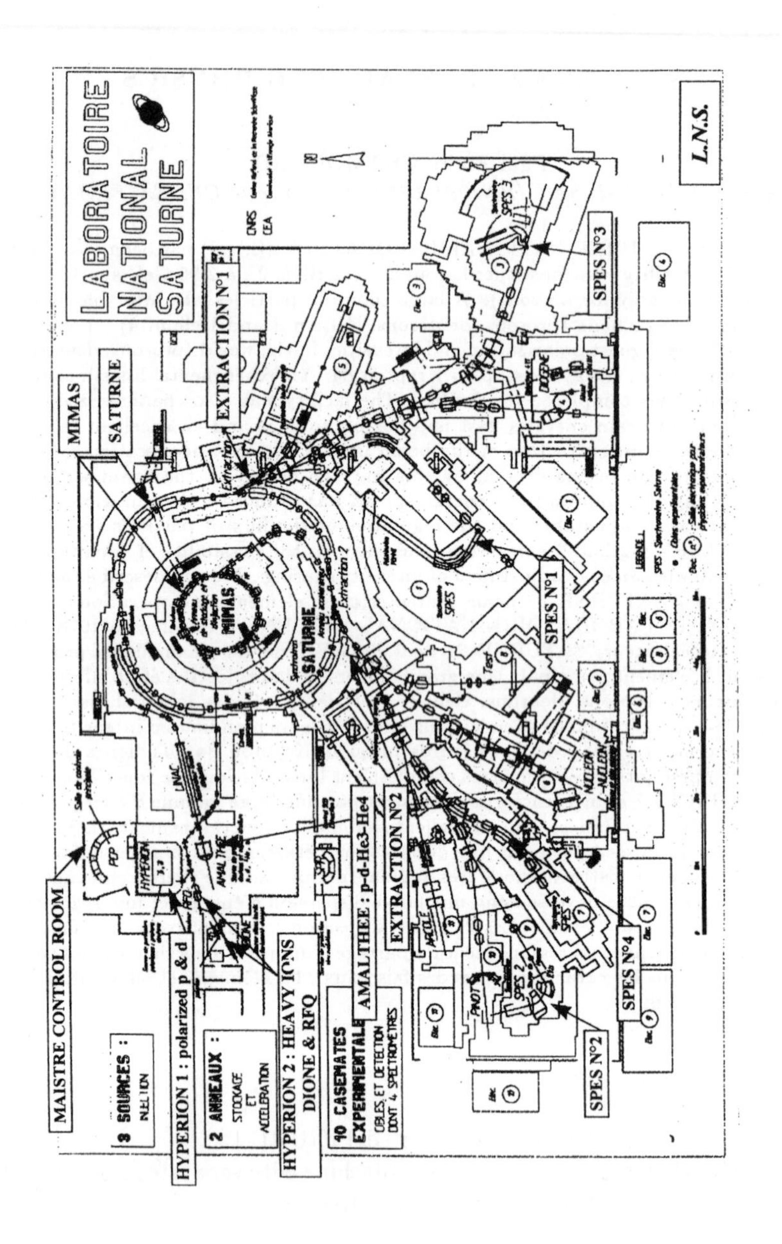

Figure 1: Map of the SATURNE NATIONAL LABORATORY.

⋆ 11 EXPERIMENTAL LINES
⋆ 4 SPECTROMETERS
⋆ 1 LINAC INJECTOR : 20 MeV (protons) able to operate on 2nd harmonic ($Q/A > 0.5$)
⋆ 1 SYNCHROTRON INJECTOR : " MIMAS " : 47 MeV (protons)
⋆ 1 LIGHT IONS PREINJECTOR : " AMALTHEE " 750 kV delivering protons, deuterons, helium 3 and 4 ions
⋆ 1 PREINJECTOR (400 kV) FOR POLARIZED PROTONS and DEUTERONS " HYPERION 1 "
⋆ 1 HEAVY IONS PREINJECTOR " HYPERION 2 " (200 keV/amu) constituted by an EBIS source and a post accelerator RFQ.

3 PRINCIPAL EVENTS DURING THE LIFE OF SATURNE 2:

1978 July	SATURNE 2	1st beam accelerated : protons with Amalthée and the linac
1978 July	SATURNE 2	1st ejected proton beam
1981 April	HYPERION 1	1st polarized protons : LINAC (2nd harmonic)
1982 March	HYPERION 1	1st polarized deuterons : LINAC (2nd harmonic)
1984 May	HYPERION 2	1st heavy ions : LINAC (2nd harmonic)
1987 October	MIMAS	1st beam for physics : polarized deuterons beam
1987 October	HYPERION 2 and MIMAS	A new source " DIONE " replaces CRYEBIS, 12C(6+), 20Ne(10+) and 40Ar(16+) are accelerated in SATURNE 2. 40Ar(16+) correspond to a ratio charge/mass Q/A = 0.4 impossible to accelerate with the linac limited to $Q/A \geq 0.5$.
1989 May	HYPERION 2	1st beam of 84Kr(28+) : Q/A = 1/3
1991 February	AMALTHEE	Connection with MIMAS, end of linac operation.

1992 October	HYPERION 2	New RFQ replacing the first one built by Physik Elektronik Technik, Darmstad G.
1993 June	DIRECTION COMMITTEE	Decision to stop the facility ! The definitive shutdown is decided for 1997 December by LNS direction.
1993 November	HYPERION 2 and MIMAS	129Xe(33+) is accelerated in MIMAS (Q/A=0.256), 197Au(50+) is injected and captured in MIMAS (Q/A = 0.254).
1997 Nov. 28th	LAST BEAM	

4 MAIN CHARACTERISTICS OF SATURNE 2:

Physical radius	:	16.8 m
super periods	:	4
cell type	:	FODO
cells/super period	:	3
wave numbers	:	$3.5 < \nu_{x,z} < 4$
momentum compaction	:	0.005
dipoles	:	16 / gap height 14 cm / radius 6.34 m / maximum induction 2T.
quadrupoles	:	24 / length 48 cm / maximum gradient 10 T/m / aperture 19.2 cm
RF cavities	:	2 / gap voltage by cavity 14 kV harmonic number 3/ frequency range 0.8 to 8.5 MHz.
vacuum	:	1.10^{-7} Pa

5 ENERGY RANGES VERSUS THE RATIO Q/A:

Q/A	MINIMUM ENERGY (MeV/A)	MAXIMUM ENERGY(GeV/A)
1.000	100.0	2.950
0.667	46.0	1.746
0.500	26.0	1.170
0.400	16.7	0.839
0.300	9.4	0.532
0.250	6.6	0.392

6 BEAMS OF SATURNE 2 REALIZED WITH MIMAS (SYNCHROTRON INJECTOR):

6.1 *LIGHT IONS WITH THE PREINJECTOR " AMALTHEE "*

Main characteristics of " Amalthée " :

- duoplasmatron source:
  - arc current 0 to 50 A pulse duration 100 to 1500μsec
  - repetition rate 1 Hz
  - maximum intensity : 100 mA at 750 keV
  - normalized emittance $E/\pi = 1.10^{-6} m * rad$
- high voltage generator:
  - Cockroft-Walton nominal voltage 750 kV
  - maximum voltage 1MV / 200 μA
  - 16 stages , frequency 1 kHz total stability $\pm 1.10^{-3}$

The preinjector is set in pressurized vessel with a mixture of N_2 (75%) and CO2 (25%) at 10 bar. The ions He3 (++) and He 4 (++) are obtained by stripping of single charged ions at 750 keV on carbon foils (12 $\mu g/cm^2$).

	INTENSITY BY BURST
PROTONS	8.10^{11}
DEUTERONS	5.10^{11}
HELIUM 3 (state of charge : 2^+)	2.10^{11}
HELIUM 4 (state of charge : 2^+)	2.10^{11}

6.2 *POLARIZED IONS WITH THE PREINJECTOR " HYPERION 1 "*

Characteristics of " HYPERION 1 " :

- atomic beam :

  ground state type , operates in pulsed mode

  (gas and RF of the dissociator)

  nozzle cooled at 80 K

  hexapole with tapered poles

- transitions:

  2 for protons

  3 for deuterons

- ionizer :

  electron beam reflex ionizer : magnetic field ~2Kg

  pulsed voltage mode (d.c. conditionning)

  extraction voltage 18 kV

  beam current at 18 keV: 700 μA (protons or deuterons).

  normalized emittance : $E/\pi = 3.10^{-6} m * rad$

- high voltage generator :

  built by Haeffely (CH) : 420 kV / 2.5 mA/ stability: $\pm 1.10^{-4}$

	INTENSITY BY BURST
PROTONS	$2\ 10^{11}$
DEUTERONS	$3.2\ 10^{11}$

6.3 POLARIZATION OF BEAMS

	POLARIZATION
PROTONS	> 90% up to 1 GeV 80% up to 2.95 GeV
DEUTERONS	> 90% of $P_Z = \pm 1/3$; $P_{ZZ} = \pm 1$

At 2.95 GeV, the proton beam has crossed about 15 depolarizing resonances in SATURNE 2. There is no depolarizing resonance for polarized deuteron beam.

6.4 EXAMPLES OF HEAVY IONS WITH " HYPERION 2 " :

This preinjector is constituted by, an Electron Beam Ion Source (EBIS) " DIONE " and a post accelerator RFQ . The originality of SATURNE 2 for the acceleration of heavy ions is the choice of an EBIS source which produces very high states of charges :

totally ionized ions	:	(Li , C, N, O, Ne)
very high states of charge	:	40Ar (16+), 84Kr (26+), 129Xe (33+)

This method presents several advantages versus others which uses strippers :

no degradation of the initial emittance

no post stripper accelerator

MIMAS operates as a storage ring during the injection of heavy ions and is able to store up to 8 pulses from HYPERION 2 . This process needs a very low pressure (10^{-9} Pa) in the chamber to avoid losses of particles by charge exchange (electron capture) .

A first RFQ called RFQ1 has been built in LNS allowing the acceleration of ions with a ratio Q/A :

$$0.35 \leq Q/A \leq 0.5$$

This RFQ1 has been the first RFQ used as injector for a synchrotron ; a debunching section was included to reduce the energy spread of the beam.

A new RFQ (RFQ2) has been ordered in 1992 to Physik Elektronik Technik (Darmstadt G.). The range of Q/A was :

$$0.25 \leq Q/A \leq 0.5$$

NUCLEUS	A	Z	Q	PARTICLE BY BURST	RFQ
CARBON	12	6	6+	$1.\ 10^9$	RFQ1
NITROGEN	14	7	7+	$7.\ 10^8$	RFQ1
NITROGEN	15	7	7+	$7.\ 10^8$	RFQ1
NEON	20	10	10+	$2.\ 10^8$	RFQ1
ARGON	40	18	16+	$1.2\ 10^8$	RFQ1
KRYPTON	84	36	28+	$2.\ 10^6$	RFQ1
KRYPTON	84	36	26+	$8.\ 10^6$	RFQ2
XENON**	129	54	33+	$2.\ 10^6$	RFQ2

** XENON ions have been accelerated only in MIMAS during machine studies. FIGURE 2 shows the signals of XENON beam in MIMAS.

A source of polarized Lithium 6 has been put in the terminal of " DIONE ". This source included an atomic beam of polarized Lithium 6 and a ionizer giving single charged ions injected in DIONE.

	INTENSITY BY BURST	POLARIZATION
POLARIZED Li6 (3+)	$1.\ 10^9$	70% of Pzz = ±1

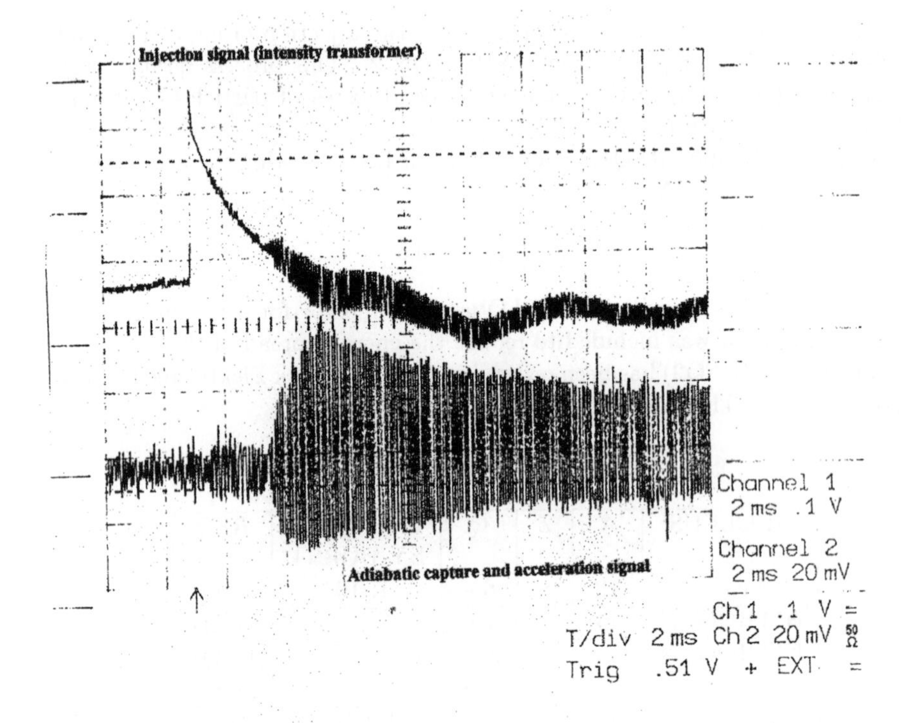

Figure 2: Acceleration by MIMAS of ^{129}Xe (33+). The intensity was 2.10^6 particle/pulse.

7 THE EXTRACTED BEAMS OF SATURNE 2:

The slow extraction uses the horizontal hexapole induced resonance:

$$\nu_x = 11/3$$

This resonance is set on the axis of the machine and the chromaticity is increased by hexapoles in order to reduce the separatrix area , by this way the energy spread DP/P and the horizontal emittance of the extracted beam are small. Before extraction the beam has been put on an external radius by the RF. The RF is switched off after. During extraction the beam is pushed to the resonance by a betatron deceleration. This method has several advantages:

- at any moment the beam is extracted on the same radius, the energy is constant
- the energy spread is small : $\Delta P/P = \pm 1.\ 10^{-4}$
- the spill duration is adjustable by the deceleration voltage
- the intensity of the spill can be regulated by a loop from a monitor set in the beam line to the betatron power supply.

TYPICAL CHARACTERISTICS OF THE EXTRACTED BEAM AT 1 GeV:

horizontal emittance	:	$E_H/\pi = 8mm * mrad$
vertical emittance	:	$E_V/\pi = 14mm * mrad$ (completely filled machine)
momentum spread	:	$\delta P/P = 1.\ 10^{-4}$

The FIGURE 3 shows an example of extraction with and without intensity

SATURNE 2 is equipped with 2 extraction channels with the followings possibilities :

single extraction in one channel , efficiency 85%

simultaneous extraction in the 2 channels , efficiency 60%, with the possibility of intensity sharing from 50%-50% to 95%-5%.

extraction in one channel at one energy and in the other channel at a different energy cycle by cycle.

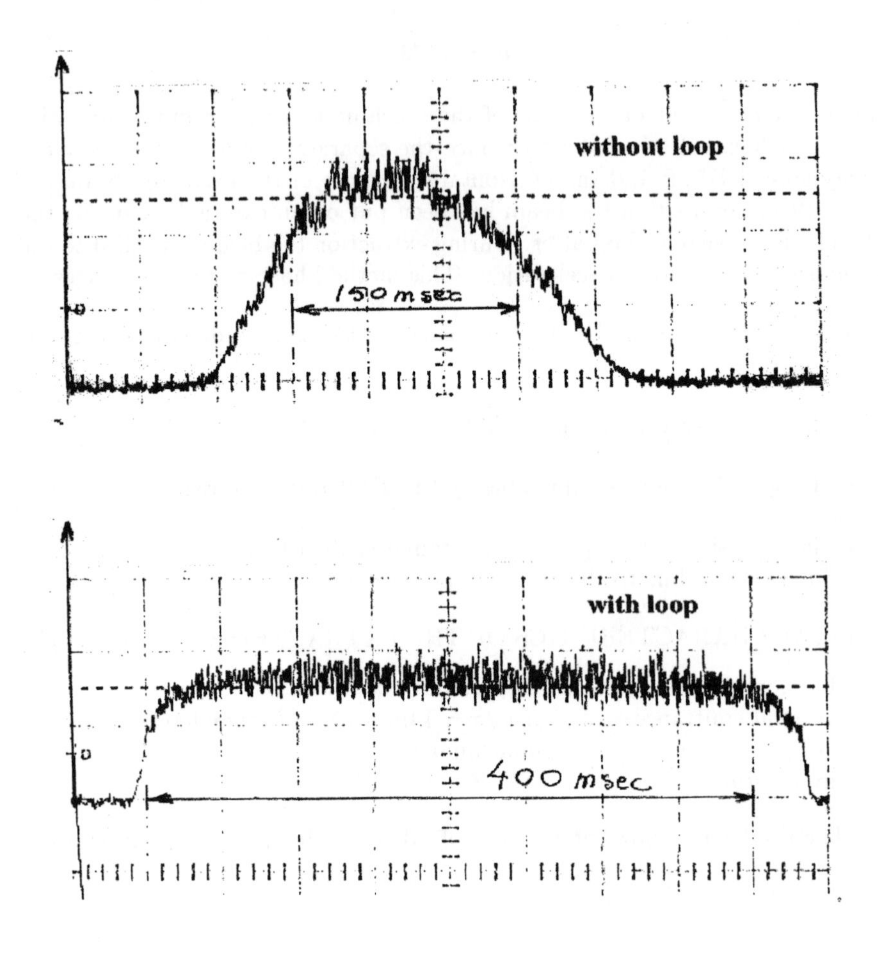

Figure 3: Example of spills with and without intensity loop on the Betatron. (beam characteristics: polarized protons 2.85 GeV, 2.10^{10} p/pulse)

8 THE BEAM LINES :

11 beam lines are connected to the extraction channels allowing a large choice of experimental possibilities.

The study of NUCLEON-NUCLEON reaction needed polarized beams (protons and deuterons) and polarized targets. A polarized frozen target was set in this line and a spin rotator to put the polarization of the incident beam in the 3 directions.

The beam lines include about 100 magnets and 45 power supplies. Each magnet can be connected to each power supply via a commutation grid.

CHARACTERISTICS OF THE 4 SPECTOMETERS :

	SPECTRO. N° 1	SPECTRO. N° 2	SPECTRO. N° 3	SPECTRO. N° 4*
SWEEPED ANGLES	-4° / 80°	0° / 60°	-5° / 80°	-9° / 17° -0.5° / 39°
SOLID ANGLE	3.8 msrd	20 msrd	10 msrd	2.4 msrd 0.4 msrd
MOMENTUM RESOLUTION	$7.\ 10^{-5}$	$3.\ 10^{-4}$	$5.\ 10^{-4}$	$3.\ 10^{-4}$ $1.\ 10^{-3}$
MAXIMUM MOMENTUM	2.00 GeV/c	0.75 GeV/c	1.4 GeV/c	3.8 GeV/c 3.8 GeV/c
MOMENTUM WIDTH	4 %	34 %	0.6 at 1.4 GeV/c	1 % 8 %
DISPERSION FOR $\delta P/P = 1\%$	15.0 cm	3.0 cm	1.4 cm	8.0 cm 7.0 cm

* SPECTROMETER N° 4 : the values depend of the target position and beam momentum.

9 THE SYNCHROTRON INJECTOR " MIMAS ":

MIMAS realizes a perfect adaptation between the 3 sources and SATURNE 2. This machine is completely described in the report of J.L. LACLARE , but it is essential to emphasize the amount of improvements made on SATURNE 2 and due to MIMAS.

polarized deuteron beam	:	intensity gain > 20 (without loss of polarization)
polarized proton beam	:	intensity gain > 50 (without loss of polarization)
heavy ions	:	possibility to accelerate ions Q/A < 0.5 (limit 0.25) (impossible with the linac)
beam qualities	:	the emittance of the beam is well controlled and reproduced due to a very easy tuning of the injection in MIMAS and transfer of the beam from MIMAS to SATURNE 2.
beam tunings	:	an efficient control system has been installed on MIMAS allowing fast tunings of MIMAS , this system has been extended to SATURNE 2 replacing the old one . A large gain of time has been obtained for startings and tunings.

MAIN CHARACTERISTICS :

injection energy	:	200 keV/A for heavy ions (1 to 8 pulses), 400 keV for polarized protons and deuterons (1 pulse) and 750 keV for protons , deuterons , He3 and He4 (1 pulse)
extraction	:	fast extraction in 1 turn at $B * \rho = 1\ T * m$ (47 MeV for protons), efficiency of the transfer to SATURNE 2 : 100%
accelerating system	:	2 RF cavities : frequency swing 0.15 to 2.5 MHz, voltage by cavity 2 kV, harmonic 1, adiabatic capture of the injected beam with an efficiency close to 90 %.
vacuum	:	pressure : 2. 10^{-9} Pa all the elements are baked "in situ" at 300° C during 1 week.
main magnets	:	8 dipoles (radius 1m), 16 quadrupoles (aperture 26cm) lattice : FODO

10 RESULTS OF SATURNE 2 OPERATION :

10.1 TUNINGS :

Constant improvements have been made up to the end of SATURNE in order to minimize the dead times due to parameter modifications and tunings. The times which are given are mean times (not shortest times) and represent the total time between the beginning of parameter modification up to the beam on target :

total starting time of the facility	:	11 h
particle change (the preinjector is changed or not)	:	4 h (8 h for heavy ions)
beam energy modification (single extraction)	:	0.5 h
beam energy modification (double extraction)	:	1 h
intensity modification (between 3. 10^8 up to 1. 10^{11} part./ burst).	:	10 min.

10.2 OPERATION RESULTS :

To characterize the results of operation we consider the following list both for the scheduled and realized times . We call " efficiency for physics " the ratio between realized and scheduled times for physics and " total efficiency " the ratio between scheduled and realized times for physics plus machine studies. The results are given for 1996.

	BEAM TIME SCHEDULED (hours)	BEAM TIME REALIZED (hours)
1 STARTING AND STOPPING TIMES	359.00	320.45
2 TUNINGS AND PARAMETER MODIFICATIONS	85.50	79.65
3 NOT SCHEDULED TUNINGS		70.60
4 BEAM FOR PHYSICS	3749.50	3518.75
5 FAILURES ON PHYSICS		195.65
6 MACHINE STUDIES	84.00	81.50
7 FAILURES ON MACHINE STUDIES		0.20
8 MAINTENANCE (during operation)	146.00	137.90
9 LOST TIMES		19.30
TOTAL OPERATION TIME (Σ 1 to 9)	4424.00	4424.00
EFFICIENCY FOR PHYSICS (3518.75/3749.50)		93.84%
TOTAL EFFICIENCY (3518.75+81.50)/(3749.50+84.00)		93.91%

We have summarized the results by a set of histograms to show the main points during the live of SATURNE 2 . We have considered the period from 1984 to 1997. Two important dates have to be retained :

October 1987 which corresponds to the beginning of MIMAS operation.

1994-1997 : the experimental areas are included in the results (not before).

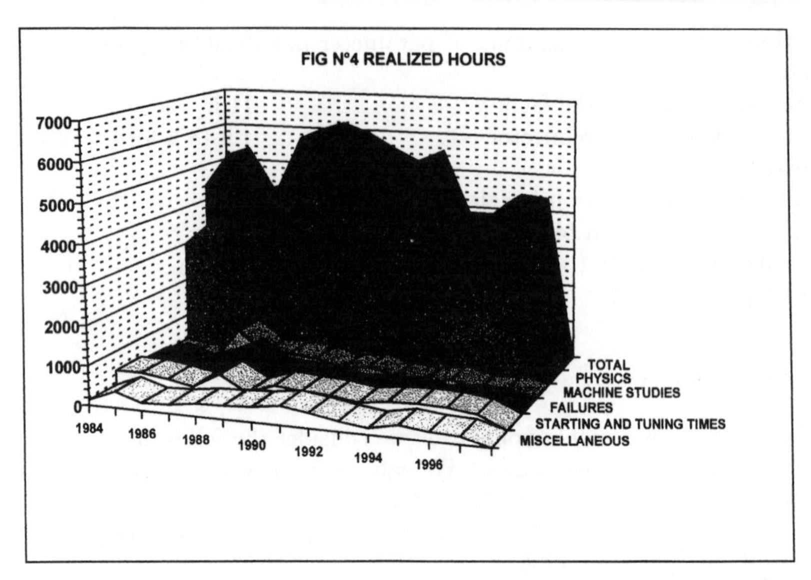

Figure 4: The realized hours for physics, starting and tuning times, failures (on physics and machine studies), machine studies, miscellaneous (maintenance, not scheduled tunings, lost times).

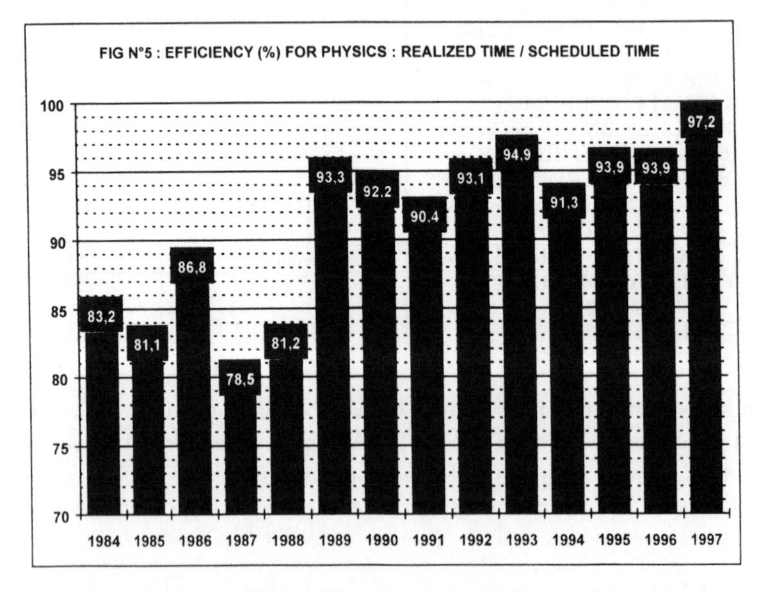

Figure 5: Efficiency for physics.

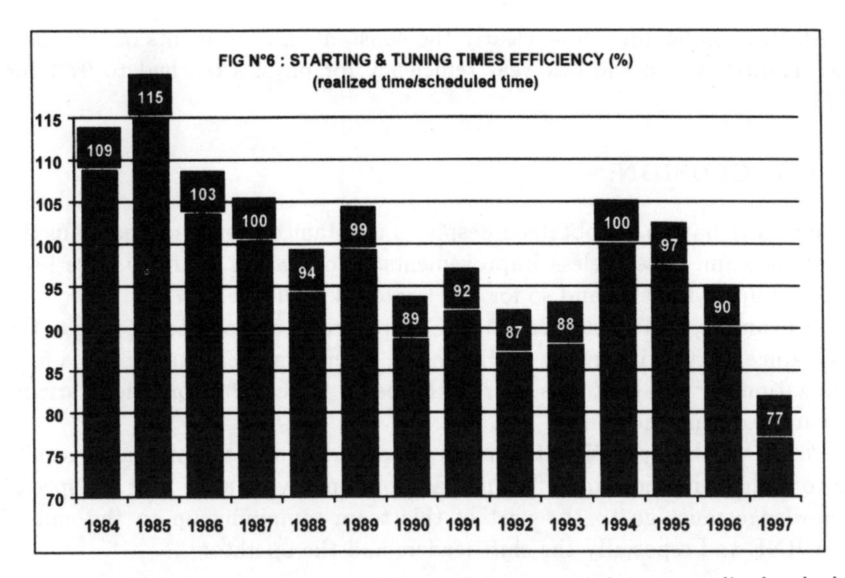

Figure 6: Scheduled starting and tuning times efficiency : ratio between realized and scheduled times. This value must be close to 100% if higher the scheduled times are under estimated or over estimated if lower.

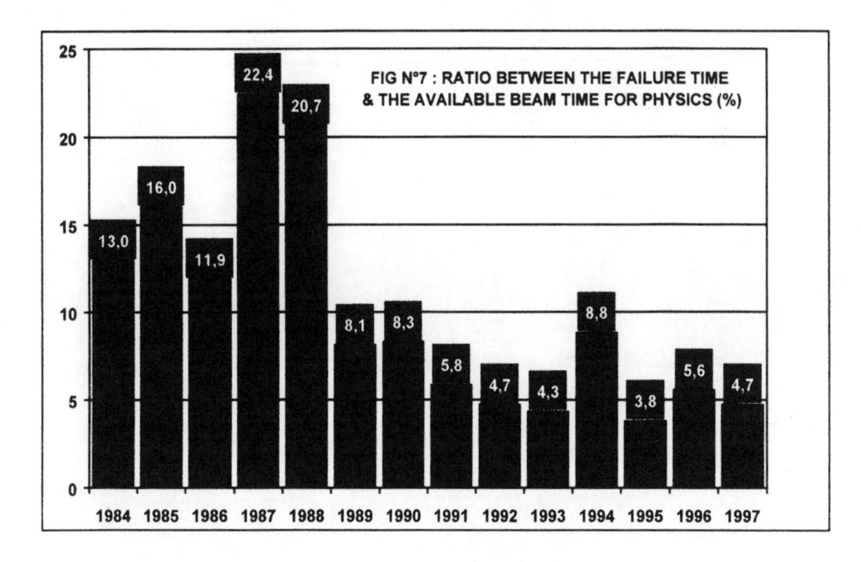

Figure 7: Ratio between failure time and available beam time.

All these indicators show clearly the constant improvements of the operation qualities up to the last year (efficiency for physics reached to 97% for 1997) .

11 CONCLUSION:

These results have been obtained despite a constant decrease of people in the operation team, nevertheless improvements on operation qualities have been improved up to the end and so for accelerators knowledge.

As example, DISTO needed a 2.85 GeV polarized proton beam (close to the maximum energy) with a very low intensity, a very small emittance and a high polarization ($\sim$ 75%); at this energy the beam is near 2 strong depolarizing resonances during the spill !

People of the operation team can be proud of these remarkable results due to their dedication and efficiency and obtained year after year. A special acknowledge must go to the part of this team remaining up to the end of SATURNE and especially the shift-leaders and the operators .

STRONG INTERACTIONS AT SATURNE

MADELEINE SOYEUR
Département d'Astrophysique, de Physique des Particules,
de Physique Nucléaire et de l'Instrumentation Associée
Service de Physique Nucléaire
Commissariat à l'Energie Atomique/Saclay
F-91191 Gif-sur-Yvette Cedex, France

The theoretical framework in which the physics of most experiments dealing with strong interactions at Saturne can be understood is reviewed and discussed.

Cet article a pour objet de présenter et de discuter le cadre théorique dans lequel s'interprètent la plupart des expériences relatives à la physique des interactions fortes réalisées à Saturne.

1 Introduction

The experiments performed at Saturne over the last twenty years dealt mostly with the physics of hadrons, the study of their interactions and the understanding of their specific role in nuclear dynamics.

The energy of the accelerator was suited to study the properties and couplings of the lowest-lying baryons (the nucleon, the $\Delta(1232)$, the $N^*(1440)$, the $N^*(1535)$, the $\Lambda(1116)$ and $\Sigma(1192)$ hyperons) and the mesons of the pseudoscalar octet and singlet (the pions, the η-meson and the kaons). These mesons are the Golstone bosons of the chiral $SU(3)$x$SU(3)$ symmetry of strong interactions [1].

The theory of strong interactions is Quantum Chromodynamics (QCD). Its degrees of freedom are the quarks and the gluons. The strong interactions of quarks and gluons are induced by their color charges [2].

For energy-momentum transfers less than or of the order of 1 GeV, the color degree of freedom appears hidden. It is of course responsible for the confinement of quarks and gluons into hadrons. However, the strong interactions of these hadrons can be understood consistently using Lagrangians whose fundamental fields are color neutral hadronic fields. The quark degree of freedom is visible in this energy range, but only in the Gell-Mann sense, through the classification of hadrons in $SU(3)$ multiplets [3].

The relevant degrees of freedom of strong interactions seem therefore to depend on the energy scale, i.e. on the wavelength at which one probes the theory. At short wavelength, the dynamics is governed by the color degree of freedom according to the Feynman rules derived from the QCD Lagrangian. In the long-wavelength limit, the colored quark and gluon fields do not play an explicit role. The low-energy structure of QCD appears equivalent to an effective theory of interacting hadronic fields. The transition from the fundamental to the effective theory occurs through a phase transition linked to the spontaneous breakdown of the chiral symmetry of massless QCD, which generates Golstone bosons. This effective hadronic theory is valid below the scale of spontaneous chiral symmetry breaking. As the energy increases and smaller wavelengths are probed, the quark and gluon degrees of freedom become relevant and must be included.

We consider in Section 2 a problem which illustrates particularly well this point and which was extensively studied at Saturne, the nucleon-nucleon interaction. The description of the nuclear force from very low energy (below the pion production threshold) until $E_p^{Lab} = 3$ GeV is reviewed and the search for the energy scale at which quarks and gluons appear explicitly in the two-nucleon problem is discussed. Section 3 is devoted to the study of pion propagation in the nuclear medium. Effective theories are used to calculate the pion spectrum at high baryon density, in particular its large self-energy due the attractive pion-nucleon p-wave coupling [4]. Charge-exchange reactions, which act as sources of virtual pions, are ideal processes to excite pionic modes in nuclei and made it possible to observe at Saturne the in-medium pion dispersion relation [5]. In Section 4, we discuss the relation between the spectral properties of hadrons and some properties of the QCD vacuum, emphasizing particularly the role of the quark condensate [6]. This relation opens very interesting perspectives to be studied experimentally mostly at GSI-Darmstadt, in nuclei and in relativistic heavy ion collisions.

2 The nucleon-nucleon interaction

- *Below the pion production threshold ($E^{Lab} < 350$ MeV)*

The meson-exchange picture of the nucleon-nucleon interaction has long been known to provide an accurate description of deuteron properties and scattering data below the pion production threshold [7].

In this picture, the long-wavelength part of the nuclear force is described by meson-exchanges involving masses up to about 1 GeV. The short-distance behavior is parametrized by form factors associated with the meson-baryon vertices. They reflect the extended structure of hadrons and effectively cut the meson-exchange contributions. The functional expression for the form factors (often monopole forms) and the value of the cutoff mass for each vertex are chosen so as to reproduce the nucleon-nucleon data. The cutoff masses are typically of the order of 1.0-1.5 GeV, i.e. values consistent with restricting exchange masses to be $\leq$ 1 GeV. We show in Figs. 1 and 2 the processes included in the Bonn meson-exchange potential [7] together with the correlated $\pi\rho$-exchange term discussed recently [8]. The single-meson exchange diagrams of Fig. 1 exhibit the four important mesonic degrees of freedom of nuclear dynamics: the $\pi(139)$ [J=0, T=1], the $\sigma(550)$ [J=T=0], the $\rho(769)$ [J=1, T=1] and the $\omega(782)$ [J=1, T=0]. The $\sigma(550)$ is an effective degree of freedom accounting for the exchange of two correlated pions in s-wave. Below the pion production threshold, only two baryons play a role in nuclear interactions, the nucleon and the $\Delta(1232)$. The box diagrams of Fig. 2 represent the two-meson exchanges of low mass, the $\pi\pi$- and the $\pi\rho$-exchanges, when they are either uncorrelated (upper graphs) or correlated (lower graphs). The tensor forces associated with the pion and the ρ-meson exchanges have opposite signs. There are therefore partial cancellations between $\pi\pi$- and the $\pi\rho$-exchanges which require that they be grouped together [9]. At higher orders (3π-exchange and beyond), cancellations become even larger and ensure the rapid convergence of the meson-exchange expansion [9].

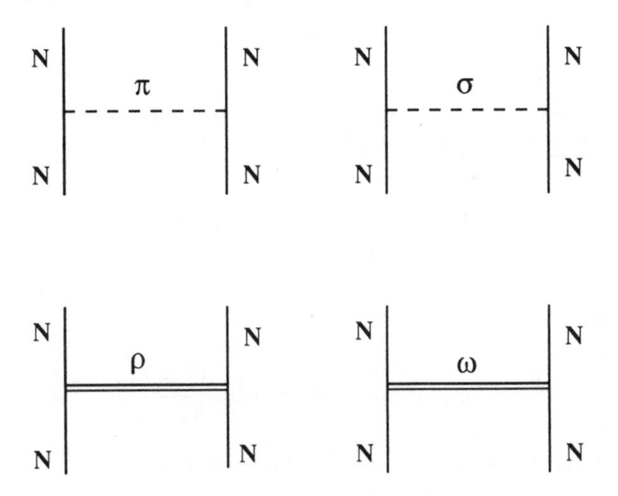

Figure 1: One-meson-exchange processes included in the Bonn potential.

It is remarkable that all recent nucleon-nucleon potentials [10,11,12], using more than 4000 data points, reproduce the two-nucleon data with a χ^2 per datum of 1.03-1.09 [10].

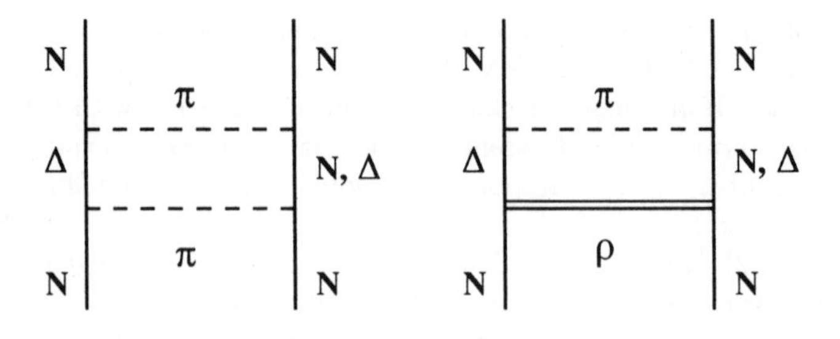

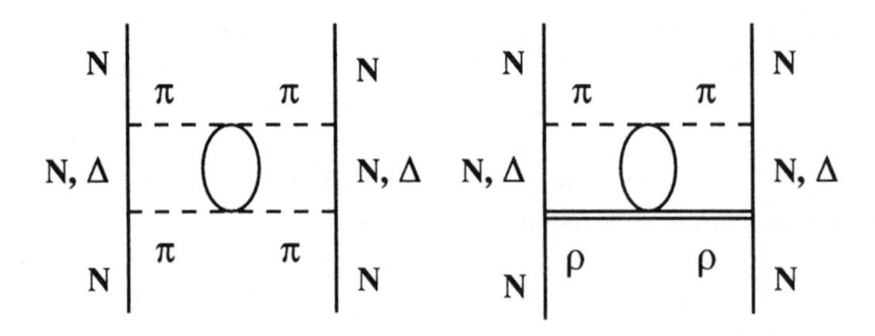

Figure 2: Box diagrams contributing to the Bonn potential. The upper graphs are uncorrelated exchanges while the lower graphs represent correlated $\pi\pi$- and $\pi\rho$-exchanges.

It is interesting to discuss in some detail the pion-nucleon vertex, which is a key quantity for the evaluation of the graphs displayed in Figs. 1 and 2. The strength of the pion-nucleon vertex (assuming pseudoscalar coupling) can be written as

$$F_{\pi NN}(q^2) = g_\pi(q^2 = m_\pi^2)\,\frac{\Lambda_\pi^2 - m_\pi^2}{\Lambda_\pi^2 - q^2}. \qquad (1)$$

All the recent high precision nucleon-nucleon potentials[10,11,12] give exactly the same value for the πNN coupling constant, i.e.

$$g_\pi^2/4\pi = 13.6. \qquad (2)$$

This number, which is of major importance for nuclear physics and low-energy chiral dynamics, is significantly lower than the previously accepted value of $g_\pi^2/4\pi = 14.6$[7]. The question of the value of g_π is not yet fully settled however. A recent np scattering measurement at 162 MeV suggested $g_\pi^2/4\pi = 14.52\pm0.26$

for the charged πNN coupling constant as a result of extrapolating the data to the pion pole [13]. The problem of the relation of this determination of g_π to the strength of the πNN vertex needed to fit the nucleon-nucleon scattering data remains to be fully understood.

A second recent result of interest on the πNN vertex is a dynamical explanation for a long-standing discrepancy between the value of Λ_π obtained in nucleon-nucleon potentials describing the two-nucleon data ($\Lambda_\pi \simeq 1.3$ GeV) [7] and the corresponding quantity for the free nucleon ($\Lambda_\pi \simeq 0.7$ GeV) [14]. The value of $\Lambda_\pi \simeq 1.3$ GeV, obtained for the Bonn potential [7] which does not include the correlated $\pi\rho$-exchange term shown in Fig. 2, was shown to effectively account for the correlated $\pi\rho$-exchange contribution [8]. The strong tensor force, needed to reproduce in particular the deuteron properties, is produced by both the single π-exchange with a rather soft form factor (comparable to what is needed to explain the pion-nucleon data) and by the correlated $\pi\rho$-exchange.

The pion-nucleon vertex as described by Eq. (1) is therefore pretty well determined, with a coupling constant $g_\pi^2/4\pi = 14.0 \pm 0.5$ and a cutoff mass Λ_π = 0.6 - 0.8 GeV.

This extended vertex characterizes the πNN coupling at short-distance and is a direct manifestation of the QCD degrees of freedom. It is therefore very interesting that the values of $g_\pi^2/4\pi$ and of the cutoff mass Λ_π could be recently calculated from QCD on the lattice (quenched approximation) [15] and that the numbers obtained for these parameters,

$$g_\pi^2/4\pi = 12.7 \pm 2.4 \tag{3}$$

and

$$\Lambda_\pi = 0.75 \pm 0.14 \; GeV, \tag{4}$$

are in remarkable agreement with the phenomenological values derived from nucleon-nucleon and pion-nucleon data.

The discussion of the πNN coupling illustrates the general theoretical approach which consists in describing strong interactions in the long-wavelength limit by effective hadronic field theories with extended vertices. These vertices are determined by QCD (probably mostly gluon) dynamics.

- *Above the pion production threshold ($E^{Lab} > 350$ MeV)*

This kinematic regime of the nucleon-nucleon interaction was extensively studied at Saturne, using polarized proton ($0.8 < E_{\vec{p}} < 2.7$ GeV) and polarized neutron ($0.8 < E_{\vec{n}} < 1.1$ GeV) beams and polarized targets [16]. Nucleon-nucleon elastic scattering amplitudes were reconstructed.

The theoretical description of the nucleon-nucleon interaction is more complicated and much less studied above the pion production threshold than below.

Above the pion production threshold, inelasticity sets in. Most pions in the $NN \to NN\pi$ reaction are produced through the excitation of the $\Delta(1232)$ resonance. The $NN \to N\Delta$ channel opens at 600 MeV. Because of the dominance of the $\Delta(1232)$ resonance in pion emission, the production of one pion is the main inelastic channel until $\sim$ 1 GeV, the threshold for double Δ excitation.

A natural extension of the nucleon-nucleon meson-exchange model discussed previously is therefore to add the $\Delta(1232)$ resonance as an explicit baryonic degree of freedom in the problem and solve coupled-channel equations for the NN, NΔ and $\Delta\Delta$ channels [7,9,17,18,19]. The overall agreement with the nucleon-nucleon phaseshifts until 1 GeV is rather good [9] eventhough some discrepancies exist, particularly for spin observables [18,19]. Further extensions of such model are in progress to fit the nucleon-nucleon data above 1 GeV [20]. Nonresonant contributions to the $NN\pi$ channel have also been studied by solving 3-body equations [21].

A completely different approach has been to start from perturbative QCD, valid at high energy, and extrapolate to lower energies by adding 'subasymptotic' corrections [22,23]. These models include the Landshoff three-gluon exchange, quark interchange and Regge terms. They fail to reproduce the nucleon-nucleon data, already at energies as high as 12 GeV [22,23], and are therefore not at all appropriate to discuss the nucleon-nucleon interaction below 3 GeV.

- *Search for narrow dibaryons*

The MIT Bag Model of hadrons [24] applied to (color neutral) six-quark configurations predicted the existence of narrow states of rather low mass [25,26]. These six-quark bags are distinct from two-baryon states in the sense that they do not factorize in two baryonic color neutral states.

The most spectacular prediction was that a strange di-Λ state (composed of uuddss quarks) could have such low mass ($M_{di-\Lambda}$=2.15 GeV) that it would be bound [25]. This state has been extensively searched for in the last twenty years and has not been identified [27]. The MIT Bag Model also predicted six-quark bags consisting of u and d quarks only [26]. In particular, such states appear as excited states of the deuteron (d') at very low excitation energies, $M'_d \simeq$ 2.1-2.2 GeV [26]. If such six-quark configurations existed, they would mean that color degrees of freedom manifest themselves at rather low excitation energy and invalidate the theoretical framework discussed above. Fortunately for effective hadronic field theories, there is no experimental evidence for such d' states.

It is tempting to search for dibaryon resonances by looking for counter-clockwise looping in Argand diagrams. However, at energies above the pion production threshold, such behavior can be produced by the opening of isobar channels [9]. A better approach is to look for narrow structures in simple spin-dependent quantities measured in the nucleon-nucleon elastic channel. A contribution of Saturne to the search for narrow, low-mass dibaryons in analyzing power excitation functions is illustrated in the lower part of Fig. 3.

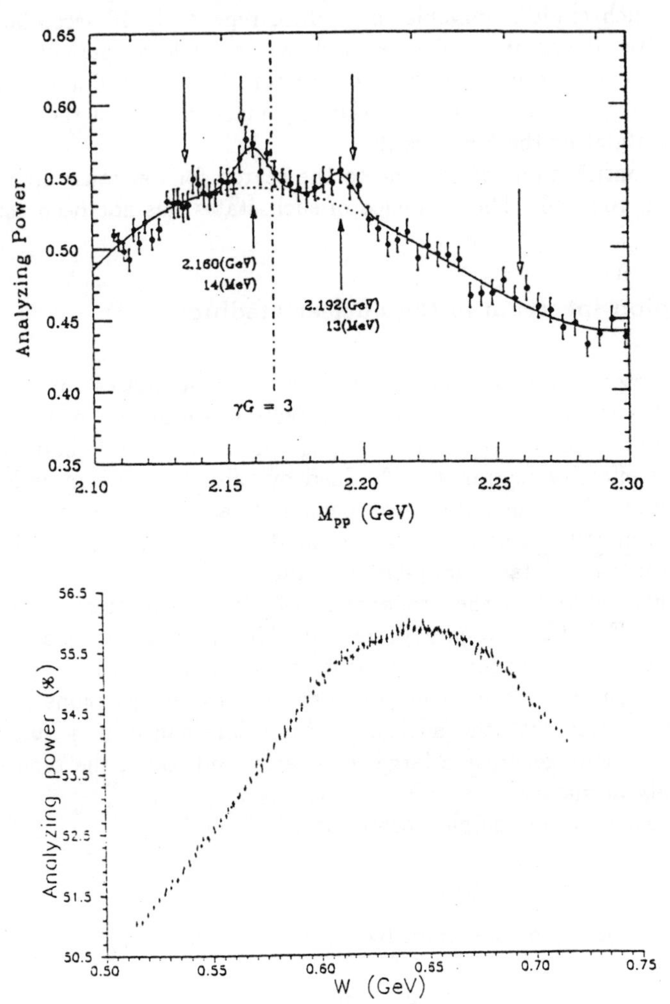

Figure 3: Proton-proton analyzing power excitation functions measured at KEK (upper figure, from Ref. 28) and at Saturne (lower figure, from Ref. 29).

The proton-proton elastic analyzing power data displayed in the upper figure were taken at KEK [28]. They exhibit narrow structures in the two-proton invariant mass distribution of the analyzing power, possibly indicating dibaryon states of low mass, compatible with the MIT Bag Model predictions [26]. The experiment was repeated at Saturne with high accuracy, using a rotating energy degrader which made it possible to produce repeatedly 16 secondary beams slightly shifted in energy [29]. The result shown in the lower part of Fig. 3 indicates that the analyzing power has no structure between 510 and 725 MeV of excitation energy (covering the invariant pp mass interval of the KEK data), thereby invalidating the KEK result.

Other contributions of Saturne to the search for low-mass dibaryons are discussed in Ref. 16. The existence of such states has not been established until now.

3 The pion spectrum in the nuclear medium

The parameters (masses, coupling constants, cutoff masses) of the effective field theories introduced in the previous section are determined by reproducing data on simple processes in free space (decay rates, two-body scattering data). The effective Lagrangian obtained by this procedure provides a theoretical framework to calculate the properties of hadrons in the nuclear medium and more generally to understand nuclear dynamics. These calculations are performed using standard many-body technics [30,31].

Of much interest is the propagation of pions in matter, as emphasized originally by Migdal [32] twenty years ago. The pion is the quantum of the nuclear field and its properties are strongly modified when it propagates in a nuclear environment rather than in free space. Low-energy pions propagating in matter interact with the nuclear medium dominantly in s- and p-waves. These interactions generate a large self-energy and soften the pion spectrum considerably in matter [4].

The free space inverse pion propagator is given by

$$D_0^\pi(\omega, \vec{q})^{-1} = \omega^2 - \vec{q}^{\,2} - m_\pi^2 \tag{5}$$

and the free space pion spectrum by the zeroes of $D_0^\pi(\omega, \vec{q})^{-1}$, i.e.

$$\omega^2 = m_\pi^2 + \vec{q}^{\,2}. \tag{6}$$

The pion mass is defined as the pion energy at vanishing 3-momentum,

$$\omega^2 = m_\pi^2. \tag{7}$$

In nuclear matter, the inverse pion propagator reflects the interaction of the pion with the medium. At baryon density ρ, it can be written as

$$D^{\pi}(\omega,\vec{q},\rho)^{-1} = \omega^2 - \vec{q}^{\,2} - m_\pi^2 - \Sigma^{\pi}(\omega,\vec{q},\rho), \tag{8}$$

in which $\Sigma^{\pi}(\omega,\vec{q},\rho)$ is the in-medium pion self-energy at density ρ. The in-medium pion spectrum obtained from $D^{\pi}(\omega,\vec{q},\rho)^{-1} = 0$ is now

$$\omega^2 = m_\pi^2 + \vec{q}^{\,2} + \Sigma^{\pi}(\omega,\vec{q},\rho) \tag{9}$$

and the pion mass is defined by the self-consistent equation

$$m_\pi^2(\rho) = m_\pi^2 + \Re e\,\Sigma^{\pi}[\omega = m_\pi(\rho), \vec{0}, \rho]. \tag{10}$$

The main contributions to the propagation of a low-energy pion in nuclear matter are illustrated in Fig. 4. The upper figure represents the s-wave pion-nuclear potential. It can be very accurately determined from low-energy theorems based on chiral symmetry [33]. The lower figure shows the p-wave contributions due to the coupling of pions to nucleon-hole (left-hand graph) and to Δ-hole (right-hand graph) states. The equations associated with these graphs are discussed for example in Ref. 4. The coupling of pions to nucleon-hole and to Δ-hole states in matter leads to the spectrum of 'pion-like' excitations [4] shown in Fig. 5. The spectrum consists of three branches, the nucleon-hole continuum, the pion branch and the Δ-hole states. The quantum mixing of pions and Δ-hole states leads to the softening of the pion mode. This means that for a given value of its 3-momentum, a pion has less energy in nuclear matter than in free space. This effect is displayed in Fig. 5. A large program was devoted to the experimental observation of the pion mode in matter at Saturne.

The pion mode in nuclei can be best studied through charge-exchange reactions at relativistic energies (E/A $\simeq$ 1 GeV) [5]. Charge-exchange reactions are processes in which a proton is changed into a neutron or a neutron into a proton. The simplest cases are the (p,n) and the (n,p) reactions. The charge-exchange reactions studied at Saturne were induced by relativistic light or heavy ions. They were mainly the $(^3He, t)$, the $(\vec{d}, 2p)$ and the $(^{12}C, {}^{12}N)$ reactions. Some processes induced by heavier nuclei were also investigated [34].

Charge-exchange reactions induced by light and heavy ions are rather peripheral processes. At small impact parameters, these reactions are strongly suppressed because the incident ion has a large probability to break into nuclear fragments [35]. The incoherent processes contributing to the cross section probe mostly the nuclear surface. Coherent processes, involving collective excitations of valence nucleons, are expected to be the most sensitive to the pion

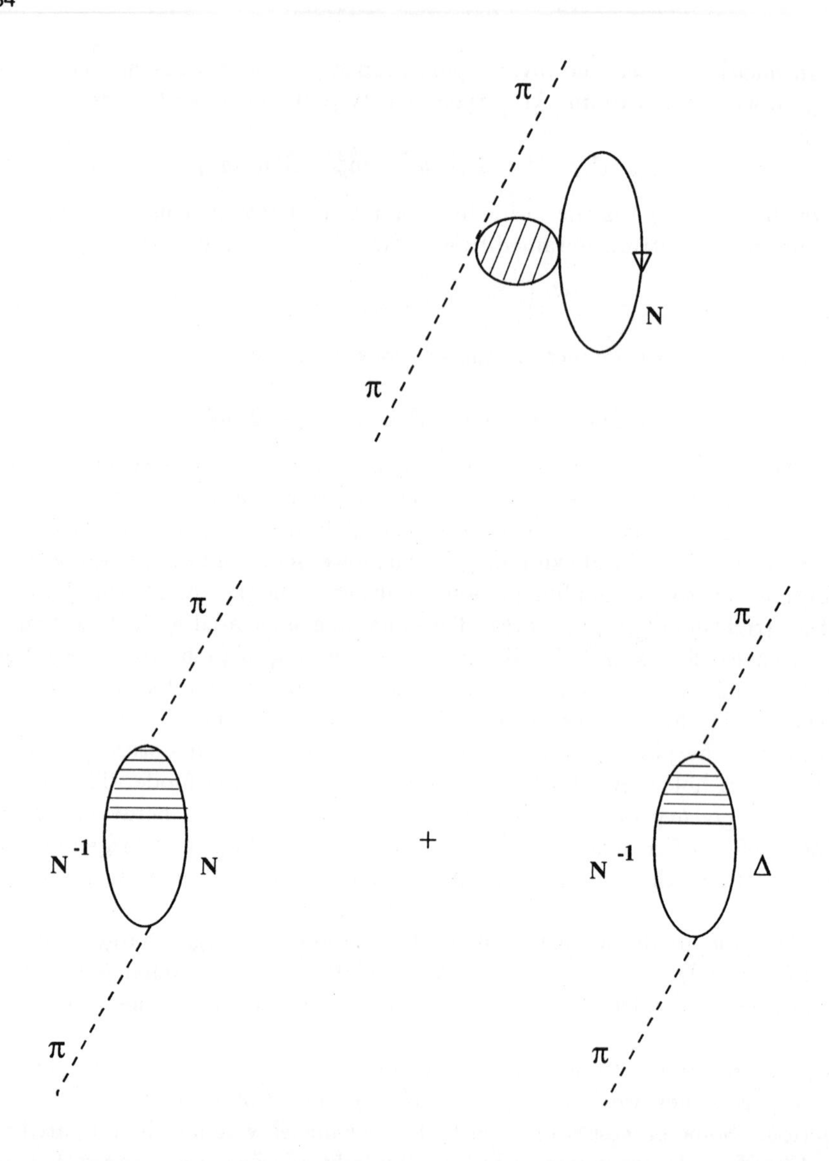

Figure 4: Contributions to the propagation of a low-energy pion in nuclear matter. The upper figure represents the s-wave pion-nuclear potential. The lower figure shows the p-wave contributions due to the coupling of pions to nucleon-hole (left-hand graph) and to Δ-hole (right-hand graph) states. The horizontally hatched areas in the lower figures indicate that the nucleon-hole and Δ-hole loops are iterated to generate coherent multiple scattering.

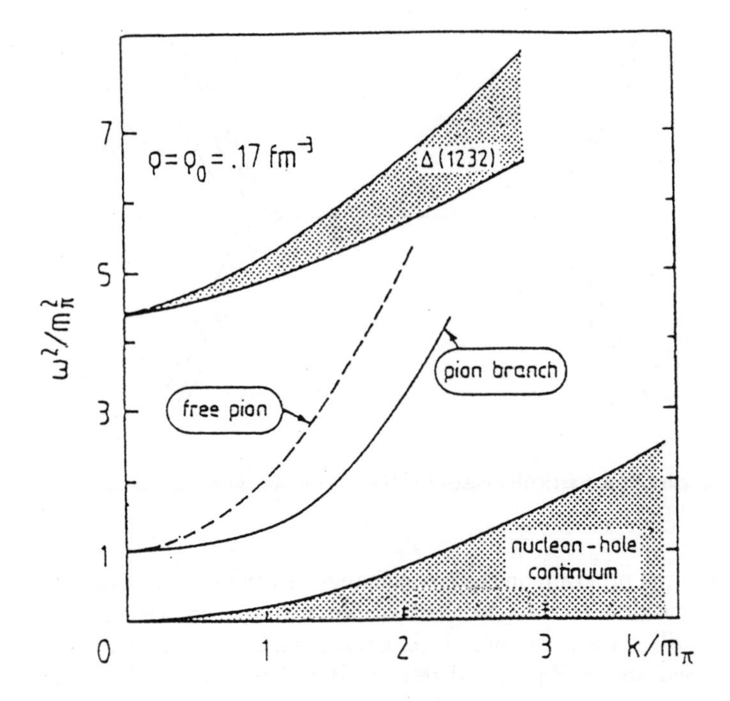

Figure 5: Pion-like spectrum in nuclear matter (from Ref. 4).

branch in nuclei. The baryon densities probed by charge-exchange reactions are in any case smaller than the saturation density of nuclear matter ($\rho_0 = 0.17\ \mathrm{fm}^{-3}$).

Charge-exchange reactions are useful tools to study the propagation of pions in the nuclear medium because they act as sources of virtual pions. This is depicted in Fig. 6. for the $(^3He, t)$ reaction. The transition from the projectile (^3He) to the ejectile (t) goes through an isovector meson, i.e. a pion or a ρ-meson. At Saturne energies ($E_{^3He}$=2 GeV), the pion-exchange dominates largely this transition. The virtual pion emitted is absorbed in the nuclear target which is excited and decays for example into a nuclear fragment, a proton and a pion as in the case illustrated in Fig. 6. The virtuality of the π^+ makes it possible to move in the (ω, k) plane of Fig. 5 by changing the kinematics of the reaction. The incident energy and the angular range studied at Saturne favor energy transfers ω around 200-300 MeV, i.e. emphasize effects associated with the Δ-hole states.

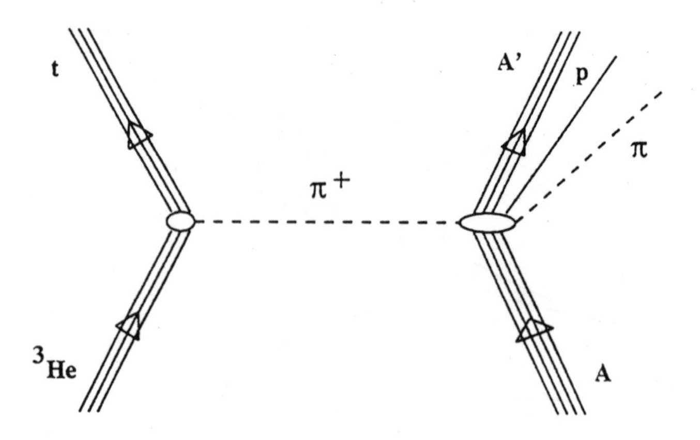

Figure 6: A particular case of (^3He, t) reaction on a nuclear target.

The Saturne program on charge-exchange reactions, aiming at determining the pion dispersion relation in the nuclear medium, proceeded in two main steps. We will state here some important results in view of the theoretical issue discussed above [Eq. 8] and refer to Ref. 5 for an actual presentation of the data and a complete review of the program.

First, a broad sample of inclusive data, where only the ejectile was detected (i.e. the triton in the (^3He, t) reaction), was taken. These experiments provided missing energy spectra. The physics of pion propagation in nuclei is revealed in the comparison of spectra taken on proton and on nuclear targets. On protons and for $E_{^3He}$=2 GeV, the main process is the excitation of the Δ^{++} resonance, which decays into a proton and a π^+. This process is represented in Fig. 7. The corresponding missing energy spectrum reflects the shape of the Δ resonance. The comparison of missing energy spectra for the (^3He, t) reaction on the proton and on nuclear targets in the Δ region ($\omega \simeq$200-300 MeV) shows that the 'Δ-peak' is shifted to a lower energy transfer in nuclei (independent of the mass number of the target) [36]. This energy shift is of the order of 70 MeV. Half of the effect is expected from target form factors and Fermi motion [37]. The rest of the energy shift should reflect many-body effects.

To understand the nature of these many-body effects, more exclusive experiments were undertaken, in which the decay products of the target in the final state, i.e. after the absorption of a virtual π^+, were measured. Two important phenomena were uncovered by these data. First, a large fraction of the dynamical shift of the 'Δ-peak' could be ascribed to pion absorption by

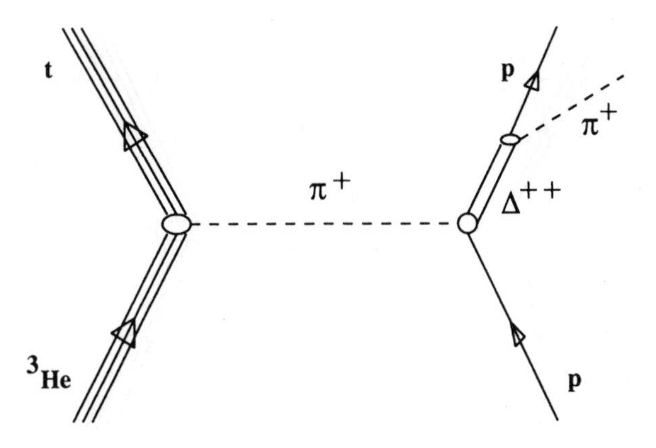

Figure 7: The (^3He, t) reaction on a proton target at $E_{^3He}$=2 GeV.

two and three nucleons [38]. This result has major consequences for the dynamics of pions in relativistic heavy ion collisions, where densities of the order of 2-3 ρ_0 can be reached. Two- and three-nucleon absorption effects go indeed like ρ^2 and ρ^3 respectively and become therefore very large at high baryon densities. They are not included in Boltzmann-like transport equations and would require relativistic transport theories, involving the quantum fluctuations of meson fields [39]. A second major result, linked directly to the issue of in-medium pion propagation, is the observation of a large number of events in which the virtual π^+ emitted in the (^3He, t) reaction scatters on the nuclear target and is reemitted on his mass shell while the target remains in the ground state [40]. This coherent pion production process is depicted in Fig. 8. The large cross section of this process indicates the collective nature of the pion branch which is singled out by selecting coherent events. The softening of the pion mode is reflected in the position of the peak in the missing energy spectrum, which is around $\omega \simeq 235$ MeV [40]. The analysis of the latest data on coherent pion production is still in progress [5].

The study of the in-medium propagation of Goldstone bosons is in general a topic of extensive experimental investigations. Comparisons of K^+ and K^- production in nucleon-nucleon, nucleon-nucleus and nucleus-nucleus collisions indicate that the production of K^-'s is very much enhanced in nucleus-nucleus collisions [41]. This suggests that their self-energy in nuclear matter could be very large. The production of η-mesons in nucleon-nucleon and nucleon-nucleus collisions has also been investigated at Saturne [42].

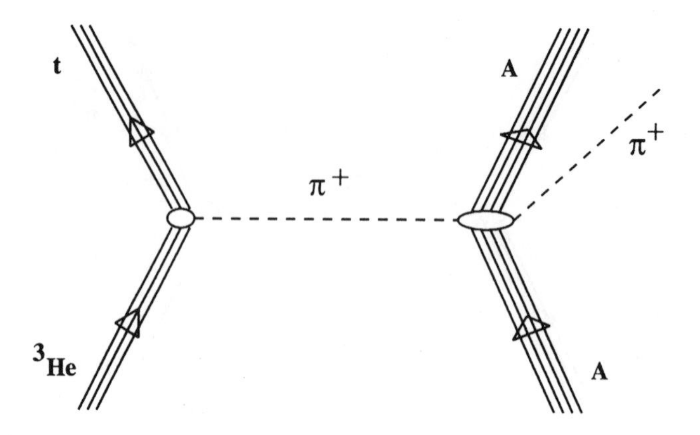

Figure 8: Coherent pion production in the (^3He, t) reaction on a nuclear target.

4 Chiral symmetry and hadron properties in nuclear matter

Quantum Chromodynamics is a chirally invariant theory in the limit where the fermion fields, the quarks, are massless. This invariance means that, in that particular limit, the helicity of left- and righthanded quarks is separately conserved by QCD dynamics. The finite masses of the quarks induce an explicit breaking of chiral symmetry. The u- and d-quark masses are very small (5-10 MeV). To a very good approximation, strong interactions are therefore chirally invariant in this sector. The strange quark mass is larger ($m_s \simeq 150$ MeV), but still small compared to other mass scales [43]. In the strange sector, chiral symmetry appears then as a good starting assumption but deviations from the chiral limit must be investigated.

Chiral symmetry is spontaneously broken. This implies that the symmetry does not manifest itself by parity multiplets in hadron spectra but rather by the presence of low-energy excitations, the Goldstone bosons. They are the pions in the u-,d-quark sector and the pions, the kaons and the η-meson in the u-,d-,s-quark sector. In the chiral limit (m_q=0), the Golstone bosons would be massless. Their finite mass is generated by the explicit breaking of chiral symmetry. As a consequence of the spontaneous breaking of chiral symmetry, the QCD vacuum has a complicated structure. It is characterized by a finite expectation value of $\langle \bar{q}q \rangle$, the quark condensate, estimated to be [43]

$$| \langle \bar{u}u \rangle + \langle \bar{d}d \rangle | \simeq 3 fm^{-3}. \tag{11}$$

Because the pions are the quanta of the nuclear field, it is expected that nuclear dynamics will be closely related to chiral symmetry and therefore to the structure of the QCD vacuum[44].

To investigate the relation between nuclear dynamics and chiral symmetry, it is important to understand first how the structure of the QCD vacuum, in particular the value of the quark condensate, is modified by the presence of nucleons.

A simpler problem is to study the structure of the QCD vacuum around a single static nucleon. Such situation has been discussed using lattice QCD methods [45]. The behavior of the quark condensate obtained in a numerical simulation on a hypercubic lattice with 8 points in the space direction and 4 points in the time direction using Wilson's action and Kogut-Susskind fermions[45] is displayed in Fig. 9. If $\langle\bar{q}q\rangle$ is interpreted as an order parameter of chiral symmetry breaking, we see from Fig. 9 that chiral symmetry is locally restored in the vicinity of the static color sources[45].

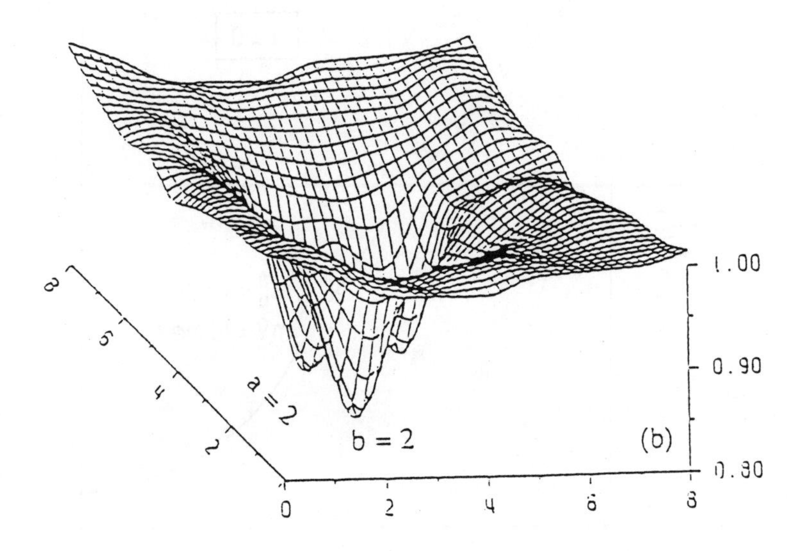

Figure 9: The quark condensate in the plane of a static three-quark system arranged in a rectangle of lengths a=$|\vec{r}_1 - \vec{r}_2|$ and b=$|\vec{r}_3 - \vec{r}_1|$, $\vec{r}_1$, $\vec{r}_2$ and $\vec{r}_3$ being the quark coordinates (from Ref. 45). The normalization is defined by Eq. (3.4) of Ref. 45.

Let us consider now the structure of the QCD vacuum in the presence of nuclear matter. We expect that the local chiral symmetry restoration in the vicinity of quarks exhibited in Fig. 9 could now become a global property of the vacuum at sufficiently high density. Lattice calculations are not available for finite baryon chemical potentials, so that the behavior of $\langle\bar{q}q\rangle$ in matter can but be studied with models. We show in Fig. 10 the prediction of a Nambu-Jona-Lasinio calculation [46]. The value of the quark condensate decreases very fast with increasing density. This decrease is linear until $\rho \simeq 2\rho_0$. This linear behavior has been shown to be a prediction of the Hellmann-Feynman theorem, and thus a model-independent result, to first order in the baryon density [47]. At nuclear matter density, the value of $\langle\bar{q}q\rangle$ has decreased by about 25% compared to its vacuum value.

How could such effect be observed experimentally?

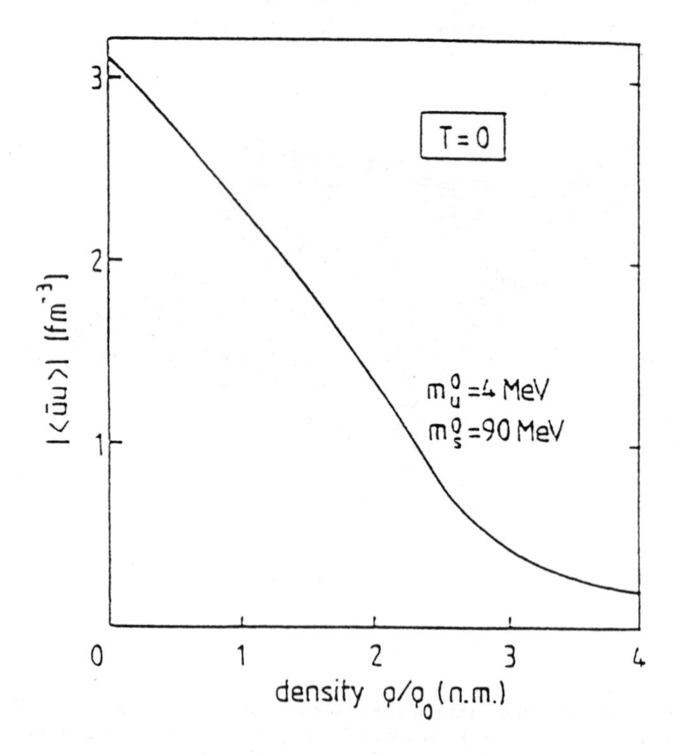

Figure 10: The quark condensate as function of baryon density in the Nambu-Jona-Lasinio model of strong interactions (from Ref. 46).

An appealing idea is that the in-medium mass of hadrons could be closely linked to the value of the quark condensate in matter [48].

In free space, such connection has been established long ago through QCD sum rules [6]. The general assumption of this approach is that the spectral properties of hadrons are determined by a small number of parameters characterizing the structure of the QCD vacuum (i.e. nonperturbative, long-distance physics) and do not involve explicitly the full complexity of the theory. To lowest order in the quark and gluon fields, these parameters are the quark condensate $\langle \bar{q}q \rangle$ and the gluon condensate $\langle G^a_{\mu\nu} G^{\mu\nu}_a \rangle$.

Let us consider the ρ-meson. The contribution of the gluon condensate to its mass being small [6], the ρ mass is largely given by the value of the quark condensate. We have [49]

$$M_\rho \simeq c \mid \langle \bar{q}q \rangle \mid^{1/3}, \tag{12}$$

in which c is a constant of the order of 3. This result is based however on a very simple expression for the ρ spectral function [49]. In general, the hadron masses predicted in the QCD sum rule approach are within 10-20 % of their experimental values [6]. The reliability of QCD sum rules depends strongly on the knowledge of spectral functions and on a precise determination of the values of the vacuum condensates.

QCD sum rule technics have been extended to nuclear matter [50].

This extension appeared particularly interesting in view of the suggestion made by Brown and Rho [48] using effective chiral Lagrangians that vector meson masses should scale like the cubic root of the quark condensate, i.e.

$$\frac{M_\rho(\rho)}{M_\rho(0)} = \frac{M_\omega(\rho)}{M_\omega(0)} = \left[\frac{\langle \bar{q}q \rangle_\rho}{\langle \bar{q}q \rangle_0}\right]^{1/3}, \tag{13}$$

where the index 0 refers to vacuum values and the ρ-dependence to quantities calculated in nuclear matter at density ρ. This scaling law implies that vector meson masses would be reduced by about 100 MeV at nuclear matter density ρ_0. QCD sum rules in matter support this relation [51].

There are however many more uncertainties involved in QCD sum rules in matter than in free space [52]. One problem is the estimate of the 4-quark condensate which plays an important role in the calculation of the mass of light quark hadrons. A second issue is the phenomenological knowledge of spectral functions in matter, which are much more uncertain there than in free space. Much work is in progress on these questions.

An other important point, also very much under discussion, is the relation between the reduction of the quark condensate at high baryon density (Fig. 10) and the restoration of chiral symmetry in matter [53]. Low-momentum pions appear to contribute to the value of the quark condensate in the nuclear medium and the modification of this contribution with increasing density should not be associated with the restoration of chiral symmetry [53]. In this perspective, the quark condensate in the nuclear medium (or vector meson masses if we accept Eq. (13)) would be no simple order parameter for chiral symmetry restoration at finite baryon density.

The experimental information on the mass of hadrons in the medium is at present quite limited but there are interesting projects for the near future. A recent experiment at GSI has led to the discovery of deeply bound π^- states in ^{207}Pb [54] and provided very detailed information on the s-wave π-nuclear potential. It is rather straightforward to derive from these data the value of the effective pion mass in (isospin symmetic) nuclear matter [33]. The effective pion mass is defined by Eq. (10). The result is $m_\pi(\rho_0) = 1.07\, m_\pi(0)$, in agreement with results from models based on chiral symmetry which predict that the pion mass is protected from in-medium changes by the symmetry [46,55]. An extensive program to study the in-medium masses (or self-energies) of vector mesons in nuclei and in heavy ion collisions through their e^+e^- decay is planned at GSI with the HADES spectrometer [56]. Very interesting developments are expected from this project in the coming years.

Acknowledgements

The author had a wonderful time during the nine years she spent at Saturne. She is very much indebted to all the Saturne users who dropped by her office with all sorts of physics questions, who submitted their ideas for new experiments to her critical comments and who discussed many preliminary data with her. She also acknowledges most rewarding personal relations with the accelerator staff and the technical groups of the Laboratory.

References

1. B.W. Lee, Chiral Dynamics (Gordon and Breach Science, New York, 1972).
2. F.E. Close, An. introduction to quarks and partons (Academic Press, London, 1979).
3. M. Gell-Mann and Y. Ne'eman, The eightfold way (Benjamin, New York, 1964).
4. T. Ericson and W. Weise, Pions and Nuclei (Clarendon Press, Oxford, 1988).
5. M. Roy-Stephan, Pion and delta dynamics, Contribution to these Proceedings.
6. M.A. Shifman, A.I. Vainshtein and V.I. Zakharov, *Nucl. Phys.* B **147**, 385, 448 (1979).
7. R. Machleidt, *Adv. Nucl. Phys.* **19**, 189 (1989) and references therein.
8. G. Janssen, K. Holinde and J. Speth, *Phys. Rev. Lett.* **73**, 1332 (1994).
9. C. Elster *et al.*, *Phys. Rev.* C **38**, 1828 (1988).
10. R. Machleidt, F. Sammarruca and Y. Song, *Phys. Rev.* C **53**, R1483 (1996).
11. V.G.J. Stoks *et al.*, *Phys. Rev.* C **49**, 2950 (1994).
12. R. B. Wiringa *et al.*, *Phys. Rev.* C **51**, 38 (1995).
13. J. Rahm *et al.*, *Phys. Rev.* C **57**, 1077 (1998).
14. A.W. Thomas and K. Holinde, *Phys. Rev. Lett.* **63**, 2025 (1989).
15. K.F. Liu *et al.*, *Phys. Rev. Lett.* **74**, 2172 (1995).
16. C. Lechanoine-Leluc and F. Lehar, The nucleon-nucleon interaction, Contribution to these Proceedings.
17. F. Sammarruca and T. Mizutani, *Phys. Rev.* C **41**, 2286 (1990).
18. F. Sammarruca, *Phys. Rev.* C **50**, 652 (1994).
19. F. Sammarruca, *Phys. Rev.* C **51**, 2823 (1995).
20. R. Machleidt, private communication.
21. J. Dubach, W.M. Kloet and R.R. Silbar, *Nucl. Phys.* A **466**, 573 (1987).
22. G.P. Ramsey and D. Sivers, *Phys. Rev.* D **45**, 79 (1992).
23. G.P. Ramsey and D. Sivers, *Phys. Rev.* D **47**, 93 (1993).
24. A. Chodos *et al.*, *Phys. Rev.* D **9**, 3471 (1974).
25. R. L. Jaffe, *Phys. Rev. Lett.* **38**, 195 (1977).
26. P.J.G. Mulders, A.T.M. Aerts and J.J. de Swart, *Phys. Rev. Lett.* **40**, 1543 (1978).
27. A.S. Carroll *et al.*, *Phys. Rev. Lett.* **41**, 777 (1978); H. Ejiri *et al.*, *Phys. Lett.* B **228**, 24 (1989); B.A. Shahbazian *et al.*, *Phys. Lett.* B **235**, 208 (1990); K. Imai, *Nucl. Phys.* A **547**, 199_c (1992); A. Rusek *et al.*, *Phys.*

Rev. C **52**, 1580 (1995); J.K. Ahn *et al.*, *Phys. Lett.* B **378**, 53 (1996).
28. H. Shimizu *et al.*, *Phys. Rev.* C **42**, R483 (1990).
29. R. Beurtey *et al.*, *Phys. Lett.* B **293**, 27 (1992).
30. A.L. Fetter and J.D. Walecka, Quantum theory of many-particle systems (McGraw-Hill Book Company, New-York, 1971).
31. C. Itzykson and J.-B. Zuber, Quantum field theory (McGraw-Hill Book Company, New-York, 1980).
32. A. Migdal, *Rev. Mod. Phys.* **50**, 107 (1978).
33. T. Waas, R. Brockmann and W. Weise, *Phys. Lett.* B **405**, 215 (1997).
34. M. Roy-Stephan, *Nucl. Phys.* A **488**, 187_c (1988).
35. C. Guet *et al.*, *Nucl. Phys.* A **494**, 558 (1989).
36. A. Brockstedt *et al.*, *Nucl. Phys.* A **530**, 571 (1991).
37. J. Delorme and P.A.M. Guichon, *Phys. Lett.* B **263**, 157 (1991).
38. T. Hennino *et al.*, *Phys. Lett.* B **283**, 42 (1992).
39. P. Siemens *et al.*, *Phys. Rev.* C **40**, 2641 (1989).
40. T. Hennino *et al.*, *Phys. Lett.* B **303**, 236 (1993).
41. R. Barth *et al.*, *Phys. Rev. Lett.* **78**, 4007 (1997).
42. G. Dellacasa, η-meson and baryonic resonance physics, Contribution to these Proceedings.
43. W. Weise, in the Proceedings of the International School of Heavy Ion Physics, 6-16 October 93 (R.A. Broglia, P. Kienle and P.F. Bortignon Editors, World Scientific, 1994).
44. M.A. Nowack, M. Rho and I. Zahed, Chiral nuclear dynamics (World Scientific Publishing Co., Singapore, 1996) and references therein.
45. M. Müller et al, *Nucl. Phys.* B **335**, 502 (1990).
46. M. Lutz, S. Klimt and W. Weise, *Nucl. Phys.* A **542**, 521 (1992).
47. T.D. Cohen, R.J. Furnstahl and D.K. Griegel, *Phys. Rev.* C **45**, 1881 (1992).
48. G.E. Brown and M. Rho, *Phys. Rev. Lett.* **66**, 2720 (1991).
49. N.V. Krasnikov, A.A. Pivovarov and N.N. Tavkhelidze, *Zeit. Phys.* C **19**, 301 (1983).
50. E.G. Drukarev and E.M. Levin, *Nucl. Phys.* A **511**, 679 (1990); *Nucl. Phys.* A **516**, 715(E) (1990); *Nucl. Phys.* A **532**, 695 (1991).
51. T. Hatsuda and S.H. Lee, *Phys. Rev.* C **46**, R34 (1992).
52. T.D. Cohen *et al.*, *Prog. Part. Nucl. Phys.* **35**, 221 (1995).
53. M.C. Birse, *Phys. Rev.* C **53**, R2048 (1996).
54. T. Yamazaki *et al.*, *Zeit. Phys.* A **355**, 219 (1996).
55. H. Yabu, S. Nakamura and K. Kubodera, *Phys. Lett.* B **317**, 269 (1993).
56. HADES, Proposal for a High-Acceptance Di-Electron Spectrometer, GSI, 1994.

HYPERION: THE SATURNE II POLARIZED PROTON AND DEUTERON SOURCE

P-Y. BEAUVAIS
Commissariat à l'Energie Atomique Saclay, DSM/DAPNIA/SEA bat. 706, 91191 Gif sur Yvette Cedex, France

HYPERION is one of the three particle preinjectors used at Saturne. Built in 1980, this source located on a high-voltage platform (400 kV), produced during 17 years proton and deuteron polarized beams with a very high reliability. Its performances were constantly improved and allowed to accelerate up to $3*10^{11}$ deuteron and $2.5*10^{11}$ proton per machine cycle with a polarization better than 90%. In this paper, we present briefly the main characteristics of the most prevailing atomic beam sources and ionizers, then we describe in detail Hyperion and the developments we carried out to achieve its current performances.

HYPERION: THE SATURNE II POLARIZED PROTON AND DEUTERON SOURCE

HYPERION est l'un des trois pré injecteurs de particules utilisés à Saturne. Construite en 1980, cette source placée sur une plate-forme haute tension (400 kV), a produit pendant 17 années des faisceaux de protons et de deutons polarisès avec une très grande fiabilité. Ses performances furent constamment améliorées et ont permis d'accélérer jusqu'à $3*10^{11}$ deutons et $2.5*10^{11}$ protons par cycle machine avec une polarisation meilleure que 90%. Dans cet article, nous présentons brièvement les différents principes de source atomique et d'ioniseur les plus courammen utilisés puis nous décrivons en détail Hypérion et les évolutions qui ont permis d'atteindre les performances actuelles.

1 Atomic beam sources and ionizers

1.1 Atomic beam type sources (ABS)

So as to produce an high-density nuclear-polarized atomic beam, a typical atomic source employs three separate systems which have completely different functions: a dissociator system, a sextupole magnet system and generally two or three radio-frequency transition systems.
Dissociator system — H_2 or D_2 gas flowing through a pyrex pipe is dissociated by a radio-frequency discharge the power of which depending of the gas flow rate (usually a few hundreds of watts). The H or D atoms, passing through a refrigerated nozzle and a series of vacuum chambers separated by skimmers, form a highly directed supersonic-jet beam.

Sextupole magnet system — The radially inhomogeneous fields produced by sextupole magnets act as lenses focusing the electron-spin-up atoms and defocusing the opposite-state ones. Sextupole magnets were originally often electromagnets, nowadays multi-element permanent magnet sextupoles become common
Radio-frequency transition system — The combination of a dissociator and one or several sextupole magnets produces an electron-spin polarized atomic beam. Since nuclear rather than electron-spin polarization is needed, atomic transitions between selected hyperfine substates are performed by passing through one to three cavities combining static magnetic fields with radio-frequency electromagnetic ones.

1.2 Optically pumped type source (OPPIS)

The development of optically pumped ion sources is relatively young. The principle of the spin-exchange optically pumped polarized hydrogen or deuterium gas source is based on the electron-spin polarization transfer of alkali atoms - (mainly Na, K or Rb) - to the hydrogen or deuterium atoms via a spin-exchange collision. The electron-spin polarization is then transferred to the nuclear-spin polarization either by hyperfine interaction in a weak magnetic field or by inducing RF transitions between hyperfine sublevels in a strong magnetic field.

1.3 Electron Beam Ionizers (EB)

In an EB ionizer, atoms ionization is obtained by impact, in a strong longitudinal magnetic field, between an electron beam (usually 1 to 3 keV) and the atomic nuclear-spin polarized beam exiting the r.f. transition cavities. The production ratio is low (0.1% to 1%) depending among other things on electron and atom densities and on the interaction region length. The polarization is very well conserved (typically 95%).

1.4 Electron Cyclotron Resonance Ionizers (ECR)

In an ECR ionizer, two solenoid coils produce a magnetic field acting as longitudinal mirror and providing the right resonance field inside the plasma chamber. Permanent hexapole surrounding this chamber confines radially the plasma. Electrons of the r.f. plasma are heated by electron-cyclotron-resonance. Ions are produced by collision between these electrons and the atomic beam. The ionization efficiency of ECR ionizers is higher than for EB ionizers (more than 1%) but the ion beam polarization does not exceed 75 to 80%.

2 The Saturne polarized beam ion source: HYPERION

2.1 The atomic beam source

The Hyperion ABS uses the main part of the classical principles described above. An auto-oscillator, magnetically coupled by a six-turn coil surrounding a bulged glass pipe, produces a r.f. discharge (1 to 2 kW at 15 MHz) in the H_2 or D_2 gas flowing inside. Both gas and r.f. power are pulsed with a 1% maximum duty factor. The atoms enter the first vacuum chamber passing through a 80 K refrigerated aluminum nozzle (3.2 mm exit aperture). The nozzle is cooled by conduction to the cold head of a closed-cycle helium refrigerator. In order to minimize atomic recombination, this nozzle is internally coated with a silicone dry film. 20 mm downstream, a 4.5 mm aperture tantalum skimmer links the first and the second chamber thus matching the supersonic-jet beam to the sextupole magnet. Each vacuum chamber is pumped by a 1150 l/s turbomolecular pump assisted by a booster. The sextupole electromagnet is 50 cm long. Its aperture radius increases from 7.4 mm up to 14 mm along the first half part, remaining constant all along the second part. The electron-spin polarized atomic beam is thus focussed 70cm downstream just at the ionizer entrance. The r.f. tansitions are located in the last chamber. Proton nuclear polarization is obtained using a 10.8 MHz weak field transition unit working at 10.7 Gauss (1-2 transition) and a 1430 MHz intermediate field unit tuned at 60 Gauss (2-4 transition). Vector and tensor deuteron polarization is obtained using the combination of a 16.2 MHz weak field transition unit working at 10.7 Gauss and two intermediate field transition units (343 MHz at 37 Gauss and 415 MHz at 60 Gauss). The right static magnetic field along the axis is produced by permanent magnets and a tilted magnetic iron circuit.

2.2 The ionizer

The Hyperion ionizer is a strong field reflex ionizer bought to the A.N.A.C. Company and later widely modified. Electrons are extracted from a direct-heated Ω-shaped tantalum ribbon filament. This filament (0.3 mm thickness) attenuates only a small fraction (about 5%) of the beam entering the interaction region. The electron beam is confined and reflected by the combination of a strong longitudinal magnetic field with potentials created by two electrodes located both sides of the ionization region (defining thus the quantization axis). The electron equivalent current inside the ionization electrode is estimated to be about 2 amperes. The magnetic field is generated by six solenoid coils surrounding the vacuum chamber. These coils are separately adjusted with respect to the electrode potentials so as to maximize the extracted beam current.

2.3 The low energy beam transport line (LEBT)

The 20keV energy ion beam extracted from the ionizer is transported through the LEBT up to the 400kV accelerator tube which links the high-voltage terminal to ground level transport line. First, the beam is guided and focussed up to the electrostatic mirror, which turn it so as to make radial the nuclear magnetic momentum. Then a spin-rotator solenoid coil makes the spin to become vertical in order to avoid a depolarization during the acceleration by the synchrotron. Finally, a set of Einzel lenses matches the ion beam to the low gradient accelerator tube.

3 Hyperion performance evolutions

3.1 First tests

The first tests of Hyperion were performed during July 1980. The A.N.A.C. ionizer was considered to be able to produce $40\mu A$ with a DC mode atomic source and up to $100\mu A$ with a pulsed mode one. So we were very disappointed, measuring only $0.8\mu A$ on the Faraday cup. Few days later, $3.1\mu A$ were obtained but we realized soon that the stainless steel ionization electrode (E3) was melted so as we were obliged to stop for repair. The same electrode melted again few months later and it is only after a precise magnetic axis tuning that we reached a current of about $20\mu A$.

3.2 first beam in Saturne

During March and April 1981, replacing the spherical deflector by an electrostatic mirror and the stainless steel ionization electrode by a tantalum one, we obtained $35\mu A$ of H^+ allowing us to accelerate 2×10^8 proton per cycle in Saturne.

3.3 Source improvements

Improvements carried out especially on the atomic source are numerous and various, particularly:
July 1981: mounting of a piezoelectric valve.
December 1981: new oscillator (auto-oscillator).
June 1982: dissociator nozzle cooled.
November 1982: new poles with larger aperture. ($195\mu A$ and $5 \times 10^9 H^+/cycle$)

1983 - 1985: improvements on glass pipe, on nozzle and skimmer shape, on pumping systems ($1.1 \times 10^{10} D^+/cycle$).
March 1986: new poles. ($340\mu A$ and $1.7 \times 10^{10} D^+/cycle$).
1987: new RF auto-oscillator. ($370\mu A$).
1990: tests at 30 Kelvin with nitrogen buffer gas.
1991 - 1992: new pole shape.

3.4 Ionizer improvements

Many decisive improvements have been carried out also on the ionizer, the main of which we can name:
1981 - 1982: mounting of new molybdenum E0 E1 and E2 electrodes.
1983: new E3 tantalum electrode.
1981 - 1988: optimization of filament shape and size.
1987: Pulsation of the ionizer electron beam.
1988 - 1989: new ionizer allowing a very precise mechanical and magnetic alignement. ($580\mu A$). New filament design with alumina support.
1991: new design of the E3 electrode and mounting of an additional E3' electrode.
1992 - 1993: Tests of ion storage by using of E2 and E3' electrodes located at both sides of E3 as potential traps ($2mA$ of peak current during a few tens of μsec).

4 Best performances achieved

We can distinguish two periods. Before 1987, the 200keV/A polarized ion beam was accelerated up to 5 MeV/A in a Drift Tube Linac (20 Mev DTL tuned in the $2\beta.\lambda$ mode) and then injected in the 3 GeV Saturne synchrotron . The overall yield was very low and the best results obtained were about $2 \times 10^{10} H^+$ or D^+ /cycle. Since 1987 up to 1997, the beam (400keV for H^+ and 200keV/A for D^+), was accelerated up to 47 MeV in a large acceptance synchrotron-injector (MIMAS) and then transferred without any losses in Saturne. With this new equipment, Intensities as high as $2.5 \times 10^{11} H^+$ /cycle and $3 \times 10^{11} D^+$ /cycle were reached.

5 Summary

Hyperion is one of the best polarized beam injector in the world. From 1981 until 1997, this injector provided widely more than the half of the beams extracted from Saturne. The association, from 1987, with the synchrotron injector MIMAS allowed to reach never equaled overall performances (beam current per cycle and polarization ratio). Up to 1993, large efforts were carried out in order to maximize the beam current delivered to the experimental areas. During the five last years, most of the beam requirements for physics evolving towards lower intensities, the main concern was to improve the reliability. Thus, the availability of this injector, even during the last year of operation, reached ratio about 95%.

SATURNE 2: POLARIZED PROTON BEAMS

Paul-André CHAMOUARD*, Jean-Michel LAGNIEL**
(*)CEA/DSM/UGP SATURNE, (**)CEA/DSM/DAPNIA/SEA

91 191 Gif sur Yvette CEDEX FRANCE

Abstract: The achievement of polarized ions in the synchrotrons requires very particular characteristics for these machines but also sophisticated tunings because resonant depolarizations occur during the cycle. Saturne is the first machine in the world for the polarized beams for several reasons:

- three possible projectiles: protons, deuterons, lithium 6,
- beam intensities: 2. 10^{11} protons by burst, 3.2 10^{11} deuterons and 1. 10^{9} Li 6 (3+),
- high polarization: 90% for protons up to 1 GeV, 80% at 2.95 GeV and more than 90% of the maximum possible for deuterons.

At 2.95 GeV the proton beam has crossed about 15 depolarizing resonances. We will describe the results for protons and the methods which are used to avoid or to cross the resonances. The hardware and software will be also described.
Résumé: L'obtention de faisceaux d'ions polarisés dans les synchrotrons nécessite des caractéristiques très particulières de ces machines, les réglages sont très délicats en raison des phénomènes résonants de dépolarisation qui interviennent pendant les différentes phases du cycle de l'accélérateur. Saturne 2 détient le record mondial au niveau des faisceaux polarisés à plusieurs titres :

- les types de projectiles disponibles (protons, deutons, Lithium 6),
- les intensités obtenues : 2. 10^{11} protons polarisés par cycle, 3.2 10^{11} deutons polarisés et 1. 10^{9} ions Li 6 (3+),
- la valeur des polarisations obtenues : typiquement pour les protons 90% à 1 GeV, 80% à 2.95 GeV, pour les deutons polarisés plus de 90% de la valeur maximale possible.

A 2.95 GeV le faisceau de protons a traversé environ 15 résonances dépolarisantes. Nous décrivons ici les résultats obtenus en protons : les méthodes utilisées pour traverser ou éviter les résonances, ainsi que les matériels et les moyens utilisés.

1 DEPOLARIZATION RESONANCES:

The theory of the mechanisms of depolarizing resonances has been developed by J.Buon [1]. We will describe the resonances met in MIMAS and SATURNE 2 and the different ways to keep the polarization.

The observed resonances for protons in MIMAS or SATURNE 2 are described by the following equations:

$\gamma * G = N_0$	(1)	with $\gamma = 1 + T/E_0$, T kinetic energy, E_0 rest energy
$\gamma * G = N_1 \pm \nu_x$	(2)	G gyromagnetic anomaly of proton: $G = 1.7928$
$\gamma * G = N_2 \pm \nu_z$	(3)	ν_x and ν_z, horizontal and vertical tunings
$\gamma * G = N_3 \pm \nu_x \pm \nu_z$	(4)	N_0 positive integer with $2 < N_0 < 7$, N_1, N_2 and N_3 integers.

(1) are called imperfection resonances and correspond to a vertical closed orbit distortion. (2) and (3) are called systematic resonances and are due to horizontal and vertical tunings. (4) are due to hexapoles exciting the betatron resonance of extraction in SATURNE 2 and hexapoles correcting the chromaticity at low energy in MIMAS.

Two situations have to be considered:

- γ varies quickly:

  This situation corresponds to SATURNE 2 acceleration ($d\gamma/dt \sim 5.4\, sec^{-1}$). The types of resonances are (1), (2) and (3).

  - Imperfection resonances (type (1)):

    These resonances are due to closed orbit distortion. Crossing is realized by the mean of dipole correctors (printed circuits set in the main quadrupoles). The orbit is corrected for $\gamma * G = 2$ (108 MeV). A fault is increased to help the spin flip with the dipole correctors for $\gamma * G = 3$ (632 MeV) and $\gamma * G = 4$ (1155 MeV). The $\gamma * G = 3$ can also be corrected (FIG.1)

    The correctors become inefficient for higher energies, and the spin flip is helped by a closed orbit distortion made by a deliberate misalignment (1 mm) of one main defocusing quadrupole. In these conditions, $\gamma * G = 5$ (1678 MeV), $\gamma * G = 6$ (2202 MeV) and $\gamma * G = 7$ (2725 MeV) are crossed with spin flip.

  - Systematic resonances (type (2) and (3)):

    The way to cross these lines has been optimized experimentally. Two very strong resonances occur because they correspond to an harmonic of the lattice of SATURNE 2 (4 super periods, $\nu_z \sim 3.6$): $\gamma * G = \nu_z$ ($\sim$ 925 MeV) and $\gamma * G = 8 - \nu_z$ ($\sim$ 1390 MeV). FIG.2 shows the evolution of the vertical tuning during the crossings of these 2 lines. This variation is made by the vertical quadrupoles power supply.

    The polarization is flipped on the 2 resonances. The crossing speed must be optimized with a very good accuracy. $\gamma * G = \nu_z$ ($\sim$

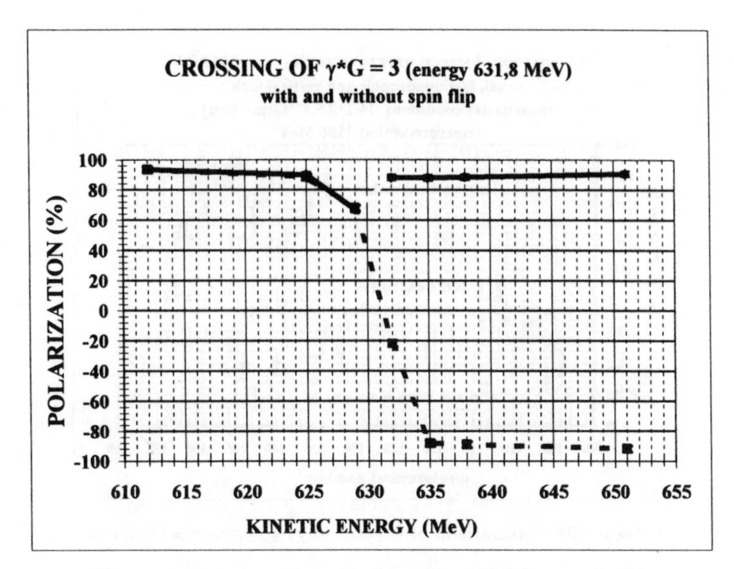

Figure 1: Crossing of $\gamma \star G = 3$ with and without spin flip.

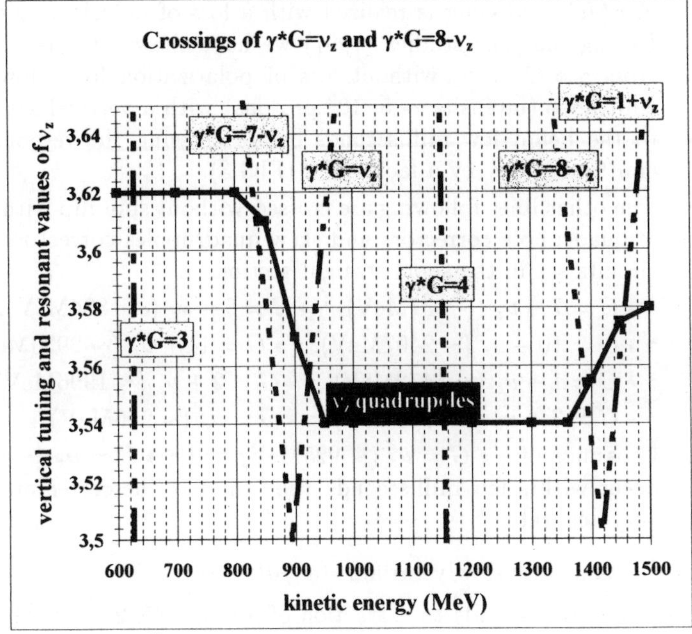

Figure 2: Crossing of $\gamma \star G = \nu_z$ and $\gamma \star G = 8 - \nu_z$.

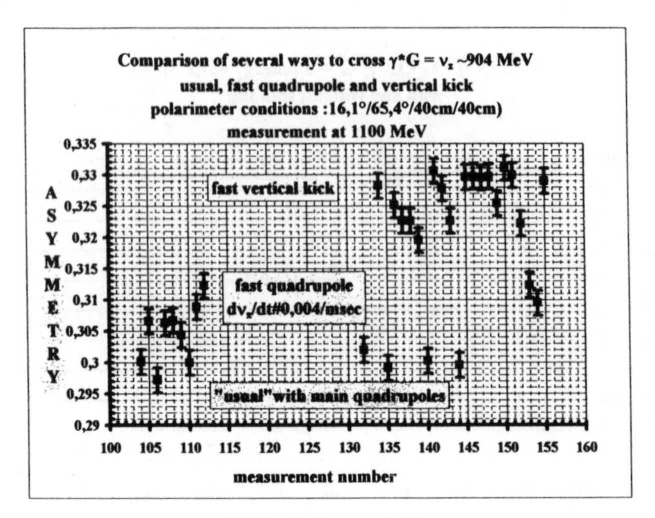

Figure 3: Comparison of several ways to cross $\gamma \star G = \nu_z$.

925 MeV) crossing is realized with a loss of polarization of $\sim 8\%$. During machine studies (1997) we have shown it was possible to cross $\gamma * G = \nu_z$ without loss of polarization for a low vertical emittance (intensity $\sim 5.\ 10^9$ protons) with a vertical kick applied to the beam few milliseconds before crossing the resonance: the emittance grew by a factor of 2 (FIG.3).

The other lines following are crossed without spin flip with the help of quadrupole correctors and skew quadrupoles correctors (printed circuits set in the short straight sections):
$\gamma * G = 7 - \nu_z$ ($\sim$ 837MeV), $\gamma * G = 9 - \nu_z$ ($\sim$ 1884MeV),
$\gamma * G = 10 - \nu_z$ ($\sim$ 2407MeV), $\gamma * G = 11 - \nu_z$ ($\sim$ 2930MeV),
$\gamma * G = 1 + \nu_z$ ($\sim$ 1472MeV), $\gamma * G = 2 + \nu_z$ ($\sim$ 1996MeV),
$\gamma * G = 3 + \nu_z$ ($\sim$ 2520MeV), $\gamma * G = \nu_x$ ($\sim$ 956MeV).
Crossings of $\gamma * G = \nu_z$ ($\sim$ 925MeV) and $\gamma * G = \nu_x$ ($\sim$ 956MeV) occur within few milliseconds, the timing of the correctors must be very accurate.

- γ is constant or is slowly variable ($d\gamma/dt \sim 0$).

γ constant corresponds to extraction of SATURNE 2. γ slowly variable corresponds to MIMAS complete cycle (from 0.400 MeV to 46.7 MeV, γ varies between 1.0004 and 1.05). In the two situations the beam will

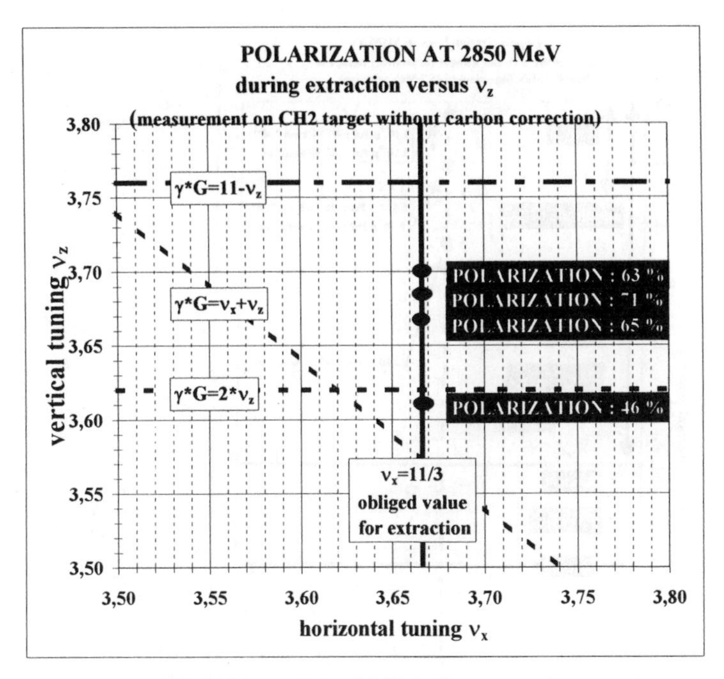

Figure 4: Polarization at 2850 MeV during extraction versus ν_z.

spend a long time above or near the resonance and multiple crossings will be made over 10^4 up to 10^6 turns: as consequence, a weak resonance can produce a very important depolarization.

- SATURNE 2: During the extraction the horizontal tuning is fixed at $\nu_x = 11/3$ (hexapole excited betatron resonance). If a type (2) depolarizing resonance occurs there is no way to avoid it. At 2550 MeV,

$$\gamma * G = 6.66 = 3 + 11/3 = 3 + \nu_x,$$

  there is no possibility to avoid this line and the polarization falls to 20%. Of course extraction on type (1) resonances is also forbidden because the correction is not possible. Type (3) or (and) (4) can be avoided because they depend of ν_z which can be adjusted (FIG.4).
- MIMAS: Two resonances have been observed: $\gamma * G = 4 - \nu_z$ and $\gamma * G = 2 + \nu_x - \nu_z$. At injection the resonant value is $\nu_z = 2.20$. If the vertical tuning is set close to this value the polarization is 70%. The correction by quadrupole correctors reach the value of 80%.

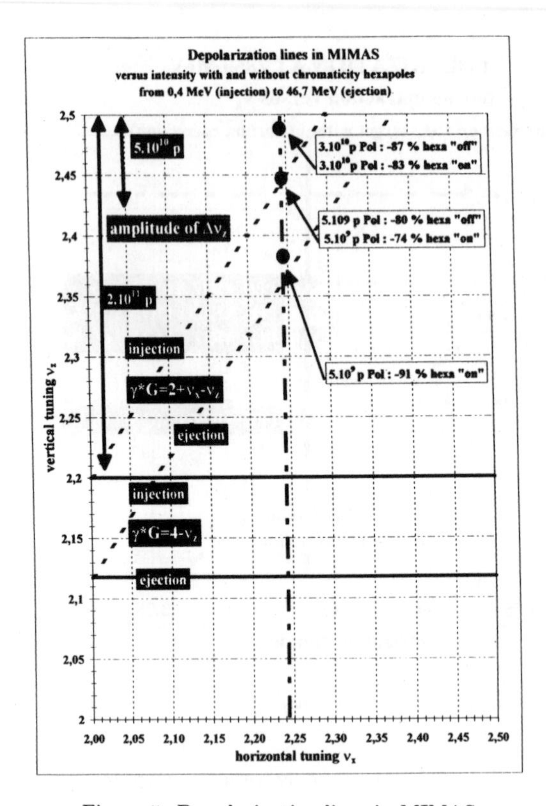

Figure 5: Depolarization lines in MIMAS.

It is necessary to change the vertical tune to get more than 90%. MIMAS is submitted to space charge effect during injection and capture: at maximum intensity (2. 10^{11} p.) it is necessary to reach the vertical tune near the half integer (2.5) to find a polarization higher than 90%. **The consequence is that the polarization is a way to measure the tune depression due to space charge !**

The FIG.5 indicates the influence of $\gamma * G = 2 + \nu_x - \nu_z$ versus the intensity, the vertical tune, and the intensity of the beam when the hexapoles used to correct the chromaticity at low energy are powered or not.

As consequence the vertical tune as to be chosen differently versus the intensity.

2 HARDWARE AND SOFTWARE:

- Horizontal and vertical tunes need a very accurate control and a fine adjustment of the current in the main quadrupoles. A vector generator "GENVEN" is used to fix the tunes at any moment during the cycle. Large bandwidth power supplies are needed.

- 12 dipoles and 24 multipoles (quadrupoles and skew quadrupoles) correctors are set around SATURNE 2: FIG.6 shows the printed circuit of a quadrupole.

  Each corrector can be pulsed 4 times during the cycle: starting time, duration, amplitude and polarity of each pulse are adjustable. Special static relays have been built (by A. Nakach) for this purpose. This system is very useful and flexible: the curves of FIG.1 show the crossing of $\gamma * G = 3$ with and without spin flip. Two sets of values have been registered for each energy during machine studies. It was possible after to change the type of crossing in few seconds.

- In order to predict what could happen during the spill, a program describing the position of the resonance versus ν_x and ν_z with beam energy as parameter has been written and allows a quick choice of the best vertical tuning (FIG.7).

- A second program has been also written to see the real variations of the horizontal and vertical tunings versus the beam energy to optimize the dynamic crossing of the resonances (FIG.8).

Several days before experiments we asked the values of the required energies to see with these programs if problems of polarization could occur during acceleration and extraction ; if a doubt appeared, machine studies were taken to solve the question. During experiments operating on the extraction channel "SD3", mainly "Nucleus-Nucleus" experiment $N°$ 225, we operate with the double extraction to follow the polarization cycle by cycle on the high energy polarimeter set on the other channel "SD2".

3 CONCLUSIONS:

One particular quality of SATURN 2 has been the flexibility of operation. Very short times were necessary for tunings and parameters modifications (energy, intensity, particle). This quality has been maintained with polarized protons despite more difficulties connected to the polarization.

Figure 6: Printed circuit quadrupole corrector of SATURNE-2.

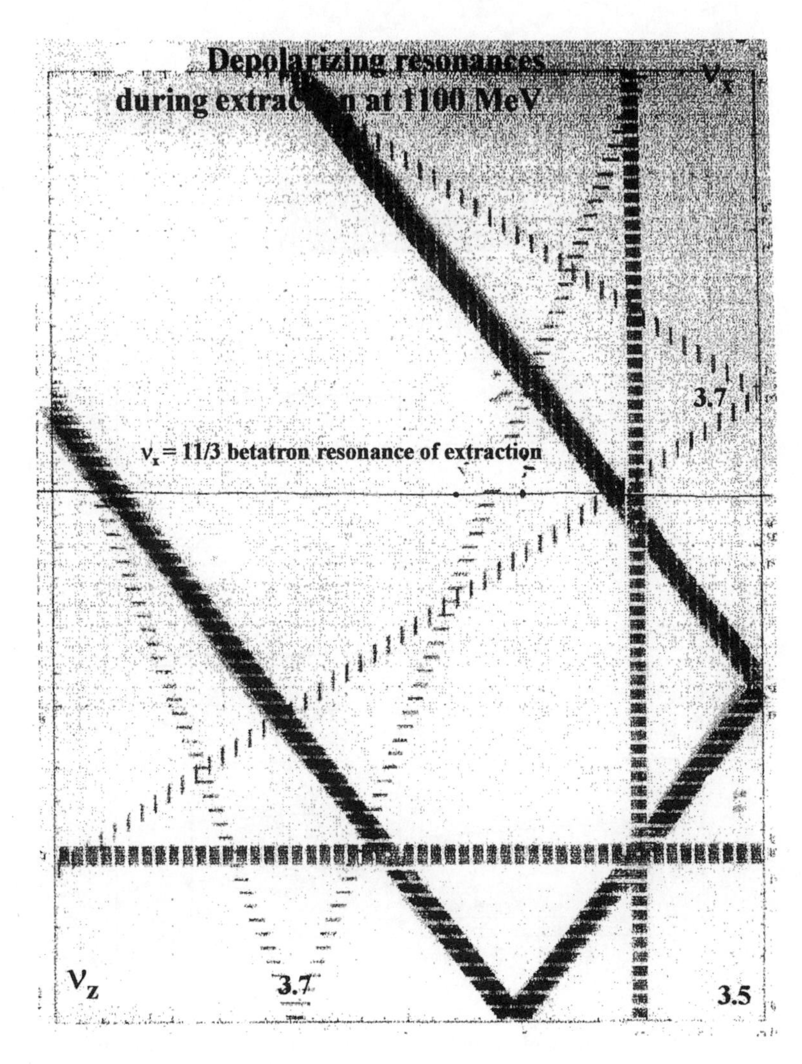

Figure 7: Depolarizing resonances during extraction at 1100 MeV.

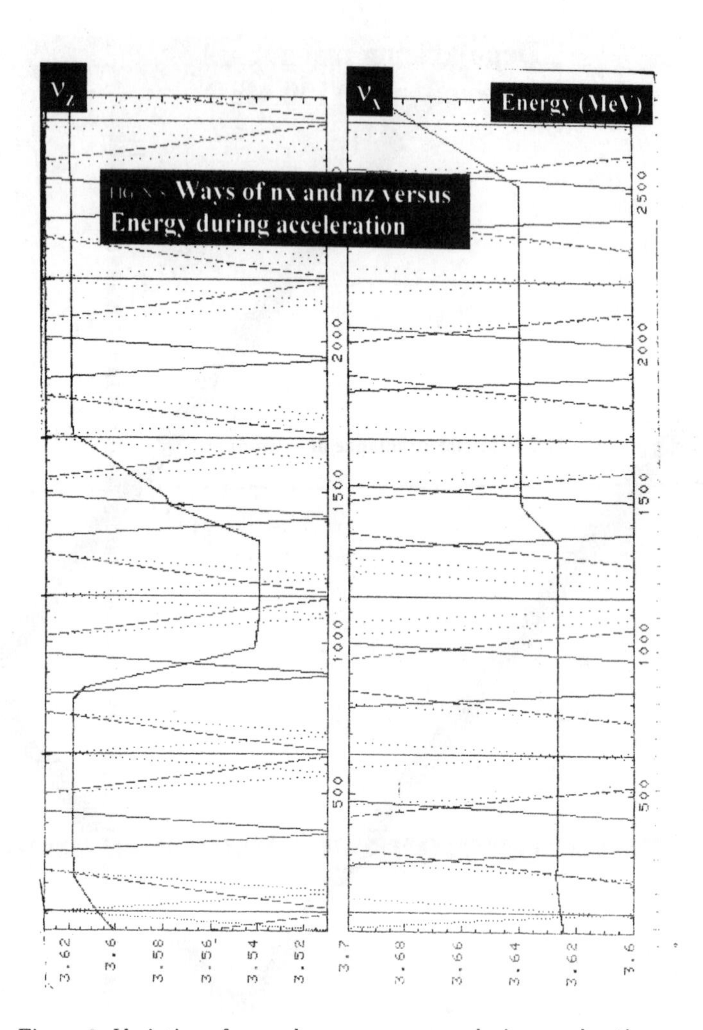

Figure 8: Variation of ν_x and ν_z versus energy during acceleration.

The intensity range was 3. 10^8 up to 2. 10^{11} polarized protons. The polarization was 80% at the maximum energy. SATURNE 2 has been the best machine up to the end for polarized beams.

We must thank F. Lehar, J.L. Boyard and J.Yonnet for polarization measurements for physics and machine studies.

Reference

1. J. BUON, Polarization in electron and proton beams, CERN ACCELERATOR SCHOOL, The Queen's College Oxford England 1985.

POLARIMETERS AND POLARIMETRY AT SATURNE

M. GARÇON
DAPNIA/SPhN, CEA-Saclay, 91191 Gif-sur-Yvette, France

The conception, calibration and use of recoil polarimeters at Saturne are reviewed, while beam polarimeters are more briefly discussed. A polarimeter based on p - C inclusive scattering was used in the standard set-up of the Nucleon-Nucleon experiment. Another one, POMME, uses the same principle, but differs in the construction and in the C thicknesses. POMME, with the addition of an absorber, is also efficient as a vector deuteron polarimeter. Three tensor deuteron polarimeters were developed or calibrated at Saturne: AHEAD (based on d-p elastic scattering between 120 and 250 MeV), POLDER (based on (d,2p) reaction between 160 and 520 MeV) and HYPOM (based on d-p elastic scattering in the 1-2 GeV region). Finally a neutron polarimeter (based on n-p elastic and quasi-elastic scattering) was calibrated.

1 Introduction

Particle and nuclear interactions depend on the spin of the interacting particles or nuclei, and result in final state particles which are in general polarized. In other words, scattering amplitudes for all processes depend on the spin of particles both in initial and final states. The full knowledge of these amplitudes may imply the measurement of the beam particles polarization and/or the scattered or recoil particles polarization. It may also imply the use of a polarized target, but this is not in the scope of this review. Depending on the physical process being studied, the complete reconstruction of the full amplitudes provides significant constraints on the physical description of the process, or sometimes is indispensible to reveal the physics at stake.

Polarimeters for nucleons or deuterons always involve a nuclear scattering, at least in the Saturne energy range.

2 Beam polarimeters:

2.1 Proton beams:

Polarimeters based on *pp* elastic scattering on a polyethylene foil were developed on both extraction lines of the accelerator [1,2]. The two final states protons are detected in small scintillators in direct view of the target, and the angular correlation identifies the *pp* elastic scattering. The remaining contribution from carbon to the asymmetry is measured and subtracted. The analyzing powers are known precisely [2], except at the very end of the energy range of Saturne.

2.2 Deuteron beams:

Due to their smaller magnetic moments, deuterons do not undergo depolarization resonances during their acceleration in the MIMAS and SATURNE rings. Their polarization is thus measured at the exit of the polarized source, at 400 keV, using the reaction $dd \rightarrow pt$ [3]. Both vector and tensor polarization can be measured with this low energy polarimeter. The subsequent depolarization during the acceleration has indeed been proven to be negligible [3].

2.3 Neutron beams:

Neutron beams are produced by break-up of extracted deuteron beams. The neutron polarization is calculated [4] and measured [5] to be nearly equal to the vector polarization of the deuteron beam, within a few % correction due to the deuteron D-state.

3 Generalities on recoil polarimeters

Polarimeters for recoil or scattered particles are composed of a target where the polarization analyzing reaction can take place, detectors upstream this target to detect (eventually identify) the incoming particles to be analyzed, detectors downstream or around the target to detect the products from the analyzing reaction. Only in the case of the neutron polarimeter was an active target used (scintillating material).

Polarimeters are caracterized by

- The analyzing reaction.
- The efficiency ε, ratio of particles which undergo the analyzing reaction (N_a) to the number of particles incident upon the polarimeter (N_i). The efficiency is an integrated cross-section over the polarimeter acceptance, multiplied by the target thickness and detection efficiencies.
- The analyzing powers, which represent a quantitative measure of the sensitivity of the analyzing reaction to the polarization of the incident particles. The vector analyzing power will be denoted A for nucleons (spin 1/2). Deuteron induced reactions have vector (iT_{11}) and tensor (T_{20}, T_{21} and T_{22}) analyzing powers. The later notation [6] is derived from the use of an irreducible tensor representation. For example, nucleons

with vertical polarization P will induce scatterings in the polarimeter, such that

$$N_a(\theta) = N_i \cdot \varepsilon(\theta) \cdot (1 + P \cdot A(\theta) \cdot \cos(\varphi)), \tag{1}$$

where θ and φ are polar and azimuthal scattering (or reaction) angles. The left-right counting asymmetry, divided by the analyzing power, yields the unknown polarization. The *systematic* errors associated with a polarization measurement are then typically given by the precision with which an asymmetry can be measured, divided by the analyzing power.

In the case of deuterons, the analyzing powers T_{kq} and the polarization components t_{kq} induce asymmetries proportional to $t_{kq}T_{kq}\cos(q\varphi)$ (and/or $\sin(q\varphi)$); after integration over φ, deuterons with alignment t_{20} will induce scatterings in the polarimeter such that

$$N_a(\theta) = N_i \cdot \varepsilon(\theta) \cdot (1 + t_{20} \cdot T_{20}(\theta)). \tag{2}$$

Note that t_{20} modifies the absolute number N_a, but does not induce any φ-asymmetry: this makes any t_{20} measurement quite more difficult.

- A figure of merit F, defined by the integration over phase-space $F^2 = \int \varepsilon \cdot A^2 \cdot d\Phi$ (or $\int \varepsilon \cdot T_{20}^2 \cdot d\Phi$ in the case of tensor polarization). From Eq. 1 (Eq. 2), one can show, in the limit of small asymmetries, that the *statistical* error associated with a measurement of P (t_{20}) is

$$\Delta P \simeq \frac{1}{F} \cdot \sqrt{\frac{2}{N_i}} \quad (\Delta t_{20} \simeq \frac{1}{F} \cdot \sqrt{\frac{1}{N_i}}). \tag{3}$$

The factor 2 in this equation comes from the fact that on average, only half of the azimuthal distribution contributes to the determination of P (in our example the up or down events do not contribute). The average effective analyzing power is then $\overline{A}$ ($\overline{T_{20}}$) $= F/\sqrt{\varepsilon}$.

- A range of particle energies where F and $\overline{A}$ ($\overline{T_{20}}$) are large enough to carry a given experiment.

4 Proton polarimeters

The first intermediate energies proton polarimeters have been developed at SIN, TRIUMF, LAMPF, Gatchina. They are all based on inclusive scattering on a carbon target: $p\ C \rightarrow 1$ *charged particle* $+ X$. The Nucleon-Nucleon group followed the same design to build its recoil polarimeter [7]. The equipment of spectrometers with focal plane polarimeters came somewhat late, due first to the fact that the high intensity polarized beams followed the operation of MIMAS and the improvement of the polarized source [8], and second to the shifting interests of various groups. In particular, in spite of the pioneering work in proton-nucleus scattering around 1 GeV at SATURNE, the Dirac phenomenology applied to this problem was studied with the additional measurements of polarization transfer at LAMPF and IUCF.

A new look at polarimetry was triggered at SATURNE by the need to measure deuteron polarization (see next section) and by the detailed study of deuteron structure. Protons from polarized deuteron break-up carry a polarization which may reveal the deuteron structure for high internal momenta, in particular the ratio of D over S waves as a function of this momentum.

The polarimeter POMME [9] was built by DPhN/ME-Saclay. The first above mentioned polarimeters used a carbon target thickness of about 6 cm (so that the probabilty of double nuclear scattering would be smaller than about 1 %). At LAMPF, a gradual increase of target thickness with energy, up to 25 cm, was successfully implemented: the C thickness was fixed by a maximal multiple scattering angle. For the design of POMME, the variation of the figure of merit F with C thickness was simulated. The calibration confirmed that the C thickness could be increased to 31 cm, gaining in efficiency without loosing in analyzing power. The proton energy range was increased up to 2.4 GeV [10]. The new generation of focal plane polarimeters at MIT-Bates and Jefferson Lab are now based on these same operating conditions.

POMME was used to measure the polarization of final state protons in experiments $p(\vec{d},\vec{p}p)n$ [11], $p(\vec{d},\vec{p})X$ [12] and $p(\vec{d},\vec{p})d$ [13]. These experiments set very significant constraints on the mechanism of deuteron break-up, since the impulse approximation is only valid up to deuteron internal momenta of about 200 MeV/c. This in turn can be used to investigate the short range structure of the deuteron, selecting the most appropriate kinematics and observables for which the reaction mechanism is under reasonable control. These studies will continue at JINR/Dubna and COSY.

5 Deuteron polarimeters

Before the work at SATURNE in the past 15 years, there was no deuteron polarimeter above 80 MeV and no identification of a good analyzing reaction, for either vector or tensor polarization measurements.

The first searches for good analyzing reactions [14] were carried out with the set-up of "Radiographie par Diffusion Nucléaire" (RDN)[15]. It was then demonstrated that any intermediate energy proton polarimeter can easily be transformed into a deuteron vector polarimeter by the addition of an absorber in front of the final particle trigger: the semi-inclusive reaction $d\,C \rightarrow 1\,charged\,particle + X$ gives a sizeable vector analyzing power provided most of the break-up protons are eliminated, in this case through range selection. This principle was implemented with success in the vector polarimeter POMME [16]. For the highest deuteron energies at Saturne, it was shown later that it is preferable not to use any absorber, since this would decrease significantly the efficiency [17]. It was also demonstrated in the RDN experiment [14] that dp elastic scattering around 200 MeV and 120° c.m. can be used for the measurement of tensor polarization. This led to the construction of the tensor polarimeter AHEAD [18] at the University of Alberta.

The next step in tensor polarimetry was triggered by theoretical calculations [19]: due to the spin structure of the charge exchange reaction $np \rightarrow pn$, the reaction $dp \rightarrow ppn$, where the final state protons are in a low relative energy $^1S_0$ state, was predicted to have sizeable tensor analyzing powers, especially in the energy ranges 200-500 MeV and 1-1.6 GeV, together with reasonable cross-sections. This was confirmed experimentally [20] and the polarimeter POLDER [21] based on this reaction was built by ISN-Grenoble, with the participation of LNS and SPhN-Saclay.

Finally, based on existing data and on simulations, it appeared that dp elastic scattering at forward angles is a good candidate for tensor polarimetry above 1 GeV. The polarimeter POMME was then transformed by using a hydrogen target and adding a detection for recoil protons in coincidence with scattered deuterons. This new polarimeter, developed by LNS, JINR/Dubna and William & Mary, was called HYPOM [22]. Contrarily to all above mentioned deuteron polarimeters, this one has both vector and tensor analyzing powers. The closing of Saturne stopped prematurely all development on this polarimeter, which is not in an optimal configuration. Data analysis concerning a calibration and the measurement of the $p(^3He, \vec{d})X$ reaction is in progress [23].

Based on all this experience, it seems that the best deuteron tensor polarimeter for the future, in the energy range 200-2000 MeV, would be a combination of POLDER and HYPOM, measuring at the same time the charge exchange reaction (d, pp) and dp elastic scattering. This is the way pursued at RIKEN with the polarimeter DPOL.

In developing these techniques, SATURNE was able to contribute decisively to

- the separation of the charge monopole and quadrupole of the deuteron (some 25 years had elapsed since the first calculations on this subject): the experimental evidence for a node in the monopole form factor came from the experiment $d(e, e'\vec{d})$ at MIT-Bates [24] using the AHEAD polarimeter. The measurement of the same t_{20} observable was continued up to a momentum transfer of 6.8 fm^{-1} at Jefferson Lab [25] using the POLDER polarimeter. Besides their own physical interests, the preliminary results from this experiment illustrate the advantage of the recoil polarimeter technique for this form factor separation at the highest possible transfers.

- the original study of isoscalar spin-flip excitations in nuclei: the inelastic scattering of polarized deuterons is very well suited to the study of $\Delta S = 1, \Delta T = 0$ transitions [26].
- the detailed structure of the $pp \to d\pi$ amplitudes [27,28].

6 Neutron polarimeters

There are indications [7] that the reaction $nC \to 1$ *charged particle* $+ X$ does not have a significant analyzing power. Therefore one cannot use a standard proton polarimeter as a neutron polarimeter. Following on their experience at lower energies, a group of Kent State University and collaborators built the polarimeter NPOL [29] and calibrated it at IUCF and SATURNE. The neutrons are incident upon a thick scintillating target and scatter into other thick scintillators at forward angles. The analyzing reaction is *np* elastic and quasi-elastic scattering at forward angles. Due to the two neutron conversion efficiencies (active target + detection) the efficiency is significantly lower than for a proton polarimeter. The analyzing powers and efficiencies were shown to decrease only very slowly up to 1 GeV, demonstrating the applicability of this technique for planned measurements of the neutron charge form factor [30].

7 Summary

In conclusion, SATURNE contributed very significantly to the advance of proton, neutron and deuteron polarimeter techniques, which in turn made it possible to address some new exciting physics. The following table contains all references to scattered or recoil particles polarimeters developed or calibrated at SATURNE, together with the references to the experiments which made use of them.

Polarimeter	Analyzing reaction	Target	Energy (MeV)	Efficiency	Analyzing power	Physics experiments
NN[7]	$p\, C \rightarrow 1\, ch$	6 cm C	200 -800	1-10%	.23-.5 $(\overline{A})$	$\vec{N}\vec{N} \rightarrow \vec{p}N$ [32]
POMME[9,10]	$p\, C \rightarrow 1\, ch$	10-31 cm C	500 -2400	13-20%	.32 - .08 $(\overline{A})$	$p(\vec{d}, \vec{p}p)n$ [11] $p(\vec{d}, \vec{p})X$ [12] $p(\vec{d}, \vec{p})d$ [13]
POMME[16,33]	$d\, C \rightarrow 1\, ch$ (+ absorber)	3-25 cm C (1-9 cm Fe)	200 -700	4-12%	.1-.35 $(\overline{iT_{11}})$	$p(\vec{p}, \vec{d})\pi$ [27] $A(\vec{d}, \vec{d}')A^*$ [26] $A(\vec{d}, \vec{d}')QE$ [31]
POMME[17,33]	$d\, C \rightarrow 1\, ch$ (no absorber)	30-35 cm C	700 -1800	14-24%	.09-.12 $(\overline{iT_{11}})$	
AHEAD[18]	$dp \rightarrow dp$	27 cm lH_2	120 -250	.23%	.26-.46 $(\lvert\overline{T_{20}}\rvert)$	$d(e, e'\vec{d})$ [24]
POLDER[21]	$p(d, pp)n$	16-20 cm lH_2	160 -520	.3-.5%	.1-.2 $(\lvert\overline{T_{20}}\rvert)$	$C(\vec{d}, \vec{d})C$ [34] $p(\vec{p}, \vec{d})\pi$ [28] $d(e, e'\vec{d})$ [25]
HYPOM[22]	$dp \rightarrow dp$	20 cm lH_2	1200 -2300	.3-1.%	0.2 $(\overline{T_{20}})$	$p(^3He, \vec{d})X$ [23]
NPOL[29]	$np \rightarrow np$	41 cm BC-408	160 -1050	.5-2.%	.12-.22 $(\overline{A})$	$d(\vec{e}, e'\vec{n})p$ [30]

References

1. J. Yonnet et al., Experiment LNS242, unpublished.
2. J. Bystricky et al., *Nucl. Instrum. Methods* **A239** (1985) 131.
3. J. Arvieux et al., *Nucl. Instrum. Methods* **A273** (1988) 48.
4. A. Boudard and C. Wilkin, J. of Phys. G**11** (1985) 583.
5. J. Bystricky et al., *Nucl. Phys.* **A444** (1985) 597.
6. The Madison convention, Proc. 3rd Int. Symp. Polarization Phenomena in Nuclear Reactions, Barschall & Haeberli eds, Wisonsin Press 1971.
7. J. Ball et al., *Nucl. Instrum. Methods* **A327** (1993) 308.
8. P.A. Chamouard, *these proceedings*, p. ??
9. B. Bonin et al., *Nucl. Instrum. Methods* **A288** (1990) 379.
10. E. Cheung et al., *Nucl. Instrum. Methods* **A363** (1995) 561.
11. S.L. Belostotsky et al., *Phys. Rev.* C**56** (1997) 160.
12. E. Cheung et al., *Phys. Lett.* B**284** (1992) 210.
13. V. Punjabi et al., *Phys. Lett.* **B350** (1995) 178.
14. M. Garçon et al., *Nucl. Phys.* **A458** (1986) 287.
15. J.-C. Duchazeaubeneix et al., IEEE Trans. Nucl. Sci. NS-30 (1983) 601.
16. B. Bonin et al., *Nucl. Instrum. Methods* **A288** (1990) 389.
17. E. Tomasi-Gustafsson et al., *Nucl. Instrum. Methods* **A366** (1995) 96.
18. J.M. Cameron et al., *Nucl. Instrum. Methods* **A305** (1991) 257; H. Witala et al., Few-Body Systems 15 (1993) 67.
19. D.V. Bugg and C. Wilkin, *Phys. Lett.* **B152** (1985) 37.
20. S. Kox et al., *Nucl. Phys.* **A556** (1993) 621.
21. S. Kox et al., *Nucl. Instrum. Methods* **A346** (1994) 527; + upgrades.
22. E. Tomasi-Gustafsson et al., *Nucl. Instrum. Methods* **A402** (1998) 361.
23. I. Sitnik, C.F. Perdrisat et al., Experiment LNS305 (1996).
24. M. Garçon et al., *Phys. Rev.* C**49** (1994) 2516.
25. S. Kox, E.J. Beise et al., CEBAF proposal 94-017.
26. F.T. Baker et al., Phys. Rep. **289** (1997) 235; M. Morlet, these proceedings, p. ??
27. B. Mayer, D. Hutcheon et al., Experiment LNS177, unpublished.
28. C. Furget et al., submitted to *Nucl. Phys.* A(1998).
29. R. Madey et al., A.I.P. Conf. Proc. **339** (1995) 47; A. Ahmidouch, private communication.
30. R. Madey et al., MIT-Bates proposal 89-04 and CEBAF proposal 93-038.
31. M.D. Holcomb et al., *Phys. Rev.* C**57** (1998) 1778.
32. C. Leluc, A review of the NN program, these proceedings, p. ??
33. V. Ladygin et al., *Nucl. Instrum. Methods* **A404** (1998) 129.
34. C. Furget et al., *Phys. Rev.* C**51** (1995) 1562.

THE PHYSICS WITH POLARIZED DEUTERONS AT SATURNE

M. MORLET
Institut de Physique Nucléaire d'Orsay,
(CNRS-IN2P3),
F91406 ORSAY CEDEX, FRANCE

Abstract: The polarized deuterons of SATURNE2 gave a conspicuous contribution in the study of few nucleon systems. The largest set of data at these energies have been obtained for meson production near threshold in the $\vec{d}+p$ system. A common mechanism was suggested in $^3\mathrm{He}\pi^0$, $^3\mathrm{He}\eta^0$ and $^3\mathrm{He}\omega^0$ channels. It fits to stress the very high cross section for the η production at threshold and the tagged η beam thus obtained. The backward elastic scattering of polarized deuterons on proton and inclusive or exclusive polarized deuteron break-up, probe the N-N interaction at very short range. It showed the disagreement with the impulse approximation prediction, may be due to subnucleonic effects. In nuclei studies, the isoscalar selectivity of deuteron was associated to the selectivity of spin observables. The response of nuclei to the isoscalar spin excitation was obtained in the polarized deuteron inelastic scattering while the distribution of neutron hole excitation in deep shell has been observed in transfer reactions induced by polarized deuterons.
Résumé: L'apport des deutons polarisés de SATURNE2 fut marquant dans l'étude des systémes à petit nombre de nucléons. La production de mésons au seuil dans le système $\vec{d}+p$ a suggéré l'existence d'un mécanisme élémentaire commun dans les trois canaux $^3\mathrm{He}\pi^0$, $^3\mathrm{He}\eta^0$ et $^3\mathrm{He}\omega^0$ et donné l'ensemble le plus complet de données actuellement disponibles. Il convient de mettre en valeur la grande section efficace de production du η au seuil qui a permis d'obtenir un faisceau de η étiqueté unique au monde. Le break-up inclusif et exclusif de deutons polarisés et la diffusion élastique à 180° sur le proton sondent l'interaction N-N à courte portée. Le désaccord avec les prédictions de l'approximation d'impulsion peut être dû à un effet subnucléonique. Dans l'étude des noyaux, c'est la sélectivité isoscalaire des deutons associée à la sélectivité des observables de spin qui ont été mises à profit. La réponse des noyaux à l'excitation de spin isoscalaire à pu être obtenue en diffusion inélastique de deutons polarisés et la distribution des excitations des trous de neutrons dans des couches profondes a été observée dans les réactions de transfert induites par des deutons polarisés.

1 Introduction

The interaction between hadrons as well as the reaction mechanisms depend on the spin and isospin of the interacting system. The measurement of cross sections leads to an average over the spin degrees of freedom in the initial and the final states, most of informations being partially lost. Spin observables, as analyzing powers or spin transfer parameters, highly contribute to recover these informations. The aim of the measurements is to obtain the largest possible number of independent parameters to constrain the models. Of course, a probe with a spin different of zero is needed. The spin 1 of the deuteron gives more spin observables than with the proton, and the deuteron alignment T_{20}, which only depends on the squared modules of amplitudes, is not too much sensitive to details of calculations. Owing to its large polarization at intermediate energy, its large intensity, and its stability on a long duration, the Saturne2 vector and tensor polarized deuteron beam was unique in the world, may be the best to perform this physics using polarized deuterons. The existence of Saturne2 was indeed determinant in the study of few nucleon systems. In this paper we try to sort the share of the experiments performed at Saturne2 in a few large subjects but this classification keep a part of arbitrary and is mainly an Ariane thread for the reader.

2 The Study of few body problems

The most important contribution of the physics using polarized deuterons at Saturne concern the study of few nucleon systems. The d-N system is a rich source of information. Data obtained may be used to study aspects of the structure of the deuteron and to probe features of N-N interaction that do not play a role in elastic N-N scattering.

2.1 The fundamental symmetries study

The deuterons have been used at Saturne to check fundamental symmetries in some experiments. The $d + d \to \alpha + \pi^0$ reaction, which is a clear violation of isospin conservation, has been observed for the first time at 1.1 GeV on *SPES4* at the LNS by the ER54 collaboration [1] with a cross section 10 times higher than predicted by the charge symmetry breaking due to $\eta^0 - \pi^0$ mixing. The cross section may be boosted by η threshold proximity where a very large η production has been observed. (*See 2.3 below*). On the other hand, the charge independence has been tested by the comparison of the deuteron tensor analyzing power T_{20} in the reactions $p(\vec{d},^3He)\pi^0$ and $p(\vec{d},^3H)\pi^+$. No first order Coulomb corrections being expected to affect T_{20}, this test depends much less of the ^3H and ^3He wave functions than the comparison of cross sections [2], leading to a better accuracy. The measurements have been done on *SPES4* at four angles with 1.1 and 1.3 GeV polarized deuterons. No charge independence violation higher than 2% has been observed.

2.2 The d-p elastic scattering amplitude

To test to which degree realistic two-nucleon interactions, adjusted to fit the 2N data, can describe the dynamic of three interacting nucleons using three-nucleon continuum calculations, a complete set of analyzing powers in the $\vec{d}+p$ scattering has been measured between 75 and 187 MeV using the polarimeter *AHEAD* on *SPES1* [3,4]. Only a few data were available before Saturne in this energy domain. Rigorous Fadeev calculations with realistic N-N potentials gave a good description of the data at all energies. This indicates that, contrarily to the lower energy case where a large discrepancy was observed in vector analyzing power, 3N force effects can be expected to be rather small for these tensor analyzing powers. Nice "complete experiments", done at the LNS, tried to obtain a sufficiently large number of independent p-d spin observables, to allow, in a model-independent manner, the reconstruction of the p-d elastic scattering amplitude. The may be nicer experiment was the measurement of vector and tensor observables in the reaction $^1\vec{H}(\vec{d},d)^1\vec{H}$ at 1.6 GeV on the line N-N [5]. It is a complete experiment because the polarization of three particles is measured. The polarization of the fourth particle can be obtained using parity conservation and time reversal invariance, the reaction being then completely determined. Twenty spin observables have been measured. Ten of these are two-spin observables and the remaining are three-spin observables. The variety and the number of observables measured in this experiment illustrates the wealth of the d-p spin structure. In the case of second-order observables, seven heretofore unmeasured observables have been obtained and can be used to place constraints on the solutions obtained in theoretical fitting of the d-p scattering data. The third-order observables, were eight of which measured for the first time, do not have as high quality as the second order observables. However they do show the feasibility of measuring such observables, alas the shut down of Saturne strongly decreases the possibility to realize such a complete experiment at the present time. A more recent experiment using the recoil tensor polarimeter *POLDER* on the line *SPES1* measured tensor polarization observables in the $H(\vec{p},\vec{d})\pi^+$ reaction between 580 and 1300 MeV [6]. Analyzing power A_y and six tensor spin observables have been measured at 132° and 151°. These data increase by a factor three the number of polarization observables available in this energy domain. Even if an overall agreement is observed between the results and the predictions of partial wave analysis, these data add new constraints to this analysis, especially for non dominant partial waves because of the sensitivity to the polarization transfer coefficients to these waves. This experiment contributes to a better understanding of the πNN system. The A_y data suggests a bump centered at $\sqrt{s} \simeq 2.37$ GeV. Although an interference of several partial amplitudes or a threshold effect could be responsible for such a structure, a dibaryon resonance can not be ruled out.

2.3 Reaction mechanism in d-N interaction

- The elastic channel

The use of the deuteron as a probe in few body problems needs the knowledge of the reaction mechanisms. Because of the relatively simple structure of the deuteron, a relatively complete theoretical treatment of the scattering process is feasible, at least for light targets. It is the reason why the information obtained using the deuterons is expected rather clean. In the first experiment using the deuteron polarized beam of Saturne2, the tensor analyzing power T_{20} in dp backward scattering has been measured at center of mass angle $\theta = 180°$ in the energy range 0.3 to 2.3 GeV [7,8]. The pd elastic scattering cross sections at intermediate energy show a backward rise explained through the one nucleon exchange (ONE) mechanism. The 180° excitation function decreases up to 400 MeV, then one observes a plateau due to the Δ excitation. In this domain the T_{20} measurement gives a new independent information helping to understand the mechanisms of pd scattering. The Saturne data showed for the first time the large negative values of T_{20}. A deep structure was observed around T_d=0.5 GeV that may be associated with one nucleon exchange mechanism. A second structure appeared around T_d=1.4 GeV not reproduced by calculations, even if Δ excitation and relativistic effects are taken into account (see fig. 9 of ref. [8]). These results are the origin of intensive search. A semi-phenomenological two step model including the $p + p \to d + \pi$ vertex proposed by Nakamura [9] improves the agreement up to 800 MeV but not beyond. The model of Boudard and Dillig [10] including re-scattering contributions in the Δ excitation leads to the same kind of agreement. Only the models including a 6 quarks component in the deuteron are in better agreement with data but the situation is not yet clear. The same kind of measurements were done more recently at Saturne as well as at Dubna. The same effect has been observed in T_{20} on proton, ^{12}C and other light targets (*See 2.4 below*). Measurements of vector and tensor spin parameters P_y, P_{yy} and P_{xx} have been done at Argonne at c.m. angles different of 180° for 600, 800, and 1000 MeV equivalent proton bombarding energy [11]. In this angular domain, a good agreement was found with predictions of a relativistic multiple scattering theory. The vector (iT_{11}) and tensor (T_{20} and T_{22}) analyzing powers for 191 and 395 MeV deuterons were also measured up to $\theta_{cm} = 150°$ at Saturne [12] for dp and dd elastic scattering as well as for inclusive scattering on Li, C, Ni, and Pb. A part of these data have been applied in the design of the

POMME recoil polarimeter.

- The inelastic channels $t\pi^+, ^3He\pi^0, ^3He\eta^0, ^3He\omega^0$

The study of exclusive production or absorption of mesons in the 3N system give useful information on reaction mechanisms at intermediate energy. Large momenta are implied giving a possible way to probe the very short range interaction. For instance broad structures appearing in the $t\pi^+, ^3He\pi^0, ^3He\eta^0$ channels for backward angles have been explained by baryonic excitations in intermediate states[13]. For these reactions, in a collinear kinematic (0° or 180°), the cross section σ and the tensor analyzing power T_{20} that depend only on two amplitudes A and B, are given by:

$$|A|^2 = (P_d/P_m)\sigma(1 - \sqrt{2}T_{20}), \ |B|^2 = (P_d/P_m)\sigma(1 + \sqrt{2}T_{20}) \qquad (1)$$

where P_d and P_m are the deuteron and meson c.m. momenta.
$t\pi^+$ ***channel:*** The first experiment probing the $t\pi^+$ channel at Saturne has been done using polarized protons at 0.9, 1., and 1.1 GeV on *SPES4*[14]. At backward angles ($\theta_{cm} > 100°$) the analyzing power varies rapidly with the incident energy around 1. GeV. This large variation has been traced back to a simple variation of the amplitudes with energy. However the origin of this phenomenon is not yet well understood since the excitation of nucleon isobars seems to be insufficient to provide a quantitative explanation. This observation was the start of one of the most important work done at Saturne on few nucleon systems.
$^3He\pi^0$ ***channel:*** In the $^3He\pi^0$ channel T_{20} do not vanish at 180° contrarily to A_y. In the reasonable hypothesis where the spin observables do not depend of isospin, it is expected that, in the domain of energy where A_y has a rapid variation, T_{20} give a robust signal associated to this variation so giving information on the origin of this phenomenon. Moreover the measurement of the cross section and T_{20} at 180° leads to the separation of the squared amplitudes $|A|^2$ and $|B|^2$. T_{20} at the threshold is related by time reversal to T_{20} obtained in π^0 absorption on ^3He at rest. If the D state of the deuteron is neglected the value of T_{20} at threshold is directly related to the ratio $|g_0/g_1|$ of the coupling constant of a pion to a nucleon pair in the isospin state T=0 (g_0) and T=1 (g_1). The first measurements were done at Saturne on *SPES4* at 0° and 180° for 19 incident energies from 0.5 up to 2.2 GeV[15,16]. At the threshold, T_{20} was found close to the prediction ($-\sqrt{2}$) of the spectator model impulse approximation (SMIA) with a pion exchange. The obtained value T_{20}=-1.31±0.039 leads to $|g_0/g_1|$=5.3 confirming the dominance of the coupling of the π with an isoscalar pair of nucleons. The results obtained for T_{20} and for the cross

section are shown in fig. 1 of ref. [15]. There is a sharp oscillation in T_{20} at 180° close to T$_d$=1.7 GeV. The modules of amplitudes A and B have been extracted showing that the T_{20} oscillation at 180° is associated with a deep minimum in $|A|^2$. This minimum leads to the observed structure in T_{20} in $^3He\pi^0$ as well as to the rapid variation of A_y close to 180° in the $t\pi^+$ channel. The large forward-backward asymmetry in the cross section is worth to be noted. A more recent experiment, specially devoted to the study of the $^3He\pi^0$ channel close the threshold [17], was performed on *SPES2* at Saturne and gave the largest set of data obtained for the pion production at threshold. It gives strong constraints on theoretical models. The P wave in $\pi - N$ system, due to the Δ formation, increases rapidly and the polarization observables, sensitive to S-P interference may give useful information on phases. Measurements of cross sections, iT_{11}, T_{22} and T_{20} have been done at 20 energies close to threshold (0.03 MeV $< T^{\pi}_{cm} <$ 10.2 MeV) with about 11 angles per energy. The mean value of T_{20} at the threshold was found T_{20}=-1.29±0.012 in agreement with the preceding measurement. At all energies, iT_{11} and T_{22} remain almost constant versus the angle and, as expected, are compatible with 0. The cross section and T_{20} vary rapidly, leading to a large backward-forward asymmetry. This behavior must be due to an interference between a non resonating S wave (sensitive to short range correlation) and the P wave associated with a virtual Δ formation in the $\pi - N$ system which appears from threshold. This combination of S and P wave suggests that the data might be simple in term of the variable $\eta cos(\theta)$ with $\eta = p_\pi/m_\pi$ and indeed, the cross section and T_{20} seems to fall on universal curves when considered in terms of this variable (see fig. 7 of ref. [17]). The model of Germond and Wilkin [18] is in good agreement with data. To explore further this channel, and obtain the phases for amplitudes A and B for instance, measurement of spin correlation parameters at 0° and 180° is essential and may improves the knowledge of short range correlation. Some predictions have been done by Ladygin [19]. This kind of measurements remains possible at Dubna, Triumf or PSI.

$^3He\eta^0$ channel: Measurements have been done on *SPES2* and *SPES4* at forward and backward angles in the $pd \rightarrow ^3He\eta^0$ for proton energies between 0.2 and 11 MeV above threshold [20], and on *SPES4* in the $\vec{d}p \rightarrow ^3He\eta^0$ reaction using polarized deuteron beam at 5 energies above threshold [13,21]. Because the $N - \eta$ system is not coupled to the Δ (except through the weak $\eta^0 - \pi^0$ mixing), no P wave is involved close by threshold, and the η production is dominated by the $S_{11}(1535)$ resonance. An isotropic cross section is then expected. Moreover, contrarily to the case of $^3He - \pi^0$, the $^3He - \eta^0$ interaction is very strong. As expected the ratio of the forward to backward cross sections remains close to 1 up to a momentum p^* of the ^3He in c.m. frame of

50 MeV/c. The forward-backward anisotropy obtained in the $pd \rightarrow^3 He\eta^0$ reaction is shown in fig. 7 of ref. [20]. A very important result is the very large experimental cross section, almost as large than in pion production, for η production and its very rapid increasing above the threshold. The maximum of cross section is reached at 2 MeV above threshold (see fig. 6 of ref. [20]). This large value was attributed to a strong final interaction (FSI) [22], suggesting the possibility of $^3He-\eta$ bound states. A two step model including the deuteron's D state proposed by Germond and Wilkin [23] is in fair agreement with data but underestimates the cross section. The more recent calculations of Fäldt and Wilkin [24] including FSI, confirm the dominant effect of FSI on the cross section behavior but underpredicts the cross section by a factor 2.4. This reaction has permitted to have a tagged η beam at Saturne with up to $10^5\eta/s$. It also allows to obtain the up to now most accurate measurement of η mass [26]. Because T_{20} is less sensitive to final state interaction, it give important information on the other part of the reaction mechanism. Like in $^3He\pi^0$ channel, if the dominant mechanism invokes only a pion exchange, the expected value is -$\sqrt{2}$. In this experiment the T_{20} value, extrapolated to the threshold, was T_{20}= -0.15$\pm$0.05 much weaker than the value obtained in $^3He\pi^0$ channel or than the value predicted by the exchange of only a pion (see fig 4 of ref. [21]). In the Germond-Wilkin model where π, η, and ρ exchange are included, T_{20} is predicted between +0.04 and +0.54 depending of choice of phases. The disagreement with data may be imputed to the large mass of η which induces large momentum transfer where 3 body processes may becomes dominant. In the Laget-Lecolley model [25] the D state of the deuteron is included and the double-scattering three-body mechanism is taken into account. The cross section has the good order of magnitude but decreases too slowly with energy close by threshold. T_{20} is predicted equal to -0.5 at the threshold and is near data indicating that the T_{20} value at threshold is a further evidence for the three-body mechanisms. The Fäldt Wilkin phenomenological two step model including FSI and 3-body processes, predicts the right variation of cross section with energy close to threshold, but its value is under-predicted by a factor 2.4 and T_{20}=+0.4 disagrees with experimental value. The $d^*(^1S_0)$ state of intermediate deuteron may perhaps explain this discrepancy.

$^3He\omega^0$ ***channel:*** In their description of $p + \vec{d} \rightarrow^3 He + \eta^0$, Fäldt and Wilkin concluded that ω or heavier meson production in the dp system should be also interpreted in term of two step model with three-body processes and, if needed, final state interaction. It is one of the motivations to study the $^3He\omega^0$ channel. Though performed using polarized proton beam, it is worth giving a few words on the experiments done at Saturne to study this channel, firstly on *SPES4* to measure the cross section [27] at 180°, then on *SPES3* to measure

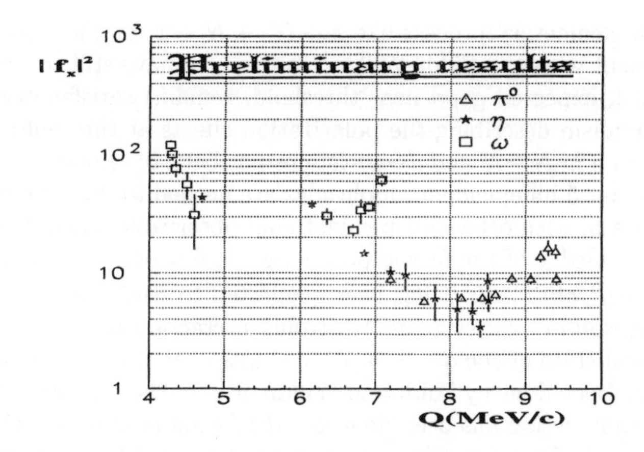

Figure 1: $|f_x|^2$ against the transfered momentum Q in ^3Hex channels.

angular distribution of cross section and analyzing power in the three channels $^3He + X$ where $X = \pi$ or η or ω [28]. The cross section increases from threshold up to 180 MeV/c then decreases. The squared amplitude defined by $|f_\omega|^2 = (p_p^*/p_\omega^*)(d\sigma/d\Omega^*)_{180^\circ}$ is shown on fig. 3 of ref. [27]. The $|f_\omega|^2$ suppression close to threshold is very similar to the effect observed by the Nimrod group in $\pi^- + p \to \omega + n$ and is not compatible with a pure S wave as in the $^3He\eta$ channel. There are two possible explanations: The presence of a P wave, or a final state interaction between one of the π coming from the ω decay and neutron. Measurement of angular distribution for spin observables is needed to go further. This is the aim of the *SPES3* experiment. The results being not yet published are still preliminary. The analyzing power A_y is very similar in the $^3He\eta$ and $^3He\pi$ channels suggesting a common elementary mechanism, like $p + p \to d + \pi^+$ for the first vertex in agreement with the Fäldt-Wilkin model where the spin observables are dominated by this vertex. In the $^3He\omega^0$ channel the results obtained at Saturne for A_y are presently the only angular distribution available in the world for this energy domain; the A_y behavior is not so similar to other channels although, in the common angular domain (100° to 180°), the zeros and extrema are at about the same angles. The cross section have a large forward-backward asymmetry indicating the presence of wave with $L > 0$. Klinger calculations (not yet published) are in fair agreement with data. If plotted against the momentum transfer $Q = p_f^* - p_i^*$ the $|f_x|^2$ amplitude do not depend of the produced x meson (see fig. 1), suggest-

ing a common elementary process like $\pi + N \to N + x$ for the second vertex also in agreement with the Fäldt-Wilkin model. One have still to understand why this model, expected good near threshold, remains satisfactory beyond. A general formalism describing the polarization effects at threshold of vector meson production in $N-N$ and $d-p$ systems has been developed by Rekalo[29] and indicates that 5 independent amplitudes are needed to describe the mechanism. More data, with other independent spin observables should be needed for a better knowledge of the meson production in few body systems.
The $d+d \to \alpha\eta^0$ channel: It has been suggested to study the coherent production of η meson in light systems[24]. Strong interaction between α and η, is expected in final state of the $d+d \to \alpha+\eta$ reaction and a two step model including FSI has been done by Fäldt and Wilkin also for the $\alpha\eta$ channel[30]. Measurement of cross section has been done on *SPES4* for $14 < p_\eta < 40 MeV/c$[31], then measurement of cross section and T_{20} were done on *SPES3* at six energies for $10 < p_\eta < 90 MeV/c$[32]. Due to Bose symmetry of the initial state, to the negative parity of the η, and to parity conservation, at threshold where L_α dominates, the cross section must be equal to 0 for 0 helicity deuterons. At threshold, we have only $\sigma_{\pm 1} \neq 0$ and $T_{20} = 1/\sqrt{2}$. In the *SPES3* data the compatibility of σ_0 with 0 at threshold indicates the validity of the background subtraction (fig 2 of ref. [32]). This is an example of using T_{20} for experimental analysis. The $|f|^2$ amplitude has been extracted and decreases more slowly with energy than in the $^3He\eta$ channel. A simultaneous fit of $^3He\eta$ and $\alpha\eta$ data using the scattering length formula leads to $a_{^3He\eta}$ = (-2.3 + 3.2 i) fm and $a_{\alpha\eta}$ = (-2.2 + 1.1 i) fm, this strongly suggests the existence of a quasi-bound state in the $\alpha\eta$ system. Strong correlation may leads to the vanishing of these state in $^3He\eta$ channel.

- The $\vec{d} \to 2p$ reaction

The spin-isospin excitation in nuclei has been intensively studied using charge exchange reaction. The reaction $(\vec{d}, 2p[^1S_0])$ is selective for these transitions and was used at Saturne to study the ΔS=1, ΔT=1 excitations in nuclei and nucleon. This reaction leads to information equivalent to that obtained in $(\vec{p}, \vec{n})$ reaction, but without the complication due to the measurement of the neutron polarization. The theory of the reaction has been given by Bugg and Wilkin[33,34]. The tensor analyzing powers are predicted different for J=L+1, L, and L-1 transitions, leading to a possible separation of 0^-, 1^-, and 2^- components of dipole isovector spin resonance for instance. The selectivity of the reaction has been checked on *SPES4* at 0° and small angles on ^1H, ^{12}C, and ^{40}Ca at 0.65 and 2 GeV[35]. Then the set of measurements was completed at 1.6

and 2 GeV on p, d, and ^{12}C targets[36]. The similarity of the spectra observed on ^{12}C and ^{40}Ca with those in (p,n) reaction, indicates that the reaction proceeds in one step. The Δ^0 excitation is clear in ^{12}C as in ^1H. The spin signal measured in these experiments is large and dominated by transverse amplitude close its maximum at q=0.8 fm^{-1}, then longitudinal at higher transfer (q$\simeq$2 fm^{-1}) (see fig 3 of ref. [35]). So, using the spin observables, it becomes possible to get information on spin structure of the excitation. The impulse approximation agrees with data up to 2 fm^{-1}, but not beyond. For all the targets, the transverse amplitude dominates the reaction mechanism at 0° in the domain of the Δ excitation (fig. 2 of ref. [36]). Due to the repulsive interaction in this channel, the dominance of transverse amplitude would leads to a shift of the Δ at higher excitation energy relative to the case of proton target in oposition with the observed shift. This observation motivated the experiments done on *DIOGENE*, *SPES4* then *SPES4π* to study the coherent meson (π, and ρ) production in inclusive then exclusive charge exchange reaction, one of the major program at Saturne, (see the contribution of M. Stephan). The $\vec{d} \rightarrow 2p$ reaction has also been used to search for a possible structure in the 2p system[37]. In that case the two protons are no longer in the $[^1S_0]$ state but are detected at an increasing relative energy in *SPES3* to observe, in A_{yy}, a signal for a possible 2p exotic resonant structure. The experiment was done at 1.7 GeV, 0° and 2 GeV, 17°. An oscillation (8 MeV width) has been observed in A_{yy} for the two energies at an invariant mass M_{pp}=1945 MeV. A weak signal only appears in A_y, more sensitive to details of mechanism. This sharp structure may be an indication for a possible dibaryon state.

2.4 Structure of the deuteron and N-N interaction

One of the largest contribution of Saturne2 in the study of few body systems concerns the structure of the deuteron. It is worthwhile to study the deuteron structure itself because the deuteron being a pure correlation between two nucleons, the study of its structures, give information on strong interaction. For instance information on short range part of the interaction, with possible subnucleonic effects or N-N* and Δ-Δ degrees of freedom, may be obtained from the contribution of large relative momenta in the deuteron wave function. The measurement of the probability to have large relative momentum in the deuteron is then a direct way to probe the N-N interaction in the domain where quark effects may appears. The inclusive and exclusive deuteron break-up, through the impulse approximation and the spectator model, is a good way to obtain this momentum distribution. The polarization observables, sensitive to small component of the wave function, are a good tool to disentangle the small effects which may be observed.

- The inclusive deuteron break-up

Inclusive deuteron break-up has been intensively studied at Saturne [38,39] and at Dubna by the same collaboration. Cross section and T_{20} were measured at Saturne at 1.25 and 2.1 GeV on ^{1}H, ^{4}He, ^{12}C, Ti, and Sn. The cross section data showed a similarity in shape when plotted against the $d_{c.m.}$ frame proton momentum q. A shoulder previously reported at higher energy for the ^{12}C (up to 7.4 GeV at Dubna) was observed in all targets at q$\simeq$ 0.30-0.35 GeV/c. Likewise the values of T_{20} were largely independent of the target's A value and almost the same at all incident energies. Associated with higher energy data on ^{12}C, these Saturne's data establish the universality of the shoulder in cross section over the energy range 1.25-7.4 GeV (fig. 4 of ref. [39]), as well as the energy independence of T_{20} (fig. 6 of ref. [39]). The shoulder in cross section as well as the abrupt turn toward less negative T_{20} values, both occurring for q>250 MeV/c, can not be explained by impulse approximation (IA) calculations even including re-scattering or a Δ-Δ component in the deuteron wave function. Calculations including multiple scattering improve the agreement with cross section data up to q=0.25 GeV/c but not beyond. The agreement with T_{20} data is improved between 0.1 and 0.2 GeV/c but not beyond [40]. Data of exclusive (p,2p) reaction, obtained at TRIUMF at lower energy, also show a shoulder in cross section at q=0.3 fm^{-1}, but it can be explained by a virtual Δ excitation [41–43]. It is not the case for the inclusive deuteron break-up measurement at Saturne and Dubna. Taking into account the rather fair agreement of the (e,e'p) data with IA predictions, the Saturne's and Dubna's data lead to a puzzle. To resolve it, relativistic effects must be taken into account, and more independent spin observables must be measured. The polarization transfer is one of these possible spin observables. The observable $\kappa_0 = P_p/P_d$ (ratio of the recoil proton polarization to the deuteron vector polarization) has been measured at Saturne in inclusive deuteron break-up and in backward elastic d+p scattering using the polarimeter *POMME* [44,45]. For measurements at $\theta_p = 0°$, $\kappa_0 = \frac{K_y^{y'}}{(1-\frac{1}{2}P_{yy}A_{yy})}$. So, in the case of the Saturne measurements where the deuterons had a pure vector polarization, $\kappa_0 = K_y^{y'}$. The IA relativistic calculations are in good agreement with data if the D state of the deuteron is included in conventional deuteron wave function. The agreement is improved up to 0.2 GeV/c if multiple scattering effects are included.

Comparison between inclusive deuteron break-up and backward dp elastic scattering: The comparison with backward elastic d+p scattering is interesting because the elastic dp scattering at 180° is a special case of

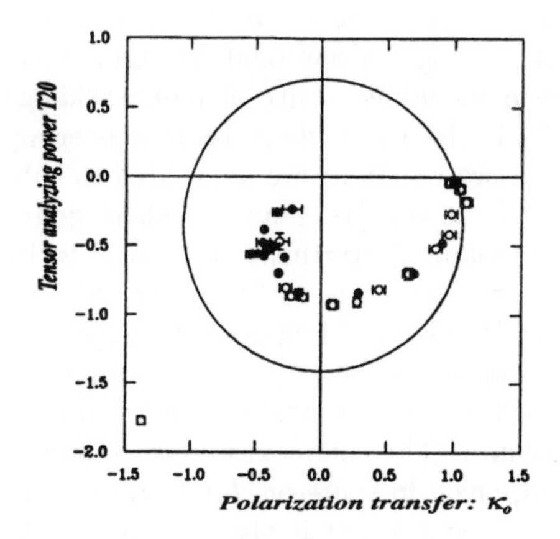

Figure 2: κ_0 versus T_{20}. Circle is the IA prediction.

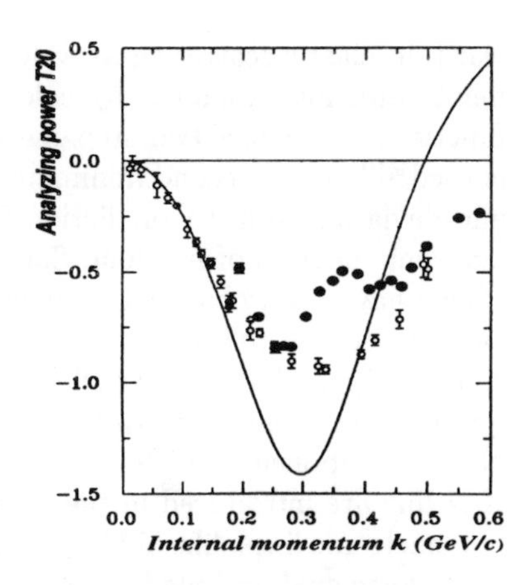

Figure 3: T_{20} for elastic backward scattering (full circles) and inclusive deuteron break-up (open circles). Line is the IA prediction.

the deuteron break-up $^1H(\vec{d},p)X$ for $X \equiv d$. At the first order the same graph dominates the two reactions but higher order processes may be different because the π is excluded in final state for the backward elastic scattering. These two reactions probe the deuteron structure but with different corrective terms. In the IA for deuteron break-up and in ONE for dp scattering, T_{20} and κ_0 are related with the S and D part of the deuteron wave function. It can be shown that, if only the S and D part of the deuteron wave function are taken into account, the Impulse approximation leads to the correlation $(T_{20} + \frac{1}{2\sqrt{2}})^2 + \kappa_0^2 = \frac{9}{8}$. So in the IA the values T_{20}, κ_0 obtained for different q values must be on a circle which don't depend on the deuteron wave function if it contains only the conventional S and D states. The $T_{20} - \kappa_0$ correlation obtained in the two reactions is the same but disagrees with IA predictions. (See data from ref. [45] in fig. 2). This similarity may indicates a common origin for the observed deviation which appears for an inter-nucleon separation lower than 0.7 fm where subnucleonic degrees of freedom may appears. The κ_0 values were found close together in the two reactions up to k=0.25 fm^{-1} but the T_{20} data (fig. 3), in agreement with preceding measurements where two minima were already observed[8], are very different in the two reactions for k>0.25 fm^{-1}. ***An evidence for quark effects at Saturne ?*** The data obtained in backward dp elastic scattering as well as in inclusive deuteron break-up measurements are not explained by calculations using conventional deuteron wave functions, even if a two step mechanism including a virtual pion exchange is used[9,10]. If the second minimum in T_{20} for backward d+p elastic scattering, and deviation from IA prediction for the two reactions are associated to subnucleonic degrees of freedom, Saturne may be the first machine where quark effects have been observed in nuclear reactions. Nevertheless one must to be careful because relativistic dynamic effects have then to be taking into account and may have a large effect on T_{20}. Dolidze and Lykasov introduce a six quarks component in the d-wave function to fit the data[46], but the complexity of this component lead to a violation of the time reversal invariance if only N-N line are introduced in the Fok's column. The complexity may be due to non nucleonic lines, like hidden color component in deuteron, *the 6q component being uncolored and made with two colored states*, but it also may be due to neglected relativistic dynamic effects. So the agreement with non conventional d-wave function is not a direct proof of the existence of exotic components, like 6q component, in the deuteron. Measurements done at higher energy at Dubna[47–51] on ^1H, and ^{12}C lead to similar conclusions. In inclusive break-up, the deviation from IA are larger for larger momentum transfer. T_{20} seems to rise to its asymptotic limit (-0.3) predicted by QCD if quark degrees of freedom are taken into account. Measurement of T_{20} at 180° in d+p elastic scattering confirm the Saturne data and a new unexpected structure appears in T_{20} at k$\simeq$0.75 GeV/c. T_{20} measurements done on ^{12}C also contain a structure at k$\simeq$0.75 GeV/c indicating that T_{20} data don't depend on incident energy and only weakly on the target.

- The exclusive deuteron break-up

It is definitely established that the inclusive deuteron break-up and backward dp elastic scattering data deviate from IA prediction at large momentum of the proton in deuteron. T_{20} is, as expected, almost the same in the two experiments up to k=0.2 GeV/c, but becomes very different beyond. T_{20} and κ_0 behavior suggest an exotic component (like 6q) in the deuteron wave function, but the introduction of such a component is not yet well mastered because the relativistic dynamic effects. On the other hand the preliminary results of the T_{20} experiment, done at TJNAF and not yet fully analyzed, are not in favor of such a non conventional wave function. The problem is still open and more complete experiments have to be done. For instance Kobushkin[52] indicates that $\vec{d}+\vec{p} \to \vec{p}+d$ give in IA, five independent spin observables with 4 correlation relations which don't depend on the deuteron wave function if only

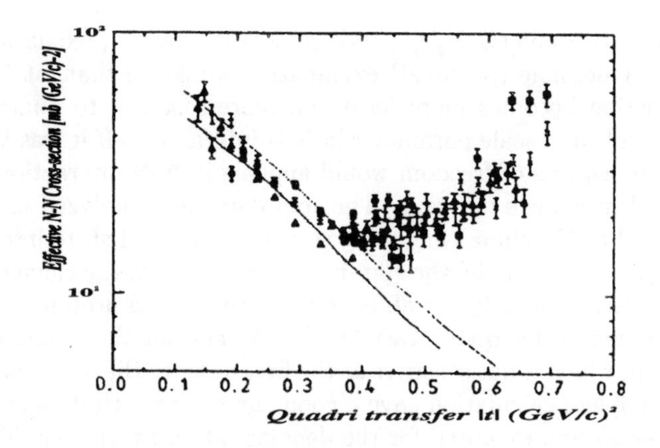

Figure 4: Effective two body cross section from experimental data against the quadri-momentum transfer. Solid lines are the free N-N cross section.

S and D states exist. They can be a useful test of the mechanism. Rekalo [53] propose a complete experiment for d-p and ^{3}He-d backward elastic scattering. The asymmetry, spin correlation, final p-polarization and d-alignment would be measured to obtain, independently of model, the four amplitudes describing the process. Spin correlation measurements are planed at Dubna. This kind of measurements may also be done at COSY, and developed at KEK, AGS, RHIC, and LISS. An other possible way to clarify the situation in d-break-up is an exclusive measurement with a complete determination of the kinematics. This detailed study has been performed at Saturne at 2 GeV on *SPES4* using the polarimeter *POMME* [54,55]. The exclusive cross section, the analyzing powers A_y and A_{yy}, and the transfer of polarization to the proton have been measured. Here again, the data are in agreement with IA prediction up to 0.2 GeV/c, then deviate beyond. The most important contribution of this experiment has been to show that the deviation from IA strongly depends on the recoil proton energy (fig. 4 of ref. [54]). The deviation appears only if T_p >0.2 GeV. A new approach is then proposed. Conventional deuteron wave function is used to calculate the momentum distribution $\rho(q)$ according to results of TJNAF. An effective N-N cross section, $d\sigma/dt$, is extracted from a fit to the data then plotted versus the quadri-momentum transfer $|t|$. For all the kinematics, the phenomenological $d\sigma/dt$ remains close to free N-N cross section up to $|t| \simeq 0.35$ $(GeV/c)^2$, then rapidly increases. (See data from ref.[54] on fig. 4). A possible (personal) explanation would be that, at large $|t|$, the N-N interaction becomes more hard and more sensitive to quark effects, $|t|$ playing the role of a scale parameter in N-N interaction. If it was the case, the subnucleonic degrees of freedom would appears in N-N interaction if they are not included in d-wave function. The spin data were analyzed using a model (already used at Gatchina) including re-scattering, final state interaction and Δ excitation. In this model the spin-isospin structure of the elementary amplitudes is treated rigorously. At all energies of the "rapid proton", A_{yy} deviates from IA prediction for q>0.2 GeV/c. For A_y and for the polarization of the rapid proton, which is not sensitive at the first order to the deuteron wave function, the complete calculation gave a good agreement with data at all energies of the recoil proton, excepted for the depolarization parameter. These results lead to the conclusion that no exotic components are revealed in deuteron wave function.

2.5 Structure of light nuclei

- The ^{3}He inclusive break-up

In this experiment, the last scheduled at Saturne, the new tensor polarimeter *HYPOM* was used to measure the tensor polarization of the deuteron coming from the ^{3}He break-up at 2.58 and 3.51 GeV in the $^3He + p \rightarrow \vec{d} + X$ reaction. Unique information on S and D states of the deuteron in ^{3}He, up to 8.4 GeV/c, may be get from these measurements. These data constitute a new approach in study of few body problems at short range. The analysis is still in progress, but theoretical predictions have been obtained for cross sections, T_{20} and $K_y^{y'}$ [56]. $K_y^{y'}$ measurements need an ^{3}He polarized beam. The possibility to have such a beam had been shown at Saturne but, alas, the shut down of Saturne stopped this program in France.

- The ^{6}Li inclusive break-up

The tensor analyzing power T_{20} in the inclusive ^{1}H(^6Li,d or α or t)X reaction with 4.5 GeV tensor polarized ^{6}Li nuclei has been measured at an angle of 0.8° [57]. The kinematic chosen favors the detection of spectator fragments, then leads directly the Lorentz boosted internal momentum of these fragments. Nonzero T_{20} values have been observed in agreement with the known non-sphericity of ^{6}Li indicated by its quadrupole moment. Owing to the polarized ^{6}Li beam of Saturne, these data sign the existence of the D state of the ^{6}Li.

3 Use of deuteron to probe the nuclei

Polarized deuterons have also been used to study the d-Nuclei interaction. In that case the reaction mechanisms become different than in few-body systems, and collective effects may appear. Owing to its isospin 0 and its spin 1, the deuteron may also probe selectively the nuclear structure using not only the cross sections, but also turning to account the specific share and the wealth of the spin observables measurement. The share of the physics using polarized deuterons at Saturne, is also consequent in the study of the nuclei, as well for the knowledge of the d-Nucleus interaction mechanisms as for the nuclear structure study.

3.1 The Rainbow effect in $(\vec{d}, d')$

Rainbow effect is observed in elastic scattering of light particles on nuclei. Namely an enhancement of the cross section at large scattering angle is associated with a damping of diffractive oscillations, the ratio σ/σ_R then exponentially decreases. Both A_y and A_{yy} rise to almost 1 in the same angular domain. This effect is due to a strong contribution of large L in partial waves analysis. In the case of deuteron, the L.S term becomes dominant and, if the tensor force may be neglected and if the classical deflection function do not change of sign, the scattering for each spin projection $(\sigma_{-1}, \sigma_0, \sigma_{+1})$ appears as three independent processes[58]. Different geometries lead to different scattering angles for the same L but for different spin projection. If large L dominate, for an attractive real L.S potential, the m=1 component is scattered at larger angle than m=0 or -1. Therefore at large scattering angles A_y and $A_{yy} \to 1$. Measurements done at Indiana on ^{58}Ni [59,60] indicate that the rainbow effect, not observed at 30 MeV and only weakly at 50 MeV, is clearly observed at 80 MeV. The role of tensor part of optical potential has been found weak. The lower limit for rainbow effect can be estimated ($\simeq$35 MeV for ^{58}Ni), but there was no estimation for the upper limit, which would corresponds to a change in the type of reaction mechanisms. Saturne has given this limit. Measurements were done on ^{16}O, ^{58}Ni at 200, 400, 700 MeV, and ^{40}Ca at 700 MeV on *SPES1.* [61,62] The rainbow effect is clearly observed on all nuclei at 200 MeV, but at 400 MeV and mainly at 700 MeV a diffractive pattern becomes dominant and is reminiscent of the behavior of the nucleon scattering at half of the energy. Using the analyzing powers the σ_{-1}, σ_0, and σ_{+1} have been extracted. At 200 MeV the rainbow effect is still present in σ_1 but vanishes in σ_0 and σ_{-1} (fig 7 of ref. [62]). This indicates that the upper limit for rainbow effect in (d,d') is at 200 MeV. Beyond the N-N elementary scattering becomes dominant. The real central potential, of Wood-Saxon type at 200 MeV, becomes in wine bottle bottom form at 700 MeV (fig. 8 of ref. [62]).

3.2 Response of nuclei to isoscalar spin excitation

Spin degrees of freedom has been intensively studied using different probes, as well electro-magnetic than hadronic. Because the weakness of the ΔS=1, ΔT=0 nucleon-nucleon interaction, only little information was get on the isoscalar spin channel. In $(\vec{p},\vec{p'})$, the ΔS=1 channel is dominated by ΔT=1 transitions. The isovector spin quadrupole and dipole excitations were localized in ^{12}C and ^{40}Ca but almost no information was obtained on the isoscalar spin strength distribution in nuclei. Using the vector and tensor polarized deuteron beam of Saturne and the vector polarimeter *POMME*, a robust signature for ΔS=1 transitions has been obtained [63]. This "vector" signature can be written as:

$$S_d^y = \frac{4}{3} + \frac{2}{3}A_{yy} - 2K_y^{y'} \tag{2}$$

It associates the ΔT=0 selectivity of the deuteron inelastic scattering with the ΔS=1 selectivity of the spin-flip probability and then gives a filter to observe ΔT=0, ΔS=1 transitions in nuclei. This selectivity is illustrated on fig. 5 for instance on the ^{12}C. After multiplication by the spin flip probability S_d^y (central part of fig. 5) the natural parity isoscalar excitations in the cross section (upper part) are washed out in spin-flip cross section (lower part).

- The low energy levels and the resonance domain:

The "vector" signature, S_d^y, is a good approximation of the spin-flip probability in (d,d') scattering: $S_1 = \frac{1}{9}(4 - A_{yy} - P^{y'y'} - 2K_{yy}^{y'y'})$, but S_d^y needs only the measure of the vector polarization of the scattered deuteron. The vector signature has been observed equal to 0 for ΔS=0 transitions. It has been tested on ^{12}C and its quality was checked by a direct measurement of S_1 on the 1^+ level at 12.7 MeV in ^{12}C and on the ground state of ^{40}Ca, using the tensor polarimeter *POLDER*. It has been found very good [64]. The isoscalar spin strength has been observed in ^{40}Ca at 400 MeV [65], in ^{12}C at 400 MeV [66] and at 600 MeV [67], then in ^{90}Zr and ^{208}Pb [68,69]. Results are compared to the predictions of microscopic DWIA calculations including finite range effect, D state of the deuteron [70], and using the transition densities obtained by F. T. Baker and G. Love in a continuum second random phase approximation (RPA) model. Most of the results obtained on ^{12}C and ^{40}Ca and the comparison with $(\vec{p},\vec{p'})$ data are given in a review article [71]. From the data, cross section σ and spin-flip probability S_d^y were directly measured then the spin-flip cross section σS_d^y (which don't depend on any model) was obtained.

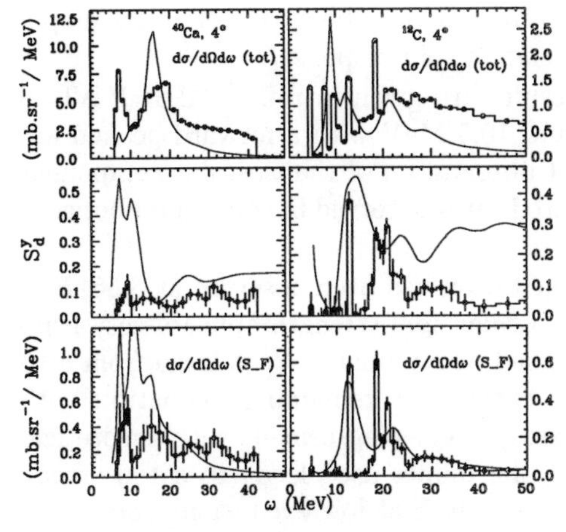

Figure 5: Directly measured quantities: cross section (upper), spin-flip probability (center), spin-flip cross section (lower) for light nuclei.

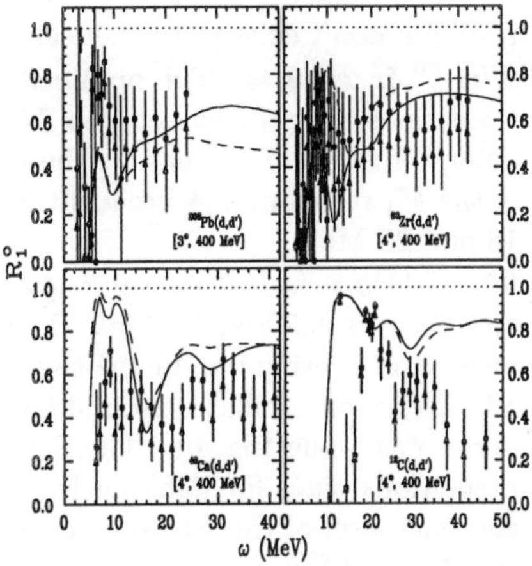

Figure 6: Isoscalar relative spin response R_1^0 for all the studied nuclei.

The derived quantities (which depend on the model used) were then extracted. There are the ΔS=1 cross section (σ_1), the ΔS=0 cross section (σ_0) and the relative isoscalar spin response (R_1^0) which gives, for each excitation energy, the ratio of ΔS=1, ΔT=0 strength relative to the total ΔT=0 strength at the same excitation energy. Many interesting results were obtained.

The ^{12}C results: A not yet known 1^+ level has been found at 20.4 MeV. Wider structures, observed at 20 and 30 MeV are compatible with a L=1 nature. The 20 MeV structure may be the isoscalar spin dipole resonance suggested by Bland at this excitation energy in pion inelastic scattering. A sample of the obtained results is given in fig. 5. The relative isoscalar spin response R_1^0 is high up to 20 MeV then decreases (fig. 6), contrarily to the predicted R_1^0 which remains high in the continuum. Measurements done at 600 MeV confirm this unexpected behavior.

The ^{40}Ca results: A part of the obtained results are also given in fig. 5. A concentration of ΔS=1, ΔT=0 excitations was observed between 9 and 15 MeV with a L=1 angular distribution. It may be the isoscalar spin dipole resonance. The comparison with $(\vec{p},\vec{p'})$ results suggests a repulsion weaker in the ΔT=0 channel than in the ΔT=1 channel. In the continuum the relative isoscalar spin response (fig. 6) is close to the Fermi gas value (0.5), suggesting a weak collectivity in this channel. Comparison with $(\vec{p},\vec{p'})$ data indicates that

the enhancement, observed in the continuum, for the isovector relative spin response don't exist in the isoscalar relative spin response.
The ^{90}***Zr results:*** The obtained results are given in ref. [68]. ΔS=1 ΔT=0 transitions are present between 7.5 and 10.5 MeV with a forward peaked angular distribution. It may be the first observation of the isoscalar component of the M_1 resonance. A broad structure is also observed in continuum between 18 and 28 MeV.
The ^{208}***Pb results:*** Isoscalar spin excitation has been observed at 3.55, 5.85 and between 6.5 and 8.9 MeV. The isoscalar spin strength appears very fragmented in continuum. In the two studied heavy nuclei, R_1^0 is large from 6 to 10 MeV, then oscillates around 0.6 in continuum, remaining compatible with Fermi gas value (fig. 6). The data are in better agreement with theoretical predictions than for light nuclei, but the model used leads to a too strong concentration of the ΔS=0 quadrupole strength at low excitation energy.
An important contribution of Saturne: The lack of enhancement in R_1^0 is an important contribution of Saturne. If the enhancement in the continuum for the isovector spin response, is due to the exhaustion of the ΔS=0 sum rules at low excitation energy, as supposed in the analysis of $(\vec{p},\vec{p'})$ data and suggested by RPA model, the lack of enhancement in the relative isoscalar spin response indicates that the ΔS=1, ΔT=0 strength is also exhausted in the same domain of excitation energy, leading only to a small shift between the ΔS=1, ΔT=0 and the ΔS=0, ΔT=0 transitions. So, from these data, the Landau parameter g_0 for the residual interaction is expected smaller than the presently admitted phenomenological value, and then closer to theoretical predictions. Non included effects, as (d,d'x) reactions with more than two particles in final state, multi-step processes, etc... may also contribute to mask a possible enhancement in R_1^0. Exclusive (very difficult) experiments would be needed to get an unambiguous explanation. The study of isoscalar spin response using the same vector signature is presently in progress at Osaka.

- Extension to the quasi-elastic region:

600 MeV polarized deuterons were also used to measure A_y, A_{yy} and the polarization transfer coefficient $K_y^{y'}$ at large transfer on H, D, C, Ca and Pb targets [72,73]. In the quasi-elastic domain (incoherent scattering), these spin observables, associated with equations obtained from DWIA, lead to isolate the isoscalar spin response for longitudinal and transverse modes. The ratio of spin longitudinal to spin transverse excitations has been obtained. More thorough analyses are in progress at Boulder and definitive results will be soon published.

3.3 Transfer reaction induced by deuterons

- Deep hole in nuclei:

The spectroscopic force is not only concentrated in low lying states, but also spreaded in continuum. Comparison with theoretical predictions need to identify the contribution of each (n, l, j)-shell. The aim of these transfer experiments is the study of simple states like the excitation corresponding to a neutron or proton hole in a valence or inner shell in medium or heavy nuclei. In ($\vec{d}$,t) or ($\vec{d}$,^{3}He) reactions a nucleon was picked in a shell, and the recoil nuclei is excited in one of these states. At the end of the eighties, the hole state fragmentation was fairly well understood for weak or medium orbital momentum L values, but not for higher L. Because the matching, the energy of Saturne was good to preferentially excite large L values. The data were obtained at Saturne on *SPES1* using the vector and tensor polarized beam at 200 and 360 MeV. At least 10 targets from ^{12}C up to ^{208}Pb were used [74–77]. At intermediate energy, the selectivity of the angular distribution of cross section is not good enough to identify the spin orbit partners $j = l \pm 1/2$. Spin observables like the analyzing powers A_y and A_{yy} lead to an unambiguous identification of these components. A_y is generally negative for the $j = l + 1/2$ and positive for the $j = l - 1/2$ partners, A_{yy} being always larger for $l - 1/2$ than for $l + 1/2$. The angular distribution of A_y and A_{yy} do not depend on the l value and only weakly of the principal quantic number n of the shell. The selectivity of A_y is illustrated on fig. 7 for the ^{142}Nd($\vec{d}$,^3He)^{141}Pr and ^{146}Nd($\vec{d}$,^3He)^{145}Pr reactions. A_y may also be used to decompose a mixing of levels with different j-natures. Data analysis was done in DWIA using exact finite range calculations including spin-orbit term and the deuteron D state. Isolated levels were decomposed using σA_y. In the continuum the unknown spectroscopic factors, ($C^2 S_i$ for the i^{eme} component), were obtained by solving, for each energy bin, the linear system:

$$\sigma^{exp} = \sum_i (C^2 S_i) \sigma_i^{th} + \sigma^b \tag{3}$$

$$\sigma^{exp} A_y^{exp} = \sum_i (C^2 S_i) \sigma_i^{th} A_{yi}^{th} + \sigma^b A_y^b \tag{4}$$

$$\sigma^{exp} A_{yy}^{exp} = \sum_i (C^2 S_i) \sigma_i^{th} A_{yyi}^{th} + \sigma^b A_{yy}^b \tag{5}$$

where σ_i^{th}, A_{yi}^{th}, and A_{yyi}^{th} are theoretically calculated. The best set of optical potential parameters was chosen to obtain the best agreement with data from ^{12}C up to ^{208}Pb. A modified optical potential, containing a large spin-orbit

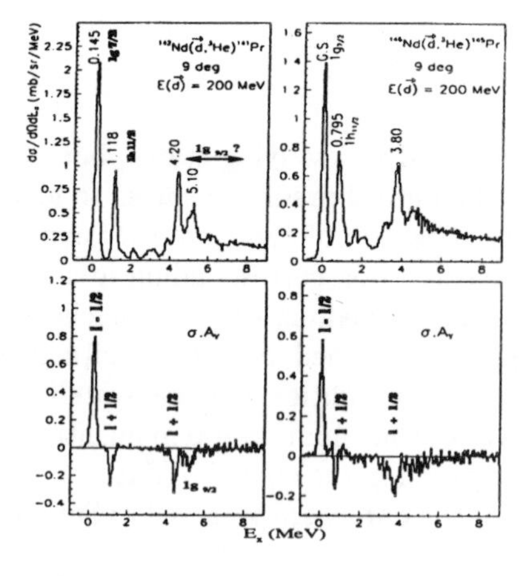

Figure 7: An example of the selectivity of A_y for $j = l \pm 1/2$ excitations.

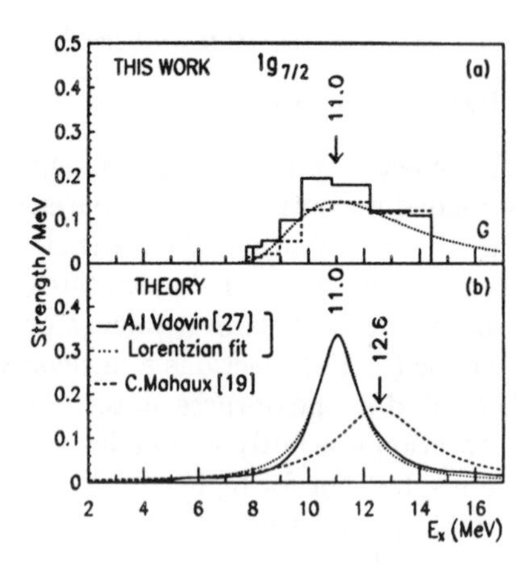

Figure 8: Distribution of n-hole excitation in the inner shell $1g_{7/2}$.

term and an imaginary tensor term was needed. The energy dependence of the radial functions, R_{nlj} has been taking into account and calculated in a quasi particle-phonon model [78]. Angular distribution of cross section and analyzing powers are predicted in good agreement with data for all the nuclei. This work, the first systematic study of $(\vec{d}$, t) reaction at intermediate energy, indicates that the shape of these angular distributions depends only weakly of the target. For the first time, the n-holes states excitation distribution in the valence shells $1i_{13/2}$ and $1h_{9/2}$ in ^{207}Pb and in the inner shells $1h_{11/2}$, $1g_{7/2}$ and $1g_{9/2}$ were obtained. It is the first observation of the deep shell $1g_{7/2}$ (fig. 8), and 85% of the $1h_{11/2}$ sum rule has been observed. A splitting spin-orbit of 5.2 MeV has been extracted for ^{208}Pb. Comparison has been done with the fragmentation predicted in different models [75]. The future of this experiment is at AGOR and GANIL.

Deeply bound pionic states: The transfer reaction (d,^{3}He) was also used to search for deeply bound pionic states in Lead. The expected signature is a sharp structure at $\simeq$140 MeV of excitation energy in ^{208}Pb(d,^{3}He) reaction. This structure was not observed at Saturne, but Saturne showed that measurements at 0° are needed. Using the fragment separator of GSI, deeply bound pionic states was recently observed at 0°. They may be the pionic $\pi^- \otimes^{207} Pb$ in $(2p)_{\pi^-}(3p_{1/2}, 3p_{3/2})_n^{-1}$ configuration.

4 Use of deuteron to probe the nucleon

The vector signature S_d^y, used to search for isoscalar spin excitation in nuclei is defined using only the spin structure of initial and final state, then it can be used to search for components ΔJ=1, ΔT=0 in baryonic resonance excitation of the nucleon. The ΔS=1 selectivity is, in principle, only for J=0 target, but it can be shown that, at 0°, where ΔL=0 excitation dominates, the ΔS=1 selectivity of S_d^y remains. It was the reason why S_d^y could be used to search for a possible ΔJ=1 contribution to the Roper excitation in $(\vec{d},\vec{d}')$. The cross section, A_y and A_{yy} have been measured at 3.7 GeV/c using the polarized deuteron beam of Saturne at 2.3 GeV in the domain of the Roper excitation in proton and of the Δ excitation in the projectile [79]. After subtraction of the Δ contribution, the Roper excitation becomes visible, but the cross section is too small for polarimetry measurements. The data were compared to the predictions of the Oset's model where ρ and π exchange as well as the interference term are included. The data for Δ excitation are in agreement with theoretical predictions at better than 15%. The Roper position is well predicted, but the experimental cross section is 30% lower than predicted and the experimental shape of the Roper seems to be wider than predicted (see fig. 9). The model is not predictive for analyzing powers. No signal was seen in A_y, which remains close to 0 at all excitation energies, even in the domain where the Roper is excited. A_{yy} presents a shoulder weakly positive, (or weakly negative for T_{20} observable), in the domain of the Roper excitation. For larger missing mass values, A_{yy} increases except for the three last points (fig. 10). These 3 last points may be questionable due to high experimental background in this missing mass domain. Measurement of T_{20} have been done at higher energy at Dubna [80] on H and ^{12}C. The data show, as for the Saturne data, a plateau weakly negative in the domain of the Roper excitation, and remains close to 0 in the domain of the Δ excitation. T_{20} in the domain of the Roper excitation, seems do not depend on the target. An exclusive ($\vec{d}$,d') experiment, to study the Roper excitation, was done on *SPES4π*. The $N^* \to \pi N$ and $N^* \to 2\pi N$ channels were identified. The analysis of this experiment is still in progress and only preliminary results are available. The A_{yy} data are compatible with the preceding Saturne data, even for the rapid decrease in A_{yy} observed at the largest missing masses. The Rekalo's model [81], including π and ρ meson exchange for the Δ excitation and η, σ and ω exchange for Roper excitation, give predictions for analyzing powers. The sensitivity to ω exchange has been stressed as well as the sensitivity of T_{20} to the isoscalar longitudinal Roper form factor. Fair agreement can be obtained with existing data [82]. On fig. 10 the 3.7 GeV/c data are the data obtained at Saturne.

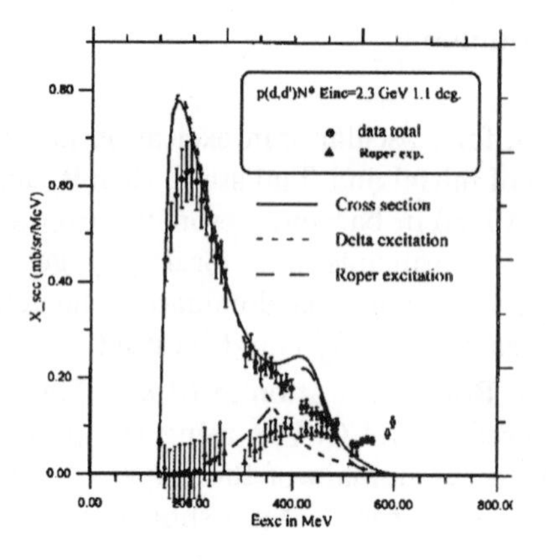

Figure 9: Cross section for Δ (in projectile) and Roper (in target) excitation. Lower data points are for Roper excitation.

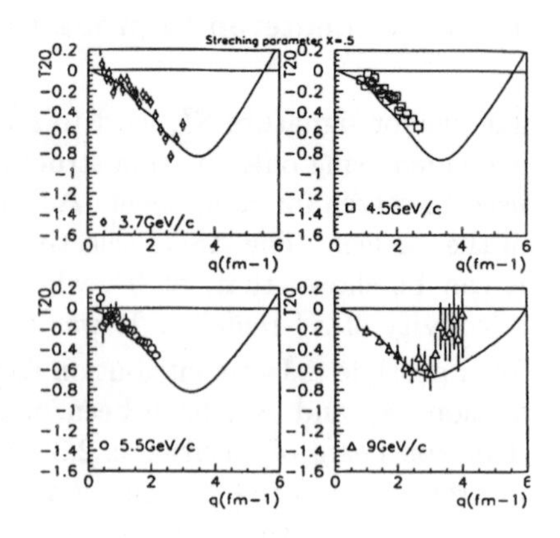

Figure 10: Analyzing power T_{20} for Roper excitation. The 3.7 GeV/c data are the Saturne data.

5 Summary and conclusion

With its highly polarized beams of high intensity and large stability, Saturne was an intermediate energy accelerator unique in the world. Its share in the physics was at the level of its capabilities and its contribution to the study of the few body problem was exceptional. Concerning the physics with polarized deuterons, though all the experiments gave a stone to the edifice of physic, let us summarize the most conspicuous contribution of Saturne on this subject. In the domain of *fundamental d-N interaction*, first complete measurements including 2nd and 3rd order spin observables gave strong constraints for models. In the *reaction mechanism study*, the upper limit for rainbow effect in (d,d') has been discovered at Saturne, but it was in few body systems that the share of Saturne was fundamental. The 3-body systems were probed at very short range where subnucleonic degrees of freedom may appears. In the $^3He\pi^0$ channel, Saturne gave the larger set of data close to threshold and stressed the P wave contribution. In the $^3He\eta^0$ channel, the contribution of Saturne was determinant. It showed the strong final state interaction leading to very large cross section at threshold. An η^0 tagged beam has been then obtained with $10^5\eta$/s, and the mass of the η measured with the best accuracy in the world. Measurements of T_{20} and cross section at threshold showed the contribution of FSI and 3-body mechanisms in a two step model, but the T_{20} values remain an open problem. In the $^3He\omega^0$ channel, the data of Saturne are the only data available in the world at intermediate energy. The two step model, used in $^3He\eta^0$ channel, remains good. A common elementary process is suggested by the data obtained in the three channels. On the other hand the ($\vec{d}$,2p) experiments were the starting point of an important work on the Δ propagation in nuclei. In the *nuclear structure study*, Saturne contributed mainly in few nucleon systems with a very large work on the inclusive and exclusive deuteron break-up, and backward elastic scattering of deuteron on the proton. Saturne showed that the predictions of ONE and IA definitively disagree with data. The Saturne and Dubna data indicate that the behavior of cross section and T_{20} don't depend on the energy and only weakly on the target. Presently no model, using conventional deuteron wave function and free N-N interaction, is able to reproduce never the shoulder in cross section nor the oscillation in T_{20}. If an hidden color component was needed in deuteron wave function, Saturne would be the the first machine to have detected this kind of effect but the problem is still open. Exclusive deuteron break-up stressed the dependence of these effects on the quadri-momentum transfered to proton, and may indicates that the quark effect is at least in the N-N interaction and not only in the deuteron wave function. The transfer reactions induced by polarized deuteron used the selectivity of A_y and A_{yy}. They gave for the first time, the distribution of n-hole in deep shell ($1h_{11/2}, 1g_{7/2}, 1g_{9/2}$) in ^{207}Pb for instance. Measurements of the spin-flip probability in ($\vec{d}, \vec{d'}$) inelastic scattering gave a robust signature of ΔS=1 transitions. The ΔT=0 selectivity of the inelastic deuteron scattering associated with the ΔS=1 selectivity of the spin-flip probability gave a filter to study the response of nuclei to the isoscalar-spin excitation. A few, still unknown, levels were found, and the isoscalar spin strength distribution in continuum was obtained in ^{12}C, ^{40}Ca, ^{90}Zr and ^{208}Pb. An important contribution was to shown that the relative isoscalar spin response remains close to its Fermi gas value in continuum for all nuclei, suggesting a Landau parameter g_0 smaller than the, until now, admitted phenomenological value. The analysis of the Roper excitation in exclusive ($\vec{d}$,d') scattering is still in progress. The obtained results will be probably also an important contribution of Saturne.

References

1. L. Goldzhal *et al.*, *Nucl. Phys.* A **533**, 675 (1991).
2. Collaboration ER54 *Nouvelles de Saturne* **13**, 37(1989).
3. J.M. Cameron *et al.*, *Nucl. Instrum. Methods* A **305**, 257 (1991).
4. H. Witala *et al.*, *Few Body Systems* **15**, 67 (1993).

5. V. Ghazikhanian *et al.*, *Phys. Rev.* C **43**, 1532 (1991).
6. C. Furget *et al.*, *Nucl. Phys.* A **631**, 747c (1998).
7. P. Berthet *et al.*, *J. Phys.* G **8**, L111 (1982).
8. J. Arvieux *et al.*, *Nucl. Phys.* A **431**, 613 (1984).
9. A. Nakamura and L. Satta, *Nucl. Phys.* A **445**, 706 (1985).
10. A. Boudard and M. Dillig, *Phys. Rev.* C **31**, 302 (1985).
11. M. Haji-Saied *et al.*, *Phys. Rev.* C **36**, 2010 (1987).
12. M. Garçon *et al.*, *Nucl. Phys.* A **458**, 287 (1986).
13. P. Berthet *et al.*, *Nucl. Phys.* A **443**, 589 (1985).
14. B. Mayer *et al.*, *Phys. Lett.* B **181**, 25 (1986).
15. C. Kerboul *et al.*, *Phys. Lett.* B **181**, 29 (1986).
16. A. Boudard *et al.*, *Phys. Lett.* B **214**, 6 (1988).
17. V. N. Nikulin *et al.*, *Phys. Rev.* C **54**, 1732 (1996).
18. J.F. Germond and C. Wilkin, *J. Phys.* G **14**, 181 (1988).
19. V. P. Ladygin, N. B. Ladygina, *Yad. Phys.* **58**, 1365 (1995).
20. B. Mayer *et al.*, *Phys. Rev.* C **53**, 2068 (1996).
21. J. Berger *et al.*, *Phys. Rev. Lett.* **61**, 919 (1988).
22. C. Wilkin *Phys. Rev.* C **47**, R938 (1993).
23. J.F. Germond and C. Wilkin, *J. Phys.* G **15**, 437 (1989).
24. G. Fäldt and C. Wilkin, *Nucl. Phys.* A **587**, 769 (1995).
25. J. M. Laget and J. F. Lecolley, *Phys. Rev. Lett.* **61**, 2069 (1988).
26. F. Plouin *et al.*, *Nucl. Phys.* B **276**, 526 (1992).
27. R. Würzinger *et al.*, *Phys. Rev.* C **51**, R443 (1995).
28. T. Kirchner *et al.*, To be published.
29. M. P. Rekalo *et al.*, LNS/Ph/95-10, To be published in *Phys. Lett.* B.
30. G. Fäldt and C. Wilkin, *Nucl. Phys.* A **596**, 488 (1996).
31. R. Frascaria *et al.*, *Phys. Rev.* C **50**, R537 (1994).
32. N. Willis *et al.*, *Phys. Lett.* B **406**, 14 (1997).
33. D. V. Bugg and C. Wilkin, *Nucl. Phys.* A **467**, 575 (1987).
34. C. Wilkin and D. V. Bugg, *Phys. Lett.* B **154**, 243 (1985).
35. C. Ellegaard *et al.*, *Phys. Rev. Lett.* **59**, 974 (1987).
36. C. Ellegaard *et al.*, *Phys. Lett.* B **231**, 365 (1989).
37. B. Tatischeff *et al.*, *Phys. Rev.* C **45**, 2005 (1992).
38. C. F. Perdrisat *et al.*, *Phys. Rev. Lett.* **59**, 2840 (1987).
39. V. Punjabi *et al.*, *Phys. Rev.* C **39**, 608 (1989).
40. C. F. Perdrisat *et al.*, *Phys. Rev.* C **42**, 1899 (1990).
41. C. F. Perdrisat *et al.*, *Phys. Lett.* B **156**, 38 (1985).
42. V. Punjabi *et al.*, *Phys. Lett.* B **179**, 207 (1986).
43. V. Punjabi *et al.*, *Phys. Rev.* C **38**, 2728 (1988).
44. E. Cheung *et al.*, *Phys. Lett.* B **284**, 210 (1992).

45. V. Punjabi *et al.*, *Phys. Lett.* B **350**, 178 (1995).
46. M. G. Dolidze and G. I. Lykasov, *Z. Phys.* A **336**, 339 (1990).
47. T. Aono *et al.*, *Phys. Rev. Lett.* **74**, 4997 (1995).
48. V. G. Ableev *et al.*, *Pis'ma Zh.Eksp. Teor. Fiz.* **47**, 558 (1988).
49. L. S. Azhgirey *et al.*, *Phys. Lett.* B **387**, 37 (1996).
50. A. A. Nomofilov *et al.*, *Phys. Lett.* B **325**, 327 (1994).
51. L. S. Azhgirey *et al.*, *Phys. Lett.* B **391**, 22 (1997).
52. A. P. Kobushkin *et al.*, *Phys. Rev.* C **50**, 2627 (1994).
53. M. P. Rekalo *et al.*, To be published in *Few Body Systems.*
54. J. Erö *et al.*, *Phys. Rev.* C **50**, 2687 (1994).
55. S. L. Belostotsky *et al.*, *Phys. Rev.* C **56**, 50 (1997).
56. E. Tomasi-Gustafsson *et al.*, *Nucl. Instrum. Methods* A **402**, 361 (1998).
57. V. Punjabi *et al.*, *Phys. Rev.* C **46**, 984 (1992).
58. R. C. Johnson and E. J. Stephenson, *Nucl. Phys.* A **371**, 381 (1981).
59. E. J. Stephenson *et al.*, *Nucl. Phys.* A **359**, 316 (1981).
60. E. J. Stephenson *et al.*, *Phys. Rev.* C **28**, 134 (1983).
61. Nguyen Van Sen *et al.*, *Phys. Lett.* B **156**, 185 (1985).
62. Nguyen Van Sen *et al.*, *Nucl. Phys.* A **464**, 717 (1987).
63. M. Morlet *et al.*, *Phys. Lett.* B **247**, 228 (1990).
64. C. Furget *et al.*, *Phys. Rev.* C **51**, 1562 (1995).
65. M. Morlet *et al.*, *Phys. Rev.* C **46**, 1008 (1992).
66. B. N. Johnson *et al.*, *Phys. Rev.* C **51**, 1726 (1995).
67. C. Djalali *et al.*, to be published in *Phys. Rev.* C.
68. M. Morlet *et al.*, Proceedings of the Varenna conference *Ed. by E. Gadioli, Milano June 1997*, 132 (1997)
69. C. Djalali *et al.*, to be published in *Phys. Rev.* C.
70. J. Van de Wiele *et al.*, *Nucl. Phys.* A **588**, 829 (1995).
71. F. T. Baker *et al.*, *Physics Report* **289**, 235 (1997).
72. R. J. Peterson *et al.*, *Nucl. Phys.* A **577**, 161c (1994).
73. M. D. Holcomb *et al.*, *Phys. Rev.* C **57**, 1778 (1998).
74. J. Van de Wiele *et al.*, *Phys. Rev.* C **46**, 1863 (1992).
75. H. Langevin-Joliot *et al.*, *Phys. Rev.* C **47**, 1571 (1993).
76. J. Van de Wiele *et al.*, *Phys. Rev.* C **50**, 2935 (1994).
77. H. Langevin-Joliot *et al.*, to be published in *Phys. Rev.* C
78. J. Van de Wiele *et al.*, *Nucl. Phys.* A **605**, 173 (1996).
79. S. Hirenzaki *et al.*, to be published.
80. L. S. Azhgirey *et al.*, *Phys. Lett.* B **361**, 21 (1995).
81. M. P. Rekalo and E. Tomasi-Gustafsson, *Phys. Rev.* C **54**, 3125 (1996).
82. E. Tomasi-Gustafsson and M. P. Rekalo, *Internal report* DAPNIA/SPhN-97-57.

THE NUCLEON-NUCLEON FORCE

C. LECHANOINE-LELUC
D.P.N.C. University of Geneva, 24 Quai E. Ansermet,
1211 Geneva 4, Switzerland

F. LEHAR
DAPNIA/SPP, CEA/Saclay, 91191 Gif-sur-Yvette Cedex, France

Nucleon-Nucleon experiments using the SATURNE II polarized beams provided a large amount of new data. The results are reviewed here and the impact of these data on theory, semi-phenomenological analyses and predictions is treated. Evidence for or against dibaryons based on SATURNE II data are discussed.

Nucleon-Nucleon (NN) interactions at intermediate energies have an impact on a wide variety of topics in nuclear and particle physics, such as understanding of the nuclear force, nucleon scattering from nuclei, and possible existence of 6 quarks states. This region covers the opening of all meson channels and the NN excited states. The total inelastic cross-section shows a fast increase and the total elastic one a slow decrease. As a result the total cross section becomes constant. In this region the duality models could be checked. The few-GeV region is thought to be wherein the quark degrees of freedom inside the nucleon core are expected to manifest themselves directly in the scattering observables. It is the realm of non-perturbative QCD. Semi-phenomenological models adjusted to fit NN scattering data are used. As these models are inputs to further calculations such as nuclear structure calculations it is important for them to fit properly the data. A pure phenomenological approach exists now for two particle reactions from NN elastic scattering. All these reasons are the principal motivations for studies of NN scattering.

1 Why study the Nucleon-Nucleon System at SATURNE II

Study of the NN interaction has always been pursued actively at SATURNE II. The reasons for this research are manyfold:

- Energy range extending up to 2.9 GeV which makes it unique among this medium energy accelerators since the closing of the ZGS machine in 1979. The energy region below 800 MeV has been studied with great care at different accelerators (PSI in Switzerland, LAMPF in the US and TRIUMF in Canada).

- Highly polarized particules are accelerated, in particular p (with 90% polarization up to 1 GeV and 75% at 2.7 GeV) and deuterons with 98.4% polarization.
- High intensity accelerator for polarized particules: $2.10^{11}\vec{p}$ and $3.10^{11}\vec{d}$.
- Polarized neutron beams available from polarized deuteron break-up. This technique is very appealing, since the deuteron vector polarization is almost totally transferred to the outgoing neutrons (59% polarized) and protons. On top of that, the full energy spread of the neutron beam is only 5%.

As well as having been for many years the only accelerator operating at energies in the range between 800 and 2900 MeV, this combination of features made SATURNE II the ideal machine for studying NN scattering.

2 NN formalism

Assuming parity conservation, time reversal and isospin invariance, the scattering matrix is written in terms of complex amplitudes a, b, c, d and e as[1]

$$\begin{aligned} M(\vec{k}_f, \vec{k}_i) &= \tfrac{1}{2}[(a+b) + (a-b)(\vec{\sigma_1}, \vec{n})(\vec{\sigma_2}, \vec{n}) + (c+d)(\vec{\sigma_1}, \vec{m})(\vec{\sigma_2}, \vec{m}) \\ &\quad +(c-d)(\vec{\sigma}_1, \vec{\ell})(\vec{\sigma}_2, \vec{\ell}) + e(\vec{\sigma}_1 + \vec{\sigma}_2, \vec{n})] \end{aligned} \tag{1}$$

where $\vec{\sigma}_1$ and $\vec{\sigma}_2$ are the Pauli 2x2 matrices, $\vec{k}_i$ and $\vec{k}_f$ are the unit vectors in the direction of the incident and scattered particles, respectively, and $\vec{n}, \vec{m}$ and $\vec{\ell}$ are the c.m basis vectors. The pp scattering matrix contains the isotriplet (M_1) matrix only while the np scattering matrix contains both the isosinglet (M_o) and isotriplet matrices.

Observables will be denoted as X_{srbt} with the subscripts referring to the polarization orientation of the scattered, recoil, beam, and target particles, respectively. For the so-called "pure experiments," the polarizations of the incident and target particles in the laboratory system are oriented along the basis unit vectors $\vec{k}$, $\vec{n}$, $\vec{s} = [\vec{n} \times \vec{k}]$. In the experiments at SATURNE II only the polarization of recoil protons was analyzed in the second scattering. The base vectors are in the directions $\vec{k}''$, $\vec{n}$, $\vec{s}\,'' = [\vec{n} \times \vec{k}'']$, where the unit vector $\vec{k}''$ is oriented along the direction of the recoil particle momentum. Taking into account fundamental conservation laws, any observable, X_{srbt}, can be expressed in terms of the five scattering amplitudes from Eq. 1 as

$$\frac{d\sigma}{d\Omega} = \frac{1}{2}[|a|^2 + |b|^2 + |c|^2 + |d|^2 + |e|^2] \tag{2}$$

$$\frac{d\sigma}{d\Omega} A_{oonn} = \frac{1}{2}[|a|^2 - |b|^2 - |c|^2 + |d|^2 + |e|^2] \tag{3}$$

If enough observables (more than 11) are measured at the same angle and at the same energy, it is possible to perform a direct reconstruction of the five scattering amplitudes. Such a reconstruction based on SATURNE II data will be presented in Section 4.2.

3 Nucleon-Nucleon experiments

In this section we present the experiments which studied NN scattering with the aim of measuring many different observables over a large angular domain and a large energy domain with high precision. The experiments dedicated to dibaryon searches will be presented later in Section 5.

3.1 Small angle experiments

Absolute *np* differential cross section has been measured at small angles by the IKAR collaboration [2,3] at 14 different energies between 378 and 1135 MeV. A ionization chamber (IKAR) filled with methane or with hydrogen [4] was used as both a gas target and recoil detector. Results are shown in Fig.1. Measurement of the analyzing power was also performed for a few angles at 5 energies (see Fig.4). A total of 585 $d\sigma/d\Omega$ data points and 30 A_{oono} data points have been measured which helped to stabilize PSA solutions. These data allowed to extract the ratio of the real to imaginary part of the forward spin independent scattering amplitudes.

An interesting measurement of the analyzing power of *pp* elastic scattering at small angles in the Coulomb-nuclear interference region was also performed in 1989 at 4 energies between 940 and 2440 MeV (34 data points) [5]. It demonstrated the feasibility of performing polarization measurements at low momentum transfer over a wide energy range.

3.2 Nucleon-Nucleon program

The main bulk of *pp* and *np* elastic spin observables has been measured by the Nucleon-Nucleon group between 1981 and 1995 in three different periods. They concentrated on spin-dependent observables requiring a polarized beam (proton and neutron) as well as a polarized target (PPT) [6]. Both the beam

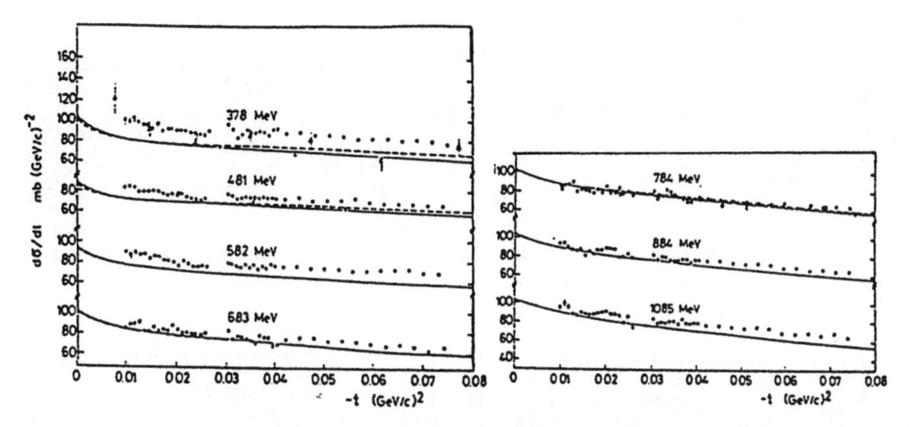

Figure 1: Differential cross sections measured by the IKAR collaboration [2,3] shown as full dots. Solid and dashed lines are 1989 phase shift predictions.

and target could have their polarizations orientated in any direction, either longitudinal, vertical or sideways. A summary of the Nucleon-Nucleon group measurements is given in Table 1 for *pp* scattering[7–36] and in Table 2 for *np* scattering[37–49].

Period 1 (1981-1986)

During this period, the *pp* total cross section differences, $\Delta\sigma_T(pp)$ and $\Delta\sigma_L(pp)$, and 11 to 15 different spin-dependent observables were measured. These amounted to a total of 2968 data points as detailed in Table 1.

A new kind of Gray code transmission counter [50] was used. About 40% of all existing $\Delta\sigma_T(pp)$ and $\Delta\sigma_L(pp)$ data were measured with this detector.

For most of the spin-dependent experiments, an apparatus consisting of a spectrometer and a polarimeter was used. It allowed measurement of single scattering observables as well as rescattering parameters for which the recoil particle transverse polarization was analyzed in a second scattering on carbon [51]. The angular dependence of the analyzing power is shown in Fig.2 at 4 energies from 0.874 to 2.70 GeV. Below 800 MeV the angular dependence has a quasi-sinusoidal behaviour. At higher energies the distributions become progressively less regular.

Some experiments were also done with a polarized deuteron beam in order to compare quasi-elastic scattering of protons (or neutrons) with free elastic *pp* (and *np*) scattering [8,26]. No significant differences between the measured

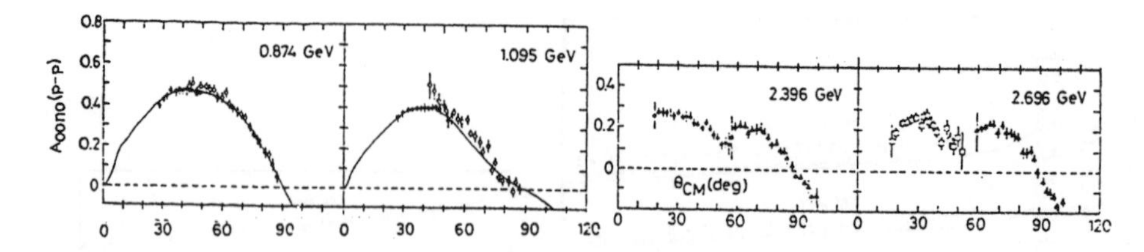

Figure 2: A_{oono} in pp elastic scattering at 4 energies between 0.874 and 2.70 GeV as measured by the Nucleon-Nucleon group [28]. Solid lines are phase shift predictions.

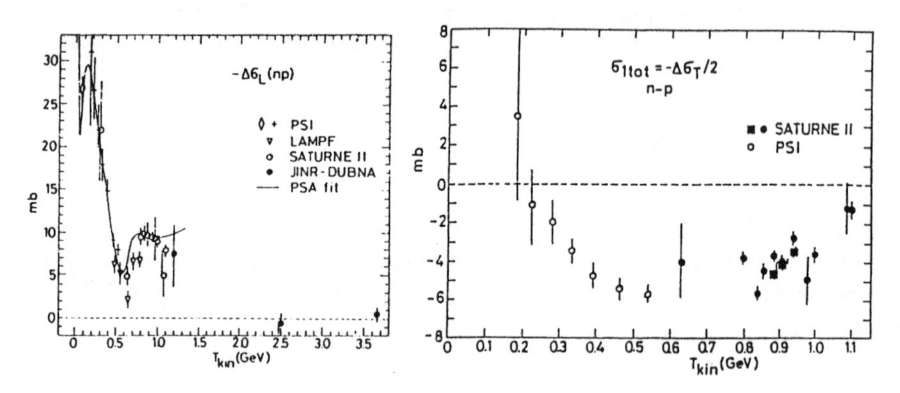

Figure 3: Energy dependence of $\Delta\sigma_L(\mathrm{np})$[47,48,54,55,56,53] and $\Delta\sigma_T(\mathrm{np})$[43,47,48,53].

spin-dependent observables were found.

Period 2 (1986-1990)

Period 2 continued with similar measurements of the np elastic spin-dependent observables with a polarized beam of free neutrons. A total of 1757 np data points, detailed in Table 2 were determined. A large neutron counter hodoscope built by the University of Geneva[52] was added to the apparatus behind the spectrometer. The MIMAS booster was installed during this period resulting in an increase of the neutron beam intensity to about 10^7 neutron/spill.

Total cross section spin-dependent differences were measured for the first time at SATURNE II using a new kind of fast detector [48,52] at zero degrees. These detectors were used later at PSI [53] and at JINR Dubna [54]. The existing np $\Delta\sigma_T$ [43,47,48,53] and $\Delta\sigma_L$ [47,48,54,55,56,53] data measured with free neutrons above 60 MeV are shown in Fig.3.

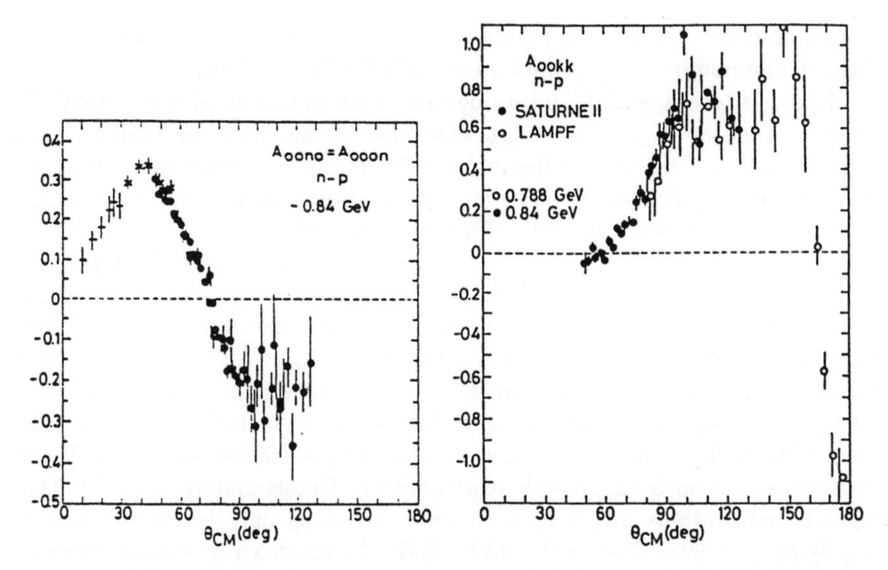

Figure 4: angular dependence of $A_{oono}(np)$ and $A_{ookk}(np)$ at 840 MeV. The full dots are the SATURNE II data[44], the open ones are LAMPF data[57] and plusses are IKAR data[2,3].

In Fig.4 the $A_{oono}(np)$ and $A_{ookk}(np)$ angular distribution at 0.84 GeV are shown. The sharp A_{ookk} variation close to 180° is a manifestation of the pion exchange.

Since 1.1 GeV is close to the maximum neutron beam energy at Saturne II, an extension to higher energy was only possible using quasielastic scattering of accelerated protons on the polarized deuteron target (PDT). The polarized ^6LiH and ^6LiD targets were developed and studied [58] by a collaboration of Saclay, Dubna, ANL, Prague and Kharkov scientists. An important part of the target materials (^6LiD and ^6LiH) was provided by the Dubna participants.

Period 3 (1991-1995)

The interesting features and structures obtained in the *pp* data (Period 1) were cross-checked (Period 3) until April 1995. A second superconducting beam solenoid was installed in order to obtain a pure sideways proton beam polarization. This polarization could be tuned using a new polarimeter constructed by the Gatchina physicists [30]. The Saclay MWPC's were considerably improved and operated by the Dubna participants. The apparatus was enhanced by use of new MWPC and electronics [52] from Argonne. An additional of 2425 new pp spin-dependent data points were obtained in Period 3 (see Table 1), bringing the number of pp points up to 5393 for Period 1 and 3.

Taking advantage of the pure sideways beam polarization orientation, observables such as K_{osso} and N_{onsk} were also measured, in particular at 1.8 and 2.1 GeV to help resolve the ambiguities found in a direct amplitude reconstruction (see Section 4). The spin correlation parameter A_{oosk} was also remeasured at 1.8 and 2.10 GeV with better statistics.

This period was also dedicated to measurements at the highest possible energies at SATURNE II, in particular to determine the A_{oono} and A_{oonn} energy dependence. Fig. 5 shows the A_{oonn} angular distribution at 2.52 GeV and 2.80 GeV [36] compared with the ANL-ZGS data at 3.17 GeV [59,60]; above 2.80 GeV the angular distribution is totally different from what is measured below. A $\sim$ 3.5σ dip is observed at 2.80 GeV close to $\theta_{CM} = 90°$. The ANL-ZGS results at 3.17 GeV show a similar angular dependence with larger errors (see Section 5 for possible explanation). Unfortunately the ANL-ZGS angular distributions up to 10 GeV were measured only below 60° and at exactly $\theta_{CM} = 90°$. Due to the SATURNE II shut-down it was not possible to investigate this interesting observation in more detail.

The measurements with the polarized beam and PPT provided 941 A_{oonn} data points [9,35,36] as well as $\sim$ 240 points for D_{onon} and K_{onno} [31]. Fig. 5 shows the D_{onno} results at 2.1 GeV as well as two phase shift predictions: The Saclay-Geneva PSA (dashed line) [61] which includes these points has a smooth angular dependence. The Virginia Polytechnic predictions (VPI PSA) [62] (solid line), which does not yet include these points, oscillates wildly and does not give a reliable representation of the data, illustrating the importance of these new D_{onno} data points.

This period was also dedicated to a careful study of the beam polarization. Although the beam polarization value increased over the years, one still had to check that the two opposite proton beam polarizations were strictly equal and that the "unpolarized" beam from the polarized ion source Hyperion was really unpolarized. The dedicated measurements carried out in 1996 [63] showed that the absolute values of the beam polarizations in the "polarized" ion source states are equal, whereas the two "unpolarized states" actually had a 6% polarization value with opposite signs.

4 Interpretation of the Data

All these pp and np SATURNE II data have been extensively used by groups performing phase shift analyses (PSA)(the VPI group [62], the japonese group [64] and the Saclay-Geneva group [65,66,67,68,61]). These data have also been used

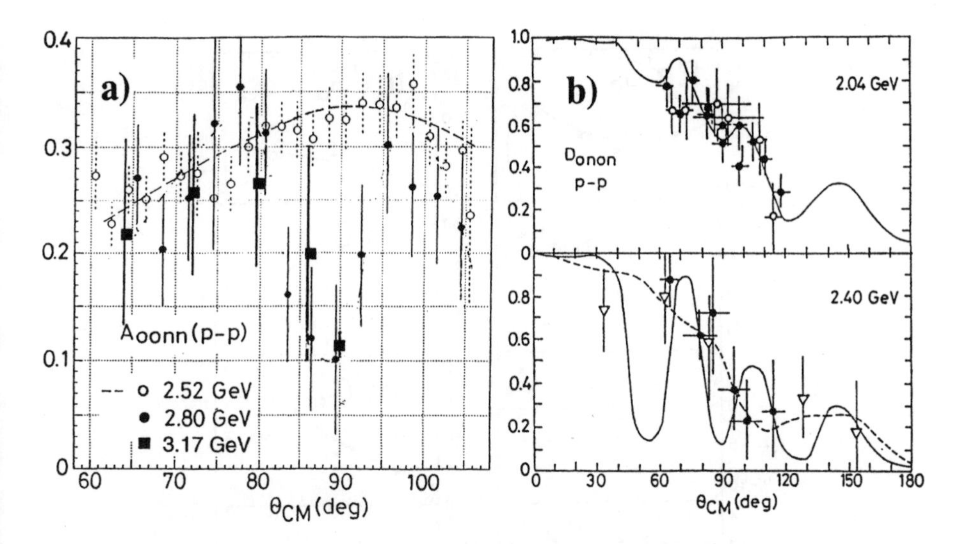

Figure 5: Angular dependence of a) $A_{oonn}(pp)$ at 2.52 and 2.8 GeV [34,36] and at 3.17 GeV [59,60]. b) D_{onon} at 2.04 and 2.4 GeV [9]. Solid and dashed lines are VPI[62] and Saclay-Geneva [61] phase shift predictions respectively.

to perform a direct amplitude reconstruction (DRSA)[23,69,61]. As illustrated in eq.1, NN observables can be expressed in terms of the scattering matrix elements. From there, two different approaches can be followed [70] – either a PSA over the entire angular range or, if enough different data are available at the same angle and the same energy, a DRSA.

4.1 *Phase shift analyses*

In the PSA approach the reconstruction of scattering amplitudes is possible even from incomplete sets of experimental quantities[71]. The lack of observables at some angles is compensated by imposed smooth functions of the angle. In the PSA the scattering amplitudes are developed as a series of Legendre polynomials and partial wave amplitudes which are independent of the scattering angle but contain phase shifts δ_{lj}. The high angular momentum phase shifts are taken from OPE in order to reduce the number of free parameters. These phase shifts are varied until a good χ^2 to the observables is obtained. The PSA below the pion production threshold for a fixed isospin state is practically model independent. At these energies only the OPE contribution may be considered as a weakly model-dependent part. With increasing energy the

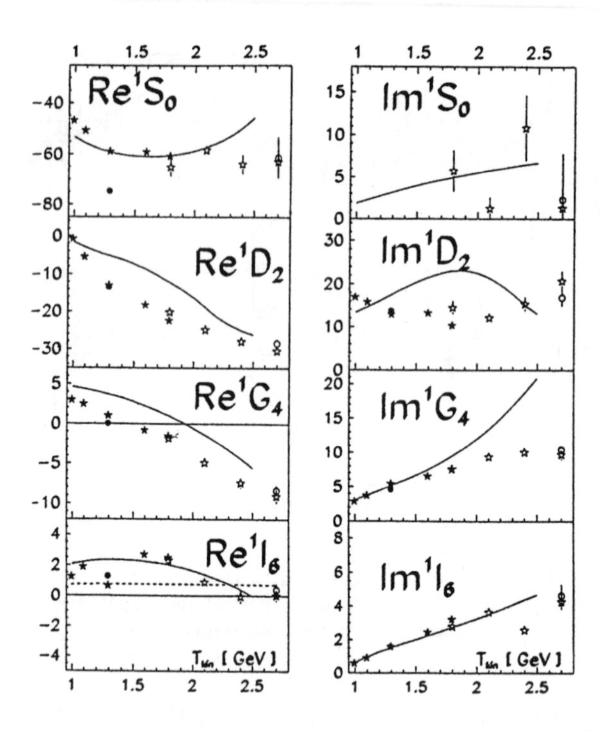

Figure 6: Spin singlet $I = 1$ phase shifts as a function of kinetic energy. Stars are Saclay-Geneva analyses[61,68,67] and solid line is VPI [62] predictions.

peripheral interaction described by OPE is introduced at progressively higher L_{max} values. Due to the increasing importance of inelastic interactions phase shifts may become complex. This increases the number of free parameters to 25 to 30 in the fixed-energy analyses. The only observable usable in the PSA which is directly related to the imaginary parts of phase shifts is the total inelastic cross section. The imaginary parts are poorly constrained for large angular momenta and must be taken from models. With increasing energy the PSA become more model dependent. An upper energy limit for the validity of the PSA method is however hard to estimate. The best way of checking the validity of model-dependent contributions to a PSA is to compare the PSA amplitude predictions with those from a DRSA.

PSA analyses above 800 MeV rely entirely on the SATURNE II data. They have allowed an extension of the *pp* PSA up to 2.7 GeV and the *np* PSA up to 1.1 GeV. Both fixed energy and energy-dependent PSA have been performed. Fig. 6 shows the lowest phase shifts above 1000 MeV as reconstructed by the Saclay-Geneva analyses (open and full stars)[67,68,61] and compares them to the VPI analysis (solid line) [62]. This later PSA is energy-dependent, the phase shifts are parametrized as functions of energy. This imposes a relatively smooth energy dependence on all parameters.

4.2 Direct amplitude reconstruction

The DRSA is performed at one energy and one angle, where it requires a complete set of measured observables [72]. For this reason it is always limited to a small number of angles and it will never be a universal tool predicting unknown quantities. The information provided by an accurately measured angular dependence of observables is lost, or may be used only indirectly to eliminate extraneous solutions. The common phase of amplitudes cannot be determined. On the other hand, a DRSA is completely model independent except for the symmetry laws. It gives a pure phenomenological description of a given interaction channel without any additional conditions and has no energy limit. Since the solution at one angle is independent of those at other angles, the amplitude analysis can reveal possible anomalies in angular or energy dependences. Using the SATURNE II data, it was possible to do this reconstruction in *pp* scattering at 11 energies from 0.834 to 2.696 GeV and several angles [23]. A similar reconstruction was possible at 6 energies between 800 and 1100 MeV and several angles in *np* scattering[69]. Fig.7 shows this *pp* reconstruction at 1.8 GeV as stars. The solid and dashed line are PSA predictions from VPI [62] and Saclay-Geneva [61] respectively. The agreement become worse at higher energies.

The best way of checking the validity of model-dependent contributions to a PSA is to compare the PSA amplitude predictions with the direct reconstruction from the amplitude analysis. In this respect, the amplitude analysis is complementary to the PSA. Agreement supports the PSA which may then be used to predict unmeasured quantities. A disagreement between the two methods either suggests a possible anomaly in the database, or casts a doubt on the PSA theoretical input. Comparison of PSA with directly-reconstructed amplitudes is also important at low energy close to the pion production threshold. The DRSA then provides a check of the validity of different potential models.

5 Search for Dibaryons

5.1 Dibaryon Resonances

Since this kind of "resonance" is often confused with resonance production (or an excited state), it is worthwhile to explain briefly this notion and its conse-

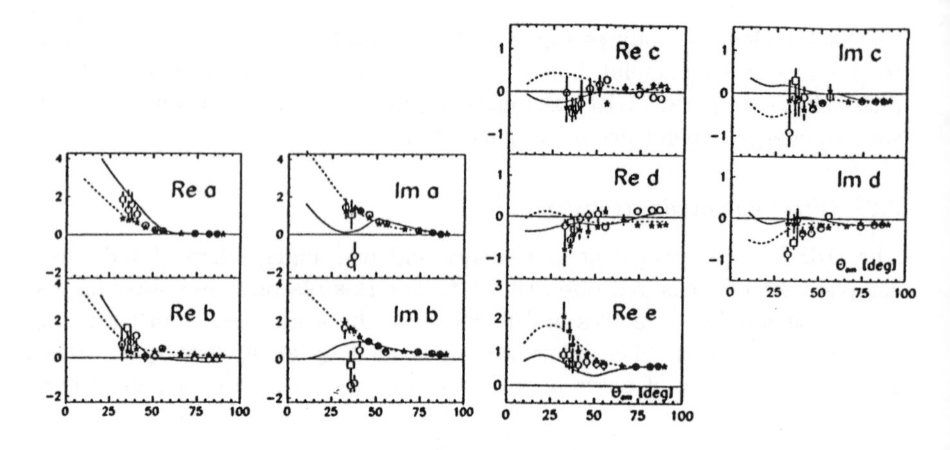

Figure 7: Direct reconstruction of the $I = 1$ scattering amplitudes at 1.8 GeV shown as points. The solid and dashed lines are predictions from PSA analyses [62,61] respectively.

quences. We restrain the discussion mainly to nucleon-nucleon interaction.

Production resonances may occur if the total CM energy of interacting baryons is larger than a production threshold of a given inelastic reaction. These are usually t-channel resonances, the Δ resonance in the $\pi - N$ system is a good exemple. Such resonances manifest itself in a fast increase of the reaction cross section. Below the production threshold only a virtual meson exchange occurs in NN scattering.

For long time it was expected that a model, containing all exchanges of possible mesons may describe energy dependence of nucleon-nucleon interaction without any free parameters. Theoreticians were encouraged by the success of the one-pion exchange (OPE), describing well peripheral elastic NN interaction. Contrary to expectation, exchange models hardly described NN spin-independent differential cross sections and, in general, failed for any spin-dependent observables.

A possible dibaryon formation corresponds to an exotic state of nucleon-nucleon system for which phenomenology of interaction corresponds to an elastic reaction. Such a formation may occur at any energy. Since the two baryons are fermions one assumes that energy dependent resonances may occur for quantified spin and orbital momenta of nucleon-nucleon states. A "resonance" will occur for possible $I = 1$ or $I = 0$ partial waves. From this point of view, a dibaryon formation may be considered as a kind of virtual exchange. The width of the resonance will depend on a given spin-state probability. However, the quark model introduce additional conditions.

Since any nucleon-nucleon interaction is described by five complex amlitudes at least, it is hard to expect that the dibaryon resonance will manifests itself mainly as an anomaly in the differential or total reaction cross sections. It is worthwhile to use spin-dependence of possible dibaryon resonance for its study. This implies to separate spin-singlet and spin-triplet contribution and apply resonant criteria on the level of scattering amplitudes or partial waves. For this reason the NN interaction, and namely the two particle reactions, remain the main tool for this purpose, although that study may be performed in arbitrary kind of baryon interaction.

A variety of QCD-based models have appeared in the literature since the 1950's[73,74,75,76,77,78,79,80,81,82]. All their mass predictions for the lowest two states for isotopic spin $I = 0$ and $I = 1$ fall within the range $\sqrt{s} = 2 - 3$ GeV although the methods used differ. This implies that interesting behavior is to be expected in NN scattering observables in the SATURNE II energy range. The oldest models, using the MIT Bag Model at the equilibrium radius that would be relevant if the multi-hadron system were confined and if there were no long range forces, give predictions around 2.2 GeV ($T_{kin}(p) \sim 0.7$ GeV)[73,74,79]. The most recent ones, taking the long range force into account [78,80,75] have predictions around 2.7 GeV ($T_{kin}(p) \sim 2.1$ GeV), i.e 500 MeV higher. Only one theory group[78] has attempted to explicitely estimate the dibaryon signature in spin observables.

5.2 NN Search for Structures

Search in missing mass experiments

Search for dibaryon resonances in nuclear reactions of nucleons or of light nuclei consists mainly in measurements of missing mass spectra. Identification of possible dibaryon resonances is more difficult and less accurate than in NN reactions. On the other hand, it is just this type of measurement which provided the major impetus to the search "dibaryon" resonances. Evidence for dibaryon resonance in reactions with light nuclei was reported in many SATURNE II papers [85,86,83,84,88] as summarized in Table.3. The resonance masses below 2.24 GeV are mentionned. Authors often referred to bubble chambers results measured at JINR Dubna. All signals are very weak, often within limits of statistical data fluctuations.

A study of spin-dependence was also carried out with polarized deuterons and/or protons, where interpretation is easier. Some dibaryon resonance signals in the lower energy region were provided by missing mass and invariant mass spectra of nucleon-nucleon inelastic reactions. For example, as indicated in Table 3, $d^2\sigma/d\Omega dM(pp)$ and the analyzing power for the rection $pp \to \pi^- X$ at $T_{kin}(p)$ 1.45, 2.1 and 2.7 GeV [87] at SATURNE II could be explained using N^*N, $N\Delta$ and $\Delta\Delta$ processes.

According to our opinion, a convincing experimental evidence for possible dibaryon resonances below 2.2 GeV mass energy is either nonexistent, or other "usual" explanation are strongly competitive.

Search in elastic *pp* scattering

Three experiments have been dedicated to the search for dibaryon resonances at SATURNE II in the mass region below 2.2 GeV: No signal for any dibaryon resonance was found.

1) Measurements of $d\sigma/d\Omega(pp)$ close to 90°_{cm} were made with the internal beam and a jet-target from 0.508 to 1.203 GeV (so-called "jets croises" experiment) [90]. The 121 data points show a smooth monotonically decreasing function without any anomaly.

2) Measurements of the $A_{oono}(pp)$ analyzing power between 0.655 and 1.017 GeV and CM angles from 39° to 82° [91] provided 1370 data points at 137 overlapping energies.

3) Measurement of the *pp* analyzing power at a fixed laboratory angle of 19.1° at energies from 0.51 to 0.725 GeV also showed no structure. This was done using a fast rotating degrader of variable thickness in conjunction with the SPES III spectrometer [92]. The experiment was triggered by a KEK result [93] claiming for narrow structures at $\sqrt{s}$=2.160 GeV and $\sqrt{s}$=2.192 GeV in a measurement performed with an internal target during the acceleration of the polarized proton beam. Results are shown in Fig. 8 and convincingly disprove the KEK experiment results.

On the other hand, structures have been found in the two following experiments: 1) Measurements of $\Delta\sigma_L$ and $\Delta\sigma_T$ have often been used as arguments for possible resonances. One can now conclude, using the SATURNE II data which studied the total cross section energy dependence, that the observed structures are most likely caused by threshold effects for the S and P wave production of $p\Delta$ final states. The threshold for the reaction $p + p \to p + \Delta^+$ is 633 MeV and the $P-$wave enters some 200 MeV higher. $-\Delta\sigma_L(pp)$ energy dependence shows two pronounced extrema : a minimum around 0.6 GeV and a maximum at 0.75 GeV. These extrema were attributed to resonances in $^1D_2$ and $^3F_3$ partial waves with masses at $\sqrt{s}$ =2.16 and 2.24 GeV, respectively. However, the $-\Delta\sigma_T(pp)$ energy dependence shows similar extrema at the same

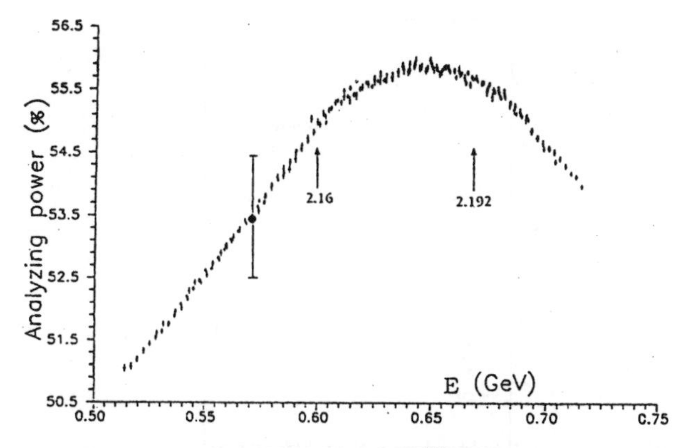

Figure 8: Energy dependence of the analyzing power [92]. The large error bar illustrates the KEK results precision [93] for comparison.

energies. The previous interpretation cannot hold here as $-\Delta\sigma_T(pp)$ contains no uncoupled triplet states, in particular no $^3F_3$ partial waves. Similar np quantities have been measured and show also a minimum at 0.6 GeV as seen in Fig.3. Using pp and np, the $I = 0$ total cross section differences can be calculated. $-\Delta\sigma_L(I = 0)$ shows a narrow minimum and $-\Delta\sigma_T(I = 0)$ a broad one, both at 0.650 GeV. There in $I = 0$ state the $^1D_2$ partial waves does not contribute and the same explanation does not apply. It is surprising that the size of the energy dependent structure seen in $-\Delta\sigma_L(I = 0)$ is comparable to that for $-\Delta\sigma_L(I = 1)$, even though the strong $NN \rightarrow N\Delta$ channel cannot contribute to $I = 0$ and the inelastic $I = 0$ cross section is much smaller than the $I = 1$ inelastic cross-section at these energies.

2) The $A_{oonn}(pp)$ measurement at 90° allows a search for possible structures in the spin-singlet scattering amplitudes. At 90°_{cm} three scattering amplitudes survive since $a = 0$ and $b = -c$. From the measured A_{oonn} observable and known pp elastic differential cross section at 90°_{cm} one can calculate the absolute value of the spin-singlet amplitude

$$|b|^2 = |c|^2 = \frac{1}{2}\frac{d\sigma}{d\Omega}(1 - A_{oonn}) \tag{4}$$

which is shown in fig 9 as measured by the Nucleon-Nucleon group[34,36]. The remaining two non-zero amplitudes d and e are pure spin-triplets. The amplitude $|b|^2 = |c|^2$ shows a break and a plateau centered around 2.1 GeV beam kinetic energy with a width $\Delta T_{kin} = \pm 0.1$ GeV (i.e $\sqrt{s} = 2.73 \pm 0.04$ GeV). A narrow maximum at $T_{kin} = 2.8$ GeV ($\sqrt{s}$= 2.96 GeV) also appears. Above

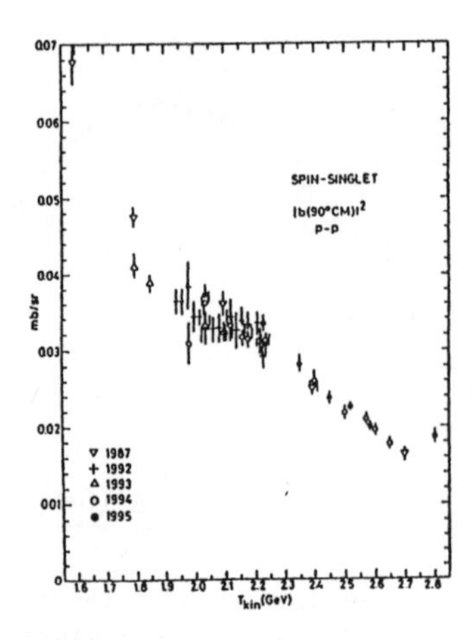

Figure 9: Amplitude $|b|^2$ at 90°_{cm} in pp elastic scattering[34,36]

this point the amplitude $|b|^2$ drops down with energy as $d\sigma/d\Omega$ due to the constant A_{oonn} values from the existing ANL-ZGS results [60]. Similar, but less pronounced behaviour of $d\sigma/d\Omega(1 - A_{oonn})$ may be observed at $80°CM$, where the spin-triplet contribution to amplitudes b and c remains small. A decreasing energy dependence can be seen at $70°CM$, where the spin-triplet contribution becomes dominant. The amplitude $|d|^2$ is a decreasing function of energy from 1.9 to 2.3 GeV. Between 2.4 and 2.6 GeV one observes constant values. The dominant amplitude $|e|^2$ is monotonicaly decreasing in the entire energy interval.

The results obtained by the NN group represent a consistent experimental indication for a possible narrow structure in spin-singlet pp elastic scattering. If dibaryon resonances are involved, their masses could be around 2.73 GeV and 2.96 GeV.

Search in pp→ πd

Some other evidence came from π production experiments. Between 725 and 1000 MeV, measurements of analyzing power and cross sections angular distributions have shown no sign of unusual structure over widths of 25 to 50 MeV in the c.m[94]. Possible evidence for a structure was suggested by similar measurements but at higher energies[95,96] and confirmed recently by new measurements of $d\sigma/d\Omega$ (45 points) and A_{no} (77 points) at 1.30, 1.60, 1.70, 1.88, 2.10 and 2.40

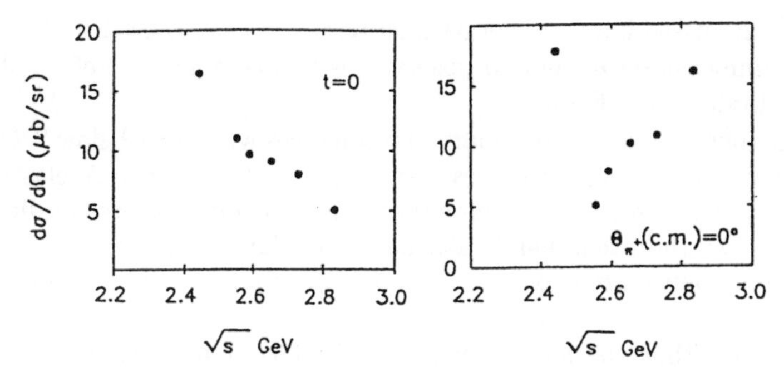

Figure 10: Excitation function of the pp $\rightarrow \pi$d cross-sections for $t = 0$ or $\theta_\pi(c.m) = 0°$

GeV [97,98]. The angular distributions of cross sections and analyzing powers were fitted simultanously with Legendre polynomials and associated Legendre polynomials. Some of the coefficients show structures around $\sqrt{s}$= 2.2, 2.4 and 2.65 GeV. Based on these fits, at each energy the values of $d\sigma/d\Omega$ and A_{no} for the Mandelstam variables $t = 0$, $u = 0$ or for $\theta_\pi(c.m) = 90°$ were obtained. The cross sections at $t = 0$ and extrapolations to 0° both show a shoulder at $\sqrt{s} \simeq$2.65 GeV as illustrated in Fig.10. The plot for the analyzing power confirmed the rapid variations suggested by earlier experiment [95,96]. The same authors mentioned that the observed structures may be also explained without invoking exotic dibaryon resonances. The peak at $\sqrt{s} = 2.2$ GeV is explained as an excitation of an $N\Delta$ state with $L_{N\Delta} = 0$. Using meson-nucleon model, it is conceivable that the structures at high energy may be provided by threshold effect due to opening of higher $N\Delta$ orbital momentum channels[99]. To distinguish pion-nucleon model explanation from the narrow dibaryon resonance it will be necessary to perform measurement of other spin dependent observables and compared it to these models predictions. Such a measurement of K_{nn} has been performed but is not fully analyzed yet .

6 Conclusions

The results obtained during the 17 years of SATURNE II activity have had an appreciable impact in the field. Systematic measurements allowed the extension of semi-phenomenological descriptions of the elastic proton-proton interaction up to 2.7 GeV and provided evidence for interesting features in different spin-dependent observables. SATURNE II in conjunction with Mimas enlarged the energy domain of np experiments with free neutrons up to 1.1 GeV. The data provided the first direct reconstruction of np scattering amplitudes. Although not discussed here, accelerated deuterons were also used to determine previously unmeasured *pd* and *dd* observables and complete set of *dp* observables was obtained at 1.6 GeV.

No convincing evidence for dibaryon resonances with mass below 2.25 GeV was obtained. Interesting structures were clearly observed in NN elastic and inelastic reactions near 2.7 GeV where theoretical calculations predict dibaryon resonances. On the other hand, alternative explanations are also possible. Unambiguous confirmation of the observed structures needs new dedicated experiments.

It is a pity that the shutdown of SATURNE II has prevented further investigation of this intriguing puzzle. One can hope that NN physics close to 3 GeV can be further explored with improvements in the new COSY accelerator.

References

1. J.Bystrický *et al.*,*J. Phys. (Paris)* **39**, 1 (1978).
2. Y. Terrien *et al.*, *Phys. Rev. Lett.* **59**, 1534 (1987).
3. B.H.Silverman *et al.*, *Nucl. Phys.* A **499**, 763 (1989).
4. A.A. Vorobyov *et al.*, *Nucl. Instrum. Methods* A **270**, 419 (1988).
5. S.Dalla Torre-Colautti *et al.*,*Nucl. Phys.* A **505**, 561 (1989).
6. R.Bernard *et al.*, *Nucl. Instrum. Methods* A **249**, 176 (1986)
7. J.Arvieux *et al.*, *Z. Phys.* C **766**, 465 (1997).
8. J.Ball *et al.*, *Nucl. Phys.* B **286**, 635 (1987).
9. J.Ball *et al.*, to be published in *Eur.Phys.J.* C , (1998).
10. J.Bystrický *et al.*, *Phys. Lett.* B **142**, 130 (1984).
11. J. Bystrický *et al.*, *Nucl. Phys.* B **262**, 715 (1985).
12. J. Bystrický *et al.*, *Nucl. Phys.* B **262**, 727 (1985).
13. J. Bystrický *et al.*, *Nucl. Phys.* B **258**, 483 (1985).
14. J. Bystrický *et al.*, *Nucl. Instrum. Methods* A **239**, 131 (1985).
15. J. Bystrický *et al.*, *Nucl. Instrum. Methods* A **234**, 412 (1985).
16. J.-M.Fontaine *et al.*, *Nucl. Phys.* B **321**, 299 (1989).
17. C.D.Lac *et al.*, *Nucl. Phys.* B **297**, 653 (1988).
18. C.D.Lac *et al.*, *Nucl. Phys.* B **315**, 269 (1989).
19. C.D.Lac *et al.*, *Nucl. Phys.* B **315**, 284 (1989).
20. C.D.Lac *et al.*, *Nucl. Phys.* B **321**, 269 (1989).
21. C.D.Lac *et al.*, *Nucl. Phys.* B **321**, 284 (1989).
22. F.Lehar *et al.*, *Nucl. Phys.* B **294**, 1013 (1987).
23. C.D.Lac *et al.*, *J. Phys. (France)* **51**, 2689 (1990).
24. F.Lehar *et al.*, *Europhys. Lett.* **3**, 1175 (1987).
25. F.Lehar *et al.*, *Nucl. Phys.* B **296**, 535 (1988).

26. A. de Lesquen *et al.*, *Nucl. Phys.* B **304**, 673 (1988).
27. F.Perrot *et al.*, *Nucl. Phys.* B **278**, 881 (1986).
28. F.Perrot *et al.*, *Nucl. Phys.* B **294**, 1001 (1987).
29. F.Perrot *et al.*, *Nucl. Phys.* B **296**, 527 (1988).
30. C.E.Allgower *et al.*, *Eur.Phys.J.* C **1**, 131 (1998).
31. C.E.Allgower *et al.*, *Nucl. Phys.* A , (in print).
32. C.E.Allgower *et al.*, submitted to *Eur.Phys.J.* C , (1998).
33. C.E.Allgower *et al.*, submitted to *Phys. Rev.* C , (1998).
34. C.E.Allgower *et al.*, submitted to *Phys. Rev.* C , (1998).
35. C.E.Allgower *et al.*, submitted to *Phys. Rev.* C , (1998).
36. C.E.Allgower *et al.*, submitted to *Phys. Rev.* C , (1998).
37. D.Adams *et al.*, *Acta Polytechnica (Prague)* **36**, 11 (1996).
38. J.Ball *et al.*, *Z. Phys.* C **40**, 193 (1988).
39. J.Ball *et al.*, *Nucl. Phys.* A **559**, 477 (1993).
40. J.Ball *et al.*, *Nucl. Phys.* A **559**, 489 (1993).
41. J.Ball *et al.*, *Nucl. Phys.* A **559**, 511 (1993).
42. J.Ball *et al.*, *Nucl. Phys.* A **576**, 640 (1994).
43. J.Ball *et al.*, *Z. Phys.* C **61**, 53 (1994).
44. J.Ball *et al.*, *Nucl. Phys.* A **574**, 697 (1994).
45. J.Ball *et al.*, *Z. Phys.* C **61**, 579 (1994).
46. J.Bystrický *et al.*,*Nucl. Phys.* A **444**, 597 (1985).
47. J.-M.Fontaine *et al.*, *Nucl. Phys.* B **358**, 297 (1991).
48. F.Lehar *et al.*, *Phys. Lett.* B **189**, 241 (1987).
49. A. de Lesquen *et al.*, *Nucl. Phys.* B **304**, 673 (1988).
50. M.Arignon *et al.*, *Nucl. Instrum. Methods* A **235**, 523 (1985).
51. M.Arignon *et al.*, *Nucl. Instrum. Methods* A **262**, 207 (1987).
52. J.Ball *et al.*, *Nucl. Instrum. Methods* A **327**, 308 (1993).
53. R. Binz *et al.*, *Nucl. Phys.* A **533**, 601 (1991).
54. B.P. Adiasevich *et al.*, *Z. Phys.* C **71**, 65 (1996).
55. M. Beddo *et al.*, *Phys. Lett.* B **258**, 24 (1991).
56. P. Hafner *et al.*, *Nucl. Phys.* A **548**, 29 (1992).
57. W.R. Dietzler *et al.*, *Phys. Rev.* D **46**, 2792 (1992).
58. J.Ball *et al.*, *Nucl. Instrum. Methods* A **381**, 4 (1996).
59. D.Miller *et al.*,*Phys. Rev.* D **16**, 2016 (1977).
60. A.Lin *et al.*,*Phys. Lett.* B **74**, 273 (1978).
61. J. Bystrický *et al.*, to be published in *Eur.Phys.J.* C , (1998).
62. D. Arndt *Phys. Rev.* D **35**, 128 (1987) SAID SM97.
63. C.E.Allgower *et al.*, *Nucl. Instrum. Methods* A **399**, 171 (1997).
64. H. Yoshino *et al.*, *Prog.Theor.Phys.* **95**, 577 (1996).
65. J. Bystrický *et al.*,*J. Phys. (Paris)* **48**, 199 (1987).

66. C. Lechanoine-Leluc *et al.*,*J. Phys. (Paris)* **48**, 985 (1987).
67. F. Lehar *et al.*,*J. Phys. (Paris)* **48**, 1273 (1987).
68. J. Bystrický *et al.*,*J. Phys. (France)* **51**, 2747 (1990).
69. J. Ball *et al.*,*Nuovo Cimento* **111**, 13 (1998).
70. C. Lechanoine-Leluc and F.Lehar, *Rev. Mod. Phys* **65**, 47 (1993).
71. N.P.Klepikov *et al.*, *Sov. Phys. JETP* **20**, 505 (1964).
72. L.D.Puzikov *et al.*,*Nucl. Phys.* **3**, 436 (1957).
73. A.Th.M.Aerts *et al.*, *Phys. Rev.* D **17**, 260 (1978).
74. C.W.Wong and K. Liu , *Phys. Rev. Lett.* **41**, 82 (1978).
75. C.W.Wong , *Proc in Part. and Nucl.Phys.* **8**, 223 (1982).
76. E.L.Lomon , *Coll. Phys. (France)* **46**, C2-329 (1985).
77. P.LaFrance *et al.*, *Phys. Rev.* D **34**, 1341 (1986).
78. P.Gonzales *et al.*, *Phys. Rev.* D **35**, 2142 (1987).
79. N. Konno *et al.*, *Phys. Rev.* D **35**, 239 (1987).
80. Yu.S.Kalashnikova *et al.*, *Sov.J.Nucl.Phys.* **46**, 689 (1987).
81. L.A.Kondratyuk *et al.*, *Nuovo Cimento* **102A**, 25 (1989).
82. E.L.Lomon , *Coll. Phys. (France)* **51**, C6-363 (1990).
83. B. Tatischeff *et al.*, *Z. Phys.* A **328**, 147 (1987).
84. B.Tatischeff *et al.*, *Europhys. Lett.* **4**, 671 (1987).
85. B. Tatischeff *et al.*,*Phys. Rev. Lett.* **52**, 2022 (1984).
86. B. Tatischeff *et al.*,*Phys. Rev.* C **36**, 1995 (1987).
87. M.P.Combes-Comets *et al.*,*Phys. Rev.* C **43**, 973 (1991).
88. B.Tatischeff *et al.*,*Phys. Rev.* C **45**, 2005 (1992).
89. M.P. Combes *et al.*, *Nucl. Phys.* A **431**, 703 (1984).
90. M.Garçon *et al.*, *Nucl. Phys.* A **445**, 669 (1985).
91. M.Garcon *et al.*, *Phys. Lett.* B **183**, 273 (1987).
92. R.Beurtey *et al.*,*Phys. Lett.* B **293**, 27 (1992).
93. H. Shimizu *et al.*, *Phys. Rev.* C **42**, 42 (1990).
94. B. Mayer *et al.*,*Nucl. Phys.* A **437**, 630 (1985).
95. R.Bertini *et al.*,*Nucl. Phys.* B **1622**, 77 (1985).
96. J.Bertini *et al.*,*Nucl. Phys.* B **203**, 18 (1988).
97. J.Yonnet *et al.*,*Coll. Phys. (France)* **51**, C6-379 (1990).
98. J.Yonnet *et al.*,*Nucl. Phys.* A **562**, 352 (1993).
99. Y. Niskanen *Phys. Lett.* B **112**, 17 (1982).

Table 1: Summary of pp observables measured by the Nucleon-Nucleon group.

Observable	T_{kin} (GeV) Interval	Range (CM deg)	Points	References
$\Delta\sigma_T$	0.494 - 2.394	0	28	27
$\Delta\sigma_L$	0.511 - 2.795	0	11	10
A_{oono}, A_{ooon}	0.424 - 2.795	19-105	1904	11,12,28,8,9,7 31,32,33,35,26
A_{oonn}	0.494 - 2.795	19-105	1468	11,12,13,9 22,34,36
A_{ookk}	0.719 - 2.696	21-103	551	13,17,25,16
A_{oosk}	0.744 - 2.696	22-102	411	13,29,16,30,26
D_{onon}	0.834 - 2.795	31-97	237	19,20,30,9
$D_{os''ok}$	0.834 - 2.696	28-93	149	18,21,30
K_{onno}	0.834 - 2.795	27-96	181	19,32,9
$K_{os''ko}$	0.834 - 2.696	28-92	72	18
$K_{os''so}$	1.795 - 2.396	66-93	12	21,30
N_{onkk}	1.796 - 2.396	66-92	15	18
$N_{os''nk}$	1.796 - 2.696	66-86	5	21
$N_{os''sn}$	1.795 - 2.095	66-93	6	32
N_{onsk}	1.795 - 2.095	66-93	6	32
N_{onkk}, $K_{ok''ko}$	0.834 - 2.696	28-85	63	18
$N_{os''nk}$, $K_{os''so}$	0.834 - 2.696	28-90	74	21
$N_{os''kn}$, $N_{os''sn}$ and $N_{ok''kn}$	0.834 - 2.096	31-84	59	20
$K_{os''ko}$, $K_{os''so}$ and $K_{ok''ko}$	0.834 - 2.096	31-85	57	20
K_{onno}, $K_{ok''so}$, N_{onsk}, $N_{ok''nk}$	0.834 - 2.696	28-90	76	21
K_{onno}, $K_{ok''so}$	1.796 - 2.696	66-91	6	21
N_{onsk}, $N_{ok''nk}$	2.396, 2.696	82,83	2	21
Total points			5393	

Table 2: Summary of np observables measured by the Nucleon-Nucleon group.

Observable	T_{kin} (GeV) interval	Range (CM deg)	Points	References
$\Delta\sigma_T$	0.630 - 1.100	0	11	48,47,43
$\Delta\sigma_L$	0.312 - 1.100	0	10	48,47
A_{oono}, A_{ooon}	0.312 - 1.100	21-150	766	13,8,39,40,14,49
A_{oonn}	0.744 - 1.100	26-150	335	41,42,49
A_{ookk}	0.312 - 1.100	29-148	256	38,44
A_{oosk}	0.744 - 1.100	26-148	219	49,44
D_{onon} and K_{onno}	0.800 - 1.100	48-86	30, 30	45
$D_{os''ok}$	0.800 - 1.100	42-77	20	45
$K_{os''ko}$ and N_{onkk}	0.800 - 1.100	42-77	20	45
$K_{os''so}$ and N_{onsk}	0.800 - 1.100	42-77	20	45
Total points			1757	

Table 3: Summary of Dibaryon searches in terms of missing and effective masses.

Reaction Studied	Apparatus	Ref.	Observations	
$p+^3He \to d+X$	SPES I	83	$M_X = 1969$ MeV	$\Gamma = 9$ MeV
		84	$M_X = 2122$ MeV	$\Gamma_{\frac{1}{2}} = 5.2$ MeV
$(^3He+p \to d+X)$			$= 2198$ MeV	$\Gamma_{\frac{1}{2}} = 8.1$ MeV
			$= 2233$ MeV	$\Gamma_{\frac{1}{2}} = 13$ MeV
		85	$M_X = 2240$ MeV	$\Gamma_{\frac{1}{2}} = 16$ MeV
		86	$M_X = 2240$ MeV	$\Gamma_{\frac{1}{2}} = 16$ MeV
			$= 2192$ MeV	$\Gamma_{\frac{1}{2}} = 25$ MeV
			$= 2121$ MeV	$\Gamma_{\frac{1}{2}} = 25$ MeV
$p+p \to \pi^- + X$	SPES III	87	No πNN bound states or 6-quark states established	
$d+p \to p+p+X$	SPES III	88	$M_{pp} = 1945$ MeV	$\Gamma_{\frac{1}{2}} = 8$ MeV
$d+d \to d+X$	SPES IV	89	No narrow peak	

NUCLEON-NUCLEON INTERACTION : INELASTIC CHANNELS

W. KÜHN
II. Physikalisches Institut
Universität Gießen
Heinrich-Buff-Ring 16, 35392 Gießen, Germany
E-mail: Wolfgang.Kuehn@exp2.physik.uni-giessen.de
http://www.physik.uni-giessen.de

A review of experimental studies of the inelastic channels in nucleon-nucleon collisions at SATURNE-2 is presented. This includes the measurement of $\pi^+, \pi^-, \pi^0, \eta, \eta', K^+, K^-, \omega, \phi$ – mesons as well as investigations of Λ–and Σ^0-hyperon production. Both unpolarized and polarized beams were employed to obtain integrated and differential cross sections and polarization observables.

1 Introduction

There are 3 fundamental interactions in nature: (i) gravitation, with the solar system or binary stars as basic bound states, (ii) the electro-weak interaction with atoms and positronium as fundamental systems, (III) the strong interaction, where the fundamental bound systems are baryons and mesons. In contrast to systems bound by gravitational or electromagnetic forces, baryons and mesons are *not well understood* in terms of QCD, the fundamental theory describing the strong interaction. Several intriguing questions remain to be answered:

- What is the origin of the confinement
- What is a "constituent quark"
- What is the structure of the QCD vacuum (this relates to the origin of the hadron masses and the role of the light pseudo-scalar mesons as Goldstone bosons)
- What is the internal structure of baryons at low Q^2
- Where are the missing bound states (glueballs, hybrids,6-quark states)

From the theroetical point of view these problems address the regime of non-perturbative QCD, where lattice calculations provide a fundamental but technically very difficult approach. Lattice calculations with realistic current quark masses and large lattice size are still not feasible.

As an alternative approach, constituent quark models provide easier access to experimental observables but the justification of these models in terms of QCD is not so evident. Theories with mesons as effective degrees of freedom such as Chiral Perturbation Theory as well as meson exchange models for the description of scattering and reactions have had considerable success in describing the phenomenological aspects of the experimental data, but can only qualitatively be justified on the basis of QCD.

From the experimental point of view, SATURNE-2 has made major contributions to the field of hadron physics in the low energy regime within the last 20 years. In this review, we shall concentrate on information obtained from studies of the inelastic nucleon- nucleon reaction channels, probed via the detection of photons and π^+, π^-, π^0, η, η', K^+, K^-, ω, ϕ –mesons and hyperons.

2 Neutron-Proton Bremsstrahlung

The reaction np->npγ is one of the most fundamental processes in nucleon-nucleon interactions, since it constitutes an inelastic channel with *only 2 hadrons* involved. Since the electromagnetic interaction is well understood, unique information on the hadron-hadron interaction part of the reaction might be obtained. Moreover, neutron-proton bremsstrahlung plays a dominant role as the fundamental process for direct photon production in nucleus-nucleus collisions in the energy regime of 20-100 AGeV.

It should be noted, that the study of np bremsstrahlung yields complementary information to experiments observing pp bremsstrahlung. Dipole radiation is suppressed in pp reactions, but plays a dominant role in np reactions. Of particular interest is the sensitivity of np->npγ experiments to contributions of meson exchange currents.

Pioneering experiments measuring for the first time the photon energy spectrum in proton-neutron bremsstrahlung have been carried out by the group of H.Nifenecker[1] and collaborators. Such experiments are very difficult, since the required neutron beams are much less intense than proton beams, and cross sections are small. The experimental setup used at SATURNE is shown in fig. 1.

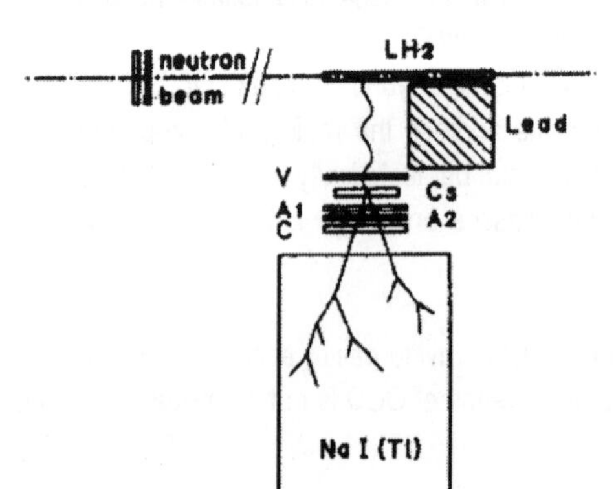

Figure 1: Experimental setup for the detection of neutron –proton bremsstrahlung

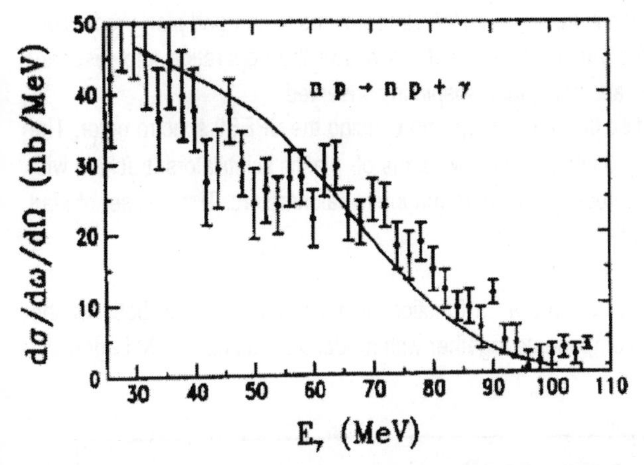

Figure 2:
Photon energy distribution in np bremsstrahlung at an incident energy of 170 MeV

Photons were detected in two identical large NaI(Tl) detectors supplemented by veto detectors for charged particle rejection. The neutron beam with an energy of about 170 MeV was produced from a 400 MeV deuterium beam incident on a beryllium target. Fig. 2 shows the resulting photon spectrum in comparison to a calculation with a one-boson exchange model[2], which is in reasonable agreement with the data.

3 Production of pseudoscalar mesons

3.1 *π^0 production*

Several experiments at SATURNE-2 were devoted to the measurement of neutral pions. Of particular interest is the role of the $\Delta(1232)$ resonance, which can be studied by

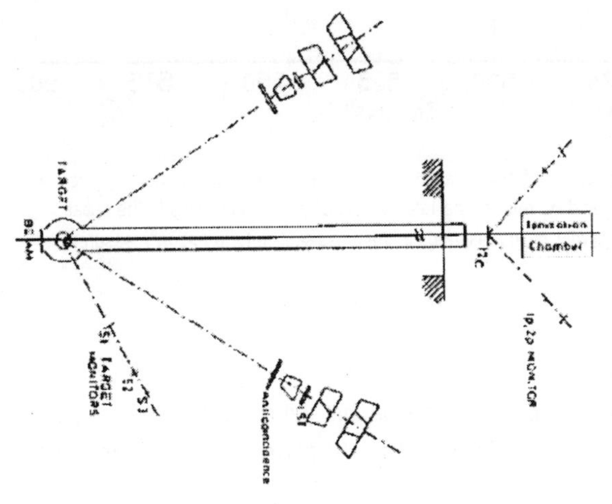

Figure 3: Experimental configuration of the SPES0 spectrometer. Photons are detected in a 2-arm setup with lead glass detectors as electromagnetic calorimeters

comparing neutral pion production in nucleon-nucleon interactions to charged pion production. An isospin decomposition of the various cross sections allows the separation into resonant and non-resonant contributions, assuming that isospin is conserved.

Didelez et al.[3] have studied the reaction pp->ppπ^0 using the SPES0 spectrometer. The experimental setup shown in fig. 3 consisted of two arms of plastic scintillators, tracking with streamer tubes and lead glass detectors as electromagnetic calorimeters. The first set of lead glass detectors were used as active converters.

Fig. 4 shows total cross sections as a function of the incident energy both for the resonant and the non-resonant components together with model predictions by J.M.Laget.

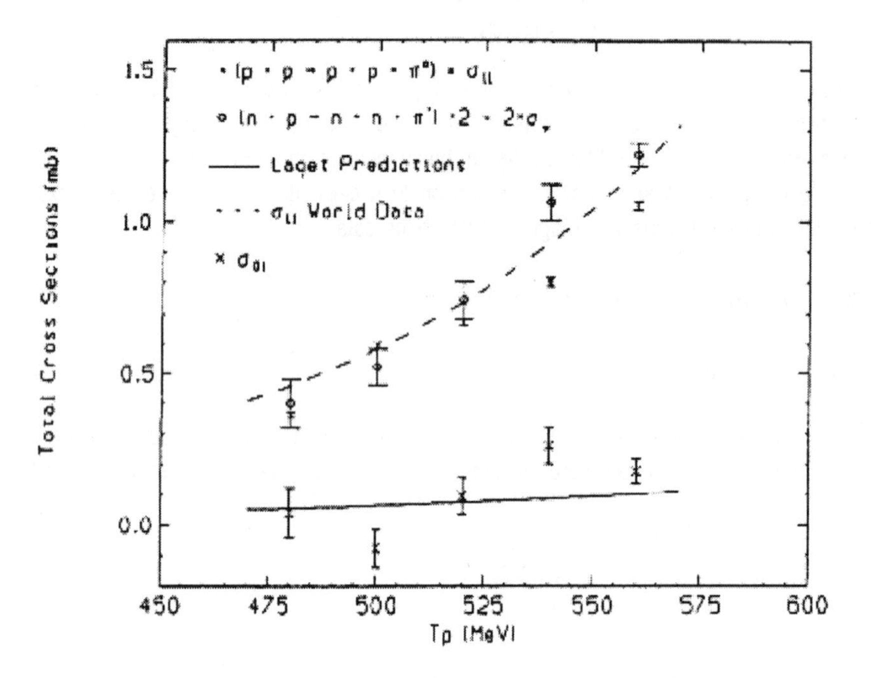

Figure 4: Total cross section of pion production as a function of incident energy. The cross sections σ_{11}, σ_{10} and σ_{01} refer to the total isospin of the nucleon pairs in the initial and final states, respectively. σ_{11} corresponds to the cross section for pp->ppπ^0 measured by Didelez et al[3].

From the analysis shown in fig. 4 it is obvious that the non-resonant part σ_{01} of the cross section is small, which emphasizes the important role played by the delta resonance in this energy regime.

3.2 Pion production with polarized beams

More detailed information on the meson production mechanism can be obtained by studies with polarized beams. Rappenecker et al.[4] have continued their SPES0 – work [3] on π^0 production in new experiments making use of polarized protons and with an improved experimental setup.

Comptour et al.[5] have employed DIOGENE to identify neutral pions via missing mass in pp reactions with polarized protons. The measured analyzing power for this reaction is shown in fig. 5. Together with measurements of differential cross sections, these experiments provide strong constraints for model calculations. Again, the dominant role of the Δ resonance was confirmed.

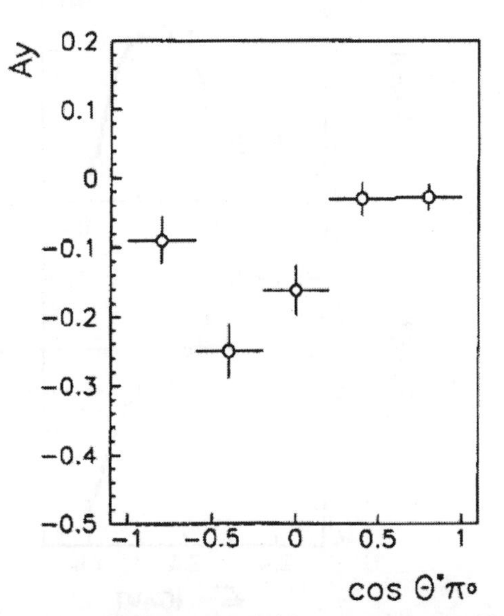

Figure 5: Analyzing power for the reaction pp->ppπ^0, taken from ref. 12.

Polarization experiments are valuable tools in the search for dibaryons. Narrow resonances in inelastic pp scattering were interpreted as evidence for exotic hadronic states[6]. Such a conjecture, however, has to be confronted with more conventional interpretations, for instance in terms of N^* resonances or threshold effects.

Complete angular distributions and measurements of polarization observables are important to test such interpretations. As an example, fig. 6 shows data obtained by Bertini et al.[7] for the reaction pp->dπ^+ with polarized beams. Experiments along these lines were also performed by J Yonnet at al.[8], making use of the SPES4 spectrometer.

Experiments studying π^- production with polarized protons were performed by Combes-Comets[9], using the SPESIII spectrometer. No evidence for T=2 dibaryons was found

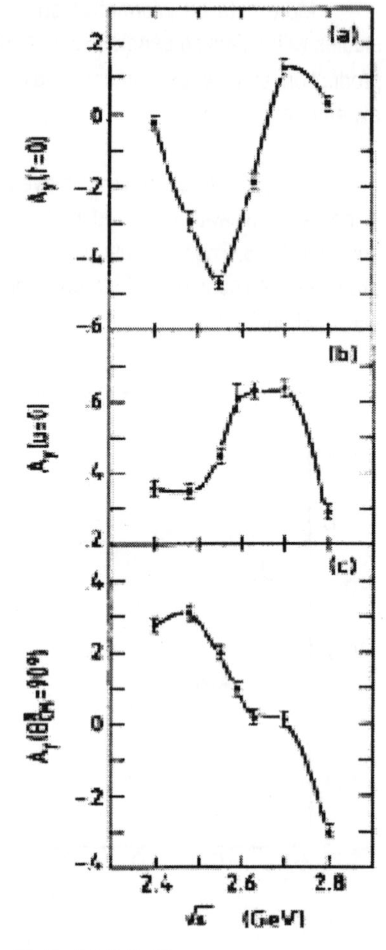

Figure 6:
Measurement of the analyzing power in pp->dπ^+ with polarized proton beams (R.Bertini et al.[14]). Shown is the energy dependence of A_{y0} at various kinematical conditions: t=0 (a), u=0(b) and θ_{cm}=90°(c). The rapid variation of the analyzing power as a function of energy indicates the existence of a resonant structure, which could, however, be consistently interpreted in terms of a N* resonance at the πN scattering vertex.

and an upper production cross section limit of 90 nb/sr MeV at 2.2 GeV resonance mass was obtained.

Very interesting data on π^- production with *polarized neutron beams* have been obtained by Y. Terrien at al.[10] employing the ARCOLE detector. Polarized neutron beams were produced at SATURNE by breakup of polarized deuterons on Be targets. More that 10^6 neutrons/s with typical polarizations of 60% were obtained. Again, models are able to predict differential cross sections but fail to describe the polarization observables in detail.

3.3 *η - Production*

Pioneering experiments on η meson production in near-threshold nucleon-nucleon collisions have been carried out at SATURNE-2. Detection of the η mesons was done both via missing mass measurement of the 2 protons in pp->ppη using the SPES3 spectrometer[11] or by detection of the decay photons using the 2-arm PINOT photon spectrometer[12,13]. As an example, fig. 5 shows the excitation function obtained by Chiavassa et al.[13] with the PINOT spectrometer.

For comparison, model calculations by Laget et al.[14] (assuming a dominant contribution of ρ-exchange) and Vetter et al.[15] (here, π exchange was assumed to be dominant) are shown, both are agreeing reasonably well with the data. All these models assume that the reaction proceeds via the formation of the $N^*(1535)$ resonance.

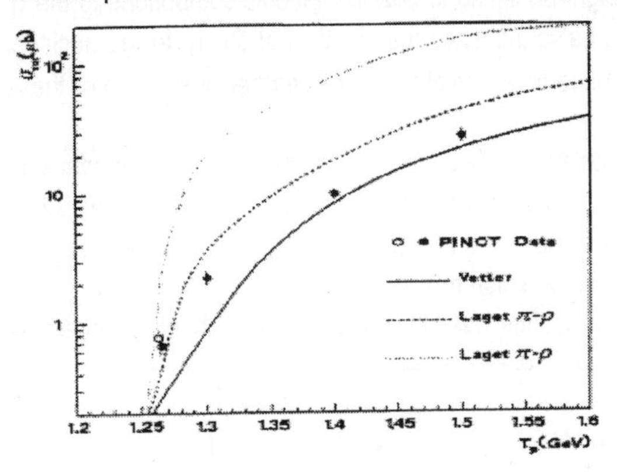

Figure7: Excitation function for the reaction pp->ppη measured with the PINOT spectrometer (see ref. 13). For comparison, calculations with meson exchange models by Laget et al.[14] and Vetter et al.[15] are shown.

A rather dramatic effect has been observed by Chiavassa et al.[13] when comparing η–meson production with proton and deuteron targets. The corresponding distributions of the

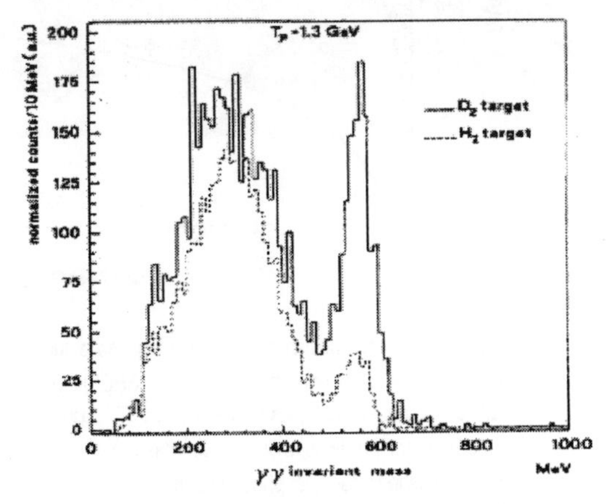

Figure 8: γγ invariant mass distribution for 1.3 GeV protons incident on proton and deuteron targets, measured with the PINOT spectrometer. The striking enhancement of the yield for the deuteron target in the invariant mass region corresponding to η production is due to the neutron contribution.

$\gamma\gamma$ invariant mass is shown in fig. 8. The experiment was performed an an incident proton beam energy of 1.3 GeV. The large pd/pp ratio in the region of the η mass indicates the dominance of the pn contribution to η production off the deuteron. This behavior has been predicted theoretically [15,16,17].

It should be pointed out that the above results are of fundamental importance for the understanding of η production in nucleus-nucleus collisions at BEVALAC/SIS energies.

3.4 η' production

The η' meson is an extremly interesting hadron, due to the fact that its mass is much larger than predicted from its quark contents and cannot be explained in a straight forward way in terms of chiral symmetry breaking The origin of the large mass gap between the η' and the other pseudoscalar mesons might be linked to possible gluonic cotributions to the η' wave function. Such contributions could also influence the coupling of the η' to the nucleon. Thus, production cross section measurements might provide further insight into these questions.

Of particular interest is a comparison to data on η production. Such a comparison, howver, has to take into account the fact, that η production near threshold is dominated by the $N^*(1535)$ resonance. So far, there is no evidence for a corresponding resonance with strong coupling to the η' in the threshold production region.

The first data on η' production in pp reactions near the threshold have been obtained by

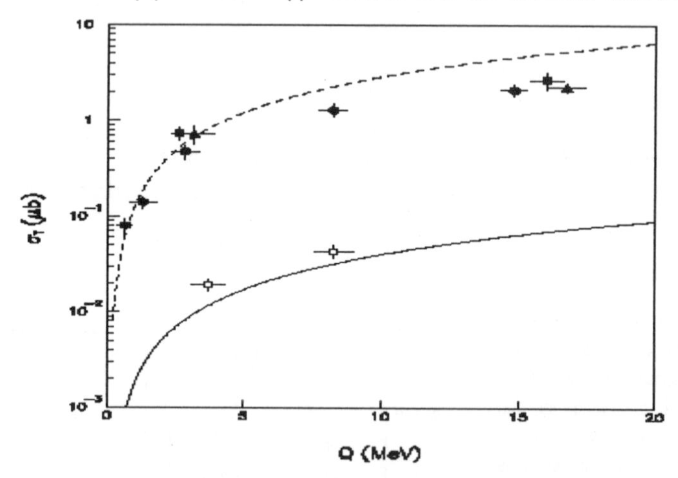

Figure 9:Total cross sections for pp -> ppη' as a function of the mean mid-target kinetic energy Q in the final state measured with SPESIII (filled squares) [18]. also shown are the CELSIUS points[19] (circles) and those of PINOT[12,13] (triangles) for η production. The data are compared to the predictions of a model (dashed curve), which reflects the proton-proton final state interaction folded with phase space.

Hibou et al.[18] employing the missing mass technique with the SPES3 spectrometer. Excitation functions for η' production in comparison to η data from SATURNE[12,13] and CELSIUS[19] is shown in fig. 9. The cross sections near threshold are strongly influenced by the final state interaction, being dominated by the S-wave pp interaction which is much weaker at higher energies. Thus, for a more accurate determination of the ppη' coupling, further data above threshold are required.

4 Production of vector mesons

First cross sections for ω production in pp reactions were obtained by Bing et al.[20] using the SPES3 spectrometer. More recently, the DISTO spectrometer has been used for extensive studies of vector mesons, covering measurements of φ/ω cross section ratios as well as angular distributions.

DISTO is a spectrometer specifically design for a kinematically complete measurement

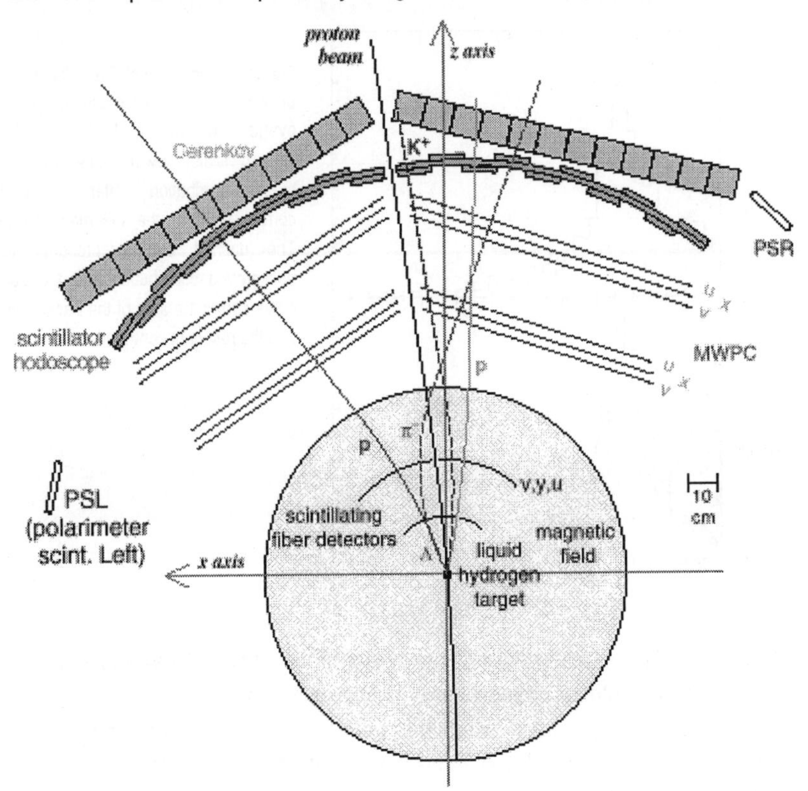

Figure 10: Schematic layout of the DISTO experimental apparatus, viewed from above, including simulated trajectories from a pp- > pK + _ event. The large shaded circle represents the effective field region. PSR and PSL are the two slabs to detect backward scattered protons in the polarimeter.

of reactions with 4 charged particles in the exit channel. Particle identification is provided by time-of-flight, energy loss measurements and water Cherenkov detectors. The latter serve for kaon identification. Momentum determination is done via tracking with scintillating fiber detectors and with sets of multi-wire proportional chambers.

DISTO has a large acceptance and is suitable both for the study of meson production and hyperon production. The hyperon part of the program will be covered in section 5. Fig. 10 shows a schematic view of the experimental setup.

A measurement of the ϕ/ω cross section ratio is of particular interest, since it might provide insight into the contribution of strange quarks to the nucleon wave function. Recently, large enhancements of this ratio in proton annihilation experiments[21] at LEAR in specific channels have been observed, which exceed predictions based on the OZI rule[22] by 2 orders of magnitude. Models based on an intrinsic strangeness component of the nucleon[23] as well as 2-step kaon exchange models [24] have been proposed to explain the annihilation data. Further insight might be provided by studying the ϕ/ω ratio in pp reactions near the absolute

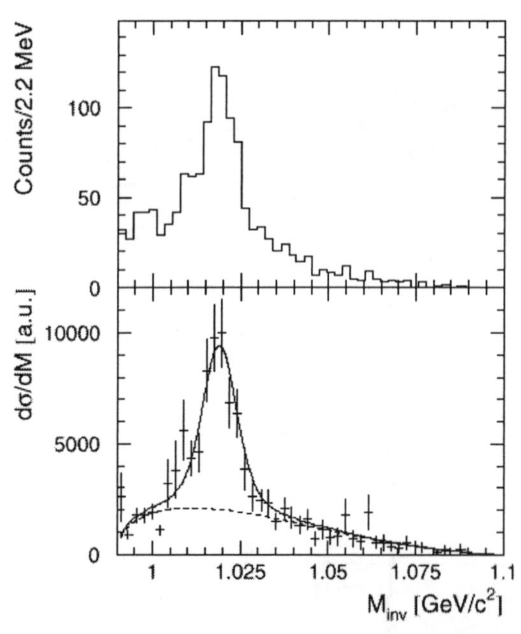

Figure 11: (top frame) Raw distribution of the K^+K^- invariant mass for ppK^+K^- events. The ϕ meson at 1.019 GeV/c^2 is clearly visible. (lower frame) Invariant mass distribution after efficiency corrections over the full mass range. The dashed curve represents the background contribution and the solid curve shows the sum of the background and the resonant contribution.

threshold, where kinematical conditions are similar to the annihilation experiments but model predictions based on intrinsic strangeness or kaon exchange might differ.

In the DISTO experiment[25] which was performed at an incident energy of 2.85 GeV (less than 100 MeV above the ϕ threshold), clean ϕ identification could be achieved employing the kaon particle id capabilities as well as kinematical conditions on the 2-body and 4-body missing mass. For ω detection, the 3-pion decay channel was used. Here, the neutral pion was not observed and ω identification was done using the pp and $pp\pi^+\pi^-$ missing mass, respectiveley. The invariant mass distribution of charged kaon pairs is shown in fig. 11.

A clear ϕ signal is obtained. Fig 12 shows the quality of the ω meson detection. The resulting

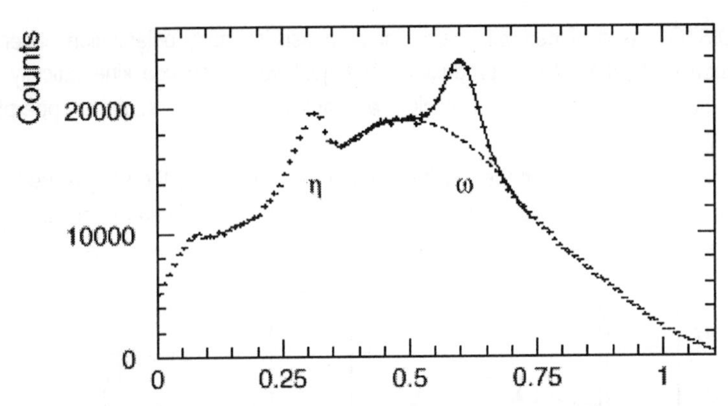

Figure 12. Raw spectrum of $(M_{pp}^{miss})^2$ for events of the type pp->pp$\pi^+\pi^-\pi^0$ after selecting a missing neutral pion via M pp$\pi^+\pi^{-miss}$. This projection shows clear signals corresponding w and h production.

ϕ/ω cross section ratio is shown in fig. 13 together with existing data[26] at higher energies. The yield ratio is observed to exceed a naive application of the OZI rule by one order of magnitude. Although more refined calculations[27] using the known $\phi\text{->}\rho\pi$ coupling are able to explain the enhancement observed at higher energies, these predictions still underestimate the near threshold behavior by a factor 3.

However, higher order mechanisms, such as two step processes with hadrons in the intermediate state, may provide an explanation for this increase without explicitly requiring a significantly enhanced contribution of strangeness to the nucleon's wave function. This observed increase of the ratio near threshold is less dramatic than the rise in antiproton

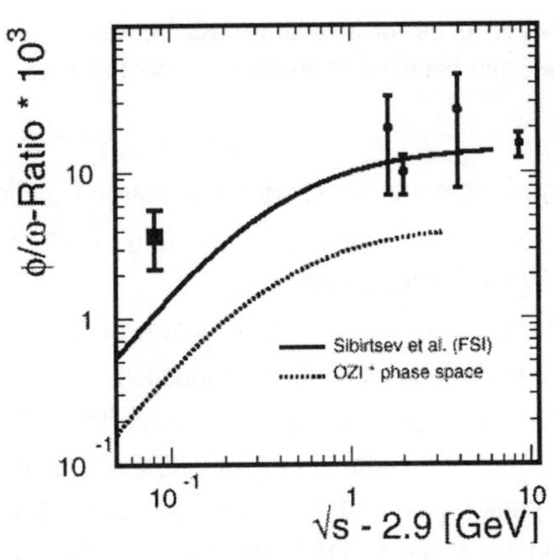

Figure 13: Ratio of the exclusive cross sections for the ϕ and ω production in pp reactions as a function of the available energy above the ϕ production threshold. Shown is the value measured in this work (square) together with data at higher energies[26] and model calculations by Sibirtsev et al.[27].

annihilation at low relative momenta. Thus it would be very useful to study the evolution of the ϕ/ω ratio to energies even closer to the production threshold.

5 Hyperon Production

DISTO was designed with particular emphasis to study polarization observables in proton induced hyperon (Λ,Σ) production. The goal was to provide kinematically complete measurements for a number of exclusive reaction channels, such as pp->p$K^+\Lambda$ and pp->p$K^+\Sigma^0$.

Tracking capabilities in the vicinity of the target provided by the scintillating fiber arrays allow the hyperon identification by observing secondary decay vertices resulting from the

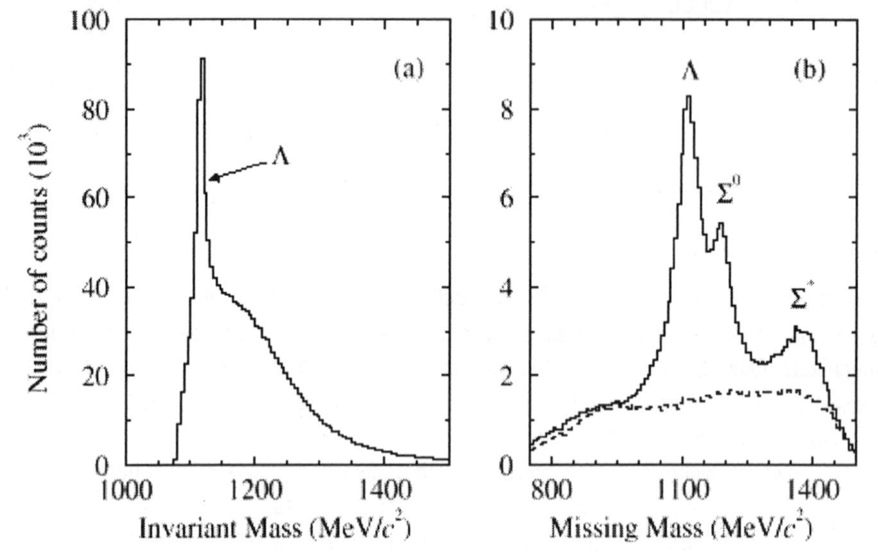

Figure 14:
(a) $p\pi^-$ invariant mass spectrum at decay vertex.
(b) (b) Missing mass spectrum from primary pK^+ pairs for invariant mass between 1100 MeV/c^2 and 1130 MeV/c^2 (solid line), and between 1150 MeV/c^2 and 1180 MeV/c^2 (dashed line)

weak Λ decays. The complete determination of momentum vectors for 4 charged particles alows also to identify reaction channels where a neutral decay daughter (e.g. a photon in case of the Σ^0(1192) or a neutral pion for the Σ^0(1385) goes undetected.

Large phase space coverage for hyperon decays is a crucial feature of the setup, thereby allowing detailed comparison to model predictions for hyperon production.

As an example, fig. 14 shows the hyperon identification capabilities of the spectrometer. The experiment has been carried out at a beam energy of 2.85 GeV. Gates on the ($p\pi^-$) invariant mass (fig.14, left) in the Λ peak region and in the background yield a pK^+ missing mass distribution (fig.14, right) showing the Λ, the Σ^0(1192) and the Σ^0(1385), or a flat background, respectively.

Fig.15 shows the measured polarization transfer D_{NN} for exclusive Λ production as a f function of transverse momentum[28]. The D_{NN} is large and negative over the entire kinematic region. This behavior is consistent with a production mechanism dominated by K meson exchange.

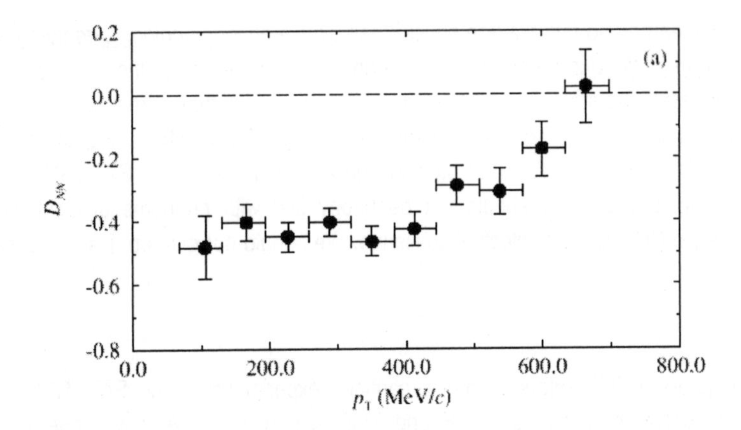

Figure 15: Measured D_{NN} values as a function of transverse momentum transfer for the exclusive pp ->pK$^+\Lambda$ reactions. The horizontal error bars reflect the width of the p_T bins analyzed. The vertical error bars reflect statistical uncertainties only.

6 Summary

The harvest of 20 Years of SATURNE-2 in the field of inelastic nucleon-nucleon interactions includes a large amount of high quality experimental information ,many channels have been investigated for the first time in detail in the near-threshold regime: η, η', ϕ, ω, Λ, Σ^0. Multi-differential cross sections and polarization observables allow much more stringent tests of phenomenological models. and theories with effective degrees of freedom Still, the death of SATURNE was premature, since much more could have been done with the new detector facilities such as DISTO and SPES4π.

What have we learned so far? Hadron physics has *simple phenomenological features,* which are *not at all understood* in terms of QCD, the underlying fundamental theory! The origin of the hadron masses, the structure of the baryons in the non-perturbative regime, the structure of the QCD vacuum, the origin of confinement, are there glueballs, hybrids, 6-quark-states? It is these questions, which need to be studied and hadron facilities like SATURN or LEAR made their contributions. We should not be happy just with phenomenological models. Non-perturbative QCD is an intellectual challenge. We should not give up, just because it is difficult.

In the near future, COSY offers a path to continue experiments in the SATURNE energy regime. New detectors such as ANKE and TOF are available. At lower energies, the CELSIUS facility plans many exciting experiments. A major new hadron facility is, however, urgently needed in Europe. Higher energies, exceeding the charm threshold as well as high quality high intensity secondary beams such as antiprotons and kaons are required to attack fundamental and challenging questions in the regime of non-perturbative QCD.

References

1. F.Malek, H.Nifenecker, J.A.Pinston, F.Schussler, S.Drissi,J.Julien, Phys.Lett. B266(1991)255
2. W.Cassing, V.Metag, U.Mosel, K.Niita, Phys.Rep. 188(1990)363
3. J.P.Didelez, T.Reposeur, M.A.Duval, R.Frascaria, G.Rappenecker, R.Siebert, E.Warde, G.Blanpied, M.M.Rigney, B.M.Preedom, G.Battistoni, C.Bloise, L.Satta, J.M.Laget, B.Sgahai, E.Bovet, J.P.Egger, F.Hinterberger, Nucl.Phys.A535(1991)445
4. G. Rappenecker, M. Rginey, J.P. Didelez, E. Hourany, L. Rosier, J. van de Wiele, G. Berrier-Ronsin, A. Elayi, R. Frascaria, P. Hoffmann-Rothe, G. Anton, J. Arends, M. Breuer, K. Büchler, G. Nöldeke, V. Zucht, G. Blanpied, B. Preedom, J.M. Laget, B.Saghai, Nucl.Phys. A590(1995)763

5. C. Comptour, J. Augerat, J. Berthod, P. Bertin, K. Bouyakhlef, H. Fonvielle, G. Audit, R. Babinet, G. Fournier, J. Gosset, M.C. Lemaire, D. L'Hote, B. Mayer, J. Poitou, B. Saghai, F. Brochard, J.M. Durand, Z. Fodor, J. Yonnet, Nucl.Phys. A579(1994)369
6. B.Tatischeff, J. Yonnet, N. Willis, M. Boivin, M.P. Comets, P. Courtat, R. Gacougnolle, Y. Le Bornec, E. Loireleux, F. Reide, Phys.Rev.Lett. 79(1997)601
7. R. Bertini, G. Roy, J.M. Durand, J. Arvieux, M. Boivin, A. Boudard, C. Kerboul, J. Yonnet, M. Bedjidian, E. Decroix, J.Y. Grossiord, A. Guichard, J.P. Pizzi, Th. Hennino, L. Antonuk. Phys.Lett. B203(1988)18
8. J. Yonnet, R. Abegg, M. Boivin, A. Boudard, G. Bruge, P. Couvert, G. Gaillard, M. Garcon, L.G. Greeniaus, D.A. Hutcheon, C. Kerboul, B. Mayer, Nucl.Phys. A562(1993)352
9. M.P. Combes-Comets, P. Courtat, R. Frascaria, Y. Le Bornec, E. Loireleux, F. Reide, B. Tatischef, N. Willis, E. Aslanides, D. Benabdelouahed, A.M. Bergdolt, G. Bergdolt, O. Bing, P. Fassnacht, F. Hibou, M. Boivin, A. Chrisholm, C. Kerboul, A.Moalem, Phys.Rev. C43(1991)973
10. Y. Terrien, P. Couvert, C. Kerboul, F. Wellers, R. Beurtex, G. Bruge, B. Fabbro, J.-C. Faivre, J.-M. Laget, M Rouger, J. Saudinos, Phys.Lett. B294(1992)40
11. A.M.Bergdolt, G.Bergdolt, O.Bing, A.Bouchakour, F.Brochard, F.Hibou, A.Moalem, A.Taleb, M.P.Combes-Comets, P.Courtat, R.Gacougnolle,Y.Le Bornec, E.Loireleux, F.Reide, B.Tatischeff, N.Willis, M.Boivin, B.M.K.Nefkens, F.Ploin, Phys.Rev.D48(1993)48
12. E.Chiavassa, G.Dellacasa, N.De Marco, C.De Oliveira Martins, M. Gallio, P. Guaita, A. Musso, A. Picotti, E. Scomparin, E. Vercellin, J.M. Durand, G.Milleret, Phys.Lett.B322(1994)270
13. E.Chiavassa, G.Dellacasa, N.De Marco, C. De Oliveira Martins, M. Gallio, P. Guaita, A. Musso, A. Picotti, E. Scomparin, E. Vercellin, J.M. Durand, G.Milleret, C.Wilkin, Phys.Lett.B337(1994)192
14. J.M.Laget, F.Wellers, J.F.Lecolley, Phys.Lett.B257(1991)254
15. T.Vetter, A.Engel, T.Biro, U.Mosel, Phys.Lett.B263(1991)153
16. J.F.Germond and C.Wilkin, Nucl.Phys.A518(1990)308
17. J.M.Laget, F.Wellers, J.F.Lecolley, Phys.Lett. B257(1991)153
18. F. Hibou, O. Bing, M. Boivin, P. Courtat, G. Fäldt, R. Gacougnolle, Y. Le Bornec, J.M. Martin, A. Moalem, F. Plouin, A. Taleb, B. Tatischeff, C. Wilkin, N. Willis, R. Wurzinger, preprint nucl-ex/9802002
19. H. Calen et al., Phys.Lett. B366(1996)39
20. O. Bing et al., Nouvelles de Saturne 19(1995)51
21. J. Reifenrither et al., Phys. Lett. B267 (1991) 299; V. G. Ableev et al., Phys. Lett. B334 (1994) 237; C. Amsler et al., Phys. Lett. B346 (1995) 363.

22. G. Zweig, CERN report 8419/Th 412 (1964); S. Okubo, Phys.Lett. B5(1965)165;I. Iizuka, Prog. Theor. Phys.Suppl. 37-38 (1966) 21; S. Okubo, Phys. Rev. D16 (1977) 2336.
23. J. Ellis et al., Phys. Lett. B353 319 (1995)
24. Ulf-G. Meißner et al., Phys. Lett. B408 (1997) 381M. P. Locher and Yang Lu, Z. Phys. A351 (1994) 83, D. Buzatu and F. M. Lev, Phys. Rev. C51 (1995) R2893, O. Gortchakov et al., Z. Phys. A353 (1996) 447.
25. F. Balestra, Y. Bedfer, R. Bertini, L.C. Bland, A. Brenschede, F. Brochard, M.P. Bussa, V. Chalyshev, Seonho Choi, M. Debowski, M. Dzemidzic, I.V. Falomkin, J.-Cl. Faivre, L.Fava, L.Ferrero, J.Foryciarz, V. Frolov, R. Garfagnini, D. Gill, A. Grasso, E. Grosse, S. Heinz, V.V. Ivanov, W.W. Jacobs, W.Kühn, A. Maggiora, M. Maggiora, A. Manara, D. Panzieri, H.-W. Pfaff, G. Piragino, G.B. Pontecorvo, A. Popov, J. Ritman, P. Salabura, P. Senger, J. Stroth, F. Tosello, S.E. Vigdor, and G. Zosi, submitted to Phys.Rev.Lett.
26. V. Blobel et al., Phys. Lett. B59 (1975) 88; R. Baldi et al., Phys. Lett. B68 (1977) 381; Arenton et al., Phys. Rev. D 25 (1982) 2241; S. V. Golovkin et al., Z. Phys. A359 (1997) 435.
27. A. Sibirtsev, Nucl. Phys. A604 (1996) 455
28. F. Balestra, Y. Bedfer, R. Bertini, L.C. Bland, A. Brenschede, F. Brochard, M.P. Bussa, V. Chalyshev, Seonho Choi, M. Debowski, M. Dzemidzic, I.V. Falomkin, J.-Cl. Faivre, L.Fava, L.Ferrero, J.Foryciarz, V. Frolov, R. Garfagnini, D. Gill, A. Grasso, E. Grosse, S. Heinz, V.V. Ivanov, W.W. Jacobs, W.Kühn, A. Maggiora, M. Maggiora, A. Manara, D. Panzieri, H.-W. Pfaff, G. Piragino, G.B. Pontecorvo, A. Popov, J. Ritman, P. Salabura, P. Senger, J. Stroth, F. Tosello, S.E. Vigdor, and G. Zosi, DISTO preprint 1998 and to be published

STUDY OF NUCLEAR MATTER DISTRIBUTION WITH 1 GeV PROTONS
SACLAY - GATCHINA COLLABORATION

A.A. VOROBYOV
Petersburg Nuclear Physics Institute,
Gatchina, Leningrad district, 188350 Russia

1 Introduction

Collaboration between Saclay and Gatchina started in 1973, and it resulted finally in 15 joint experiments successfully carried out both at Saclay and at Gatchina. Such fruitful cooperation was to a great extent due to common fields of interest. Both laboratories had constructed proton accelerators and high resolution magnetic spectrometers aimed at studies of proton-nuclear scattering at 1 GeV proton energy. The first contacts between our groups happened to be at the Uppsala conference where both laboratories presented their first results on pA-scattering and expressed their wills for cooperation. At the same time I received an invitation from J.Thirion to visit Saclay and discuss the prospects of our collaboration.During this visit in summer 1973, we decided to address to our Governments for formal approval of our collaboration and for financial support. Fortunately, just at that time a high level Agreement between our countries was signed which included also the scientific cooperation (L'accord Franco-Sovietique entre les Ministeres des Affraires Etrangeres. "La Petite Commission"). So our proposal was excepted, and the first joint experiment was carried out at Saclay already in 1974. Until 1980 our collaboration was continued in the frame of this inter-governmental Agreement. After that we signed the direct PNPI-Saclay Agreement which, with several prolongations, remains valid up today.

The first joint experiments were carried out in 1974-1975 at Saturne-1 using the high resolution spectrometer SPES-1. These experiments were:
— 1 GeV proton scattering on ^{40,42,44,48}Ca, ^{48}Ti;
— 1 GeV proton scattering on ^{58,60,62,64}Ni;
— α-particle scattering on ^{40}Ca–^{48}Ca.

When Saturne-1 was closed up for reconstruction, our collaboration was continued at Gatchina where the following experiments have been performed during 1976-1977:

— Polarization in pA-scattering at 1 GeV;
— 1 GeV proton scattering on ^{3}He;
— Small angle pp-scattering at $T_p = 0.6$–1.0 GeV;
— Small angle $pd-$ and p^4He-scattering at 1 GeV.

Since 1978, when Saturne-2 was put into operation, the collaborative experiments were restarted at Saclay exploiting favourable experimental conditions provided by the new accelerator. This activity continued until the last moment of the Saturne's life and resulted in many successful experiments:
— Small angle $\vec{n}p$-scattering at $T_n = 0.4$–1.1 GeV;
— Polarization parameters in $\vec{d}p \to ppn$ reaction at $T_d = 2$ GeV;
— T_{20} in inclusive break up of 4.5 GeV polarized ^{6}Li;
— Rare η-decay;
— Reaction $dp \to^3 He\pi^o$ at threshold;
— Nucleon-Nucleon program;
— Polarization observables in elastic backward $dp \to pd$ scattering;
— Roper resonance in αp-scattering.

While the goal of our earlier experiments was studying of the nuclear structure, the latest experiments were devoted to elementary particle physics.

In this report, I shall touch only the results of the studies of nuclear matter distribution in the proton-nuclei scattering experiments. This choice is dictated partly by my personal involvement in these experiments but mostly because these results constitute a unique and self-consistent base of experimental data which is of great importance for nuclear physics.

2 Proton-nuclei diffraction scattering at 1 GeV

Both at Gatchina and at Saclay, the pA-scattering experiments were triggered by the pioneering work at Brookhaven performed by H.Palevsky et al. [1] in 1967. In this experiment, the differential cross sections for elastic scattering of 1 GeV protons on D, ^{4}He, ^{12}C, and ^{16}O have been measured, and it was shown [2] that the experimental data could be well described by the multiple scattering theory developed by R.Glauber (also, by A.Sitenko). Why this observation produced such a wide-spread resonance in the scientific community? The answer is clear. The results of all previous experiments performed at lower energies could be analyzed only using some phenomenological models (like the optical model) with a set of arbitrary parameters. So the reliability of the nuclear matter distributions extracted from such analyses was poor. On the contrary, the Glauber theory has no arbitrary parameters at all. It expresses the pA-scattering amplitude via proton-nucleon scattering amplitudes at small angles thus opening a possibility for unambiguous determination of the nuclear

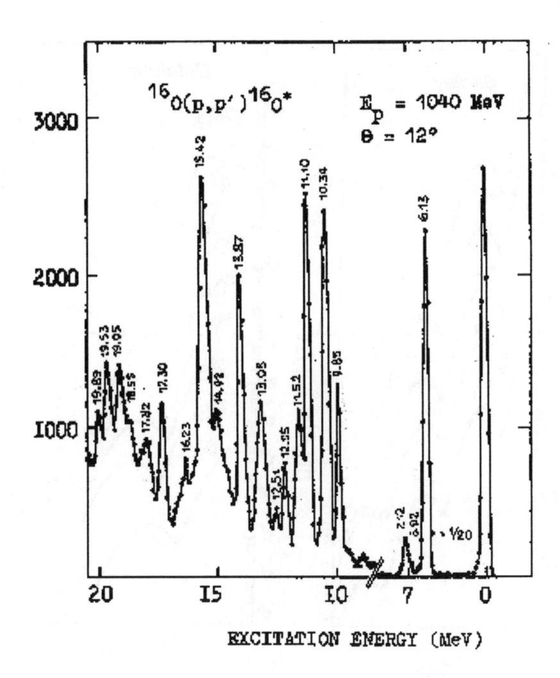

Figure 1: Energy spectrum of protons scattered from ^{16}O measured with SPES-1 spectrometer at Saclay [3].

density parameters. And the 1 GeV proton energy happened to be just the optimal energy for application of the Glauber theory.

To perform the pA-scattering experiments at T_p= 1 GeV, one should have high resolution spectrometers capable to separate the elastic scattering from the scattering with excitation of the nuclear levels. The energy resolution in the Brookhaven experiment was 3.0 MeV (FWHM). A spectrometer with 1.5 MeV (FWHM) resolution was constructed at Gatchina, and the regular studies of pA elastic scattering were started in 1972. The first results (scattering on ^{12}C, ^{28}Si, ^{32}S, ^{39}K, ^{40}Ca, ^{48}Ca) were presented at the Uppsala Conference in 1973. A unique facility SPES-1 has been built at Saclay in 1973. Its energy resolution was as high as 0.15 MeV (FWHM), so it became possible to study not only the elastic scattering but also the scattering with excitation of selected nuclear levels. Fig.1 demonstrates the energy resolution of SPES-1. With this spectrometer, the first measurements of elastic and inelastic scattering on ^{4}He, ^{12}C, and ^{16}O were performed in 1973.

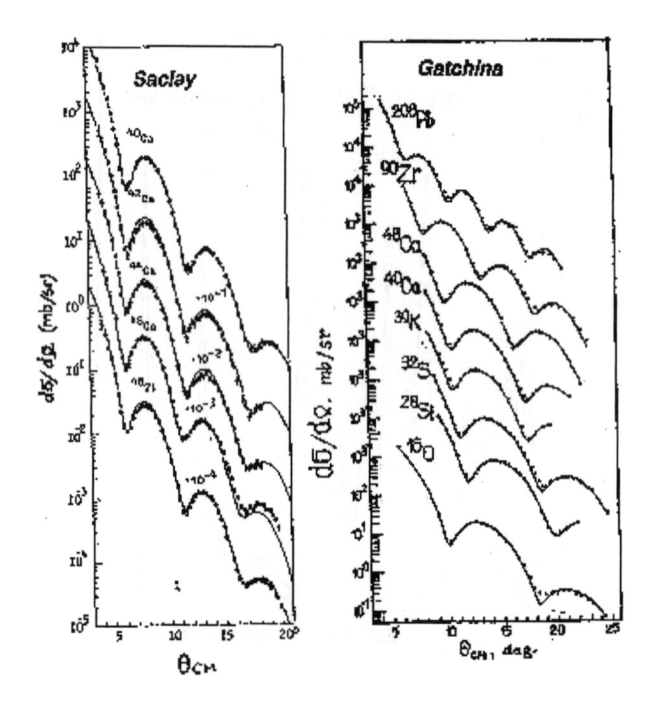

Figure 2: Differential cross sections of 1 GeV proton elastic scattering on nuclei. Left side – results of the first joint experiment at Saclay. Right side – results from Gatchina experiments. Solid lines – calculations in the independent-particle approximation of the Glauber theory with two free parameters: R_n and a_n (see section 4).

As I have already mentioned, the collaboration between the Saclay and Gatchina groups was created in 1974, and since then we were working on the pA-scattering program together. The first joint experiment was carried out at Saclay in 1974. This was the pA-scattering experiment on ^{40,42,44,48}Ca and ^{48}Ti isotopes (Fig.2). One of the goals of this experiment was comparison with the Gatchina measurements on ^{40}Ca and ^{48}Ca. The very good agreement achieved between these two measurements of both the relative and the absolute differential cross sections was a solid prove of the reliability of our experimental methods. Since then the systematic studies of pA-scattering on variety of the nuclei over the whole periodic table were continued at Saclay and at Gatchina. Below is the list of the studied nuclei [4]:
D, ^{3}He, ^{4}He, ^{6}Li, ^{9}Be, ^{11}B, ^{12}C, ^{13}C,^{14}N, ^{16}O, ^{28}Si, ^{32}S, ^{34}S, ^{39}K, ^{40}Ca, ^{42}Ca, ^{44}Ca, ^{48}Ca, ^{48}Ti, ^{58}Ni, ^{60}Ni, ^{62}Ni, ^{64}Ni, ^{90}Zr, ^{208}Pb.
Some results of Gatchina experiments are presented in Fig.2.

The measured differential cross sections were analyzed using the Glauber theory with the goal to determine the nuclear matter distribution. However, in order to apply quantitatively the Glauber theory, the proton-proton and proton-neutron scattering amplitudes at small angles were needed. The available phase shift analyses were not able to provide these amplitudes. Therefore, we have undertaken special studies of the pN small angle scattering as described in the next section.

3 Study of proton-proton and proton-neutron small angle elastic scattering

In general, pp- and pn-scattering amplitudes are rather complicated functions containing five terms each:

$$f_{pp}(q) = A_p(q) + B_p(q) + C_p(q) + D_p(q) + E_p(q) \tag{1}$$

$$f_{pn}(q) = A_n(q) + B_n(q) + C_n(q) + D_n(q) + E_n(q) \tag{2}$$

where A are the central amplitudes, B are the spin-flip terms, C, D, and E are the spin-spin interaction terms.

Fortunately, at small angles the structure of the amplitude is greatly simplified. The spin-flip amplitude $B(q) \to 0$ when $q \to 0$. Also, as it was shown in our experiments, the contribution of the spin-spin terms at $q \to 0$ is negligible at the energies $T_p \simeq 1$ GeV. So we have very simple expressions

$$f_{pp}(q) = \frac{k}{4\pi}\sigma_{pp}(\rho_{pp} + i)exp(-b_{pp}\frac{q^2}{2}) \tag{3}$$

$$f_{pn}(q) = \frac{k}{4\pi}\sigma_{pn}(\rho_{pn} + i)exp(-b_{pn}\frac{q^2}{2}) \tag{4}$$

where σ_{pp}, σ_{pn} are the total pp and pn cross sections, b_{pp} and b_{pn} are the slope parameters, and ρ_{pp} and ρ_{pn} are the real to imaginary part ratios of the central amplitudes at $q = 0$.

The total cross sections σ_{pp} and σ_{pn} being known from the world data, we had to determine only 4 parameters: $\rho_{pp}, b_{pp}, \rho_{pn}, b_{pn}$. This we were able to do using the experimental method developed at Gatchina [5]. In this method, a hydrogen filled ionization chamber was used as an active target and detector of the recoiled protons (recoil detector IKAR). The detector allowed to measure the absolute differential cross sections in the q^2-range from $2{\cdot}10^{-3}$ (GeV/c)^2 to $5{\cdot}10^{-2}$ (GeV/c)^2 that covers the Coulomb interference region (which determines the parameter ρ_{pp}) as well as the region suitable for determination of the slope parameters b_{pp} and b_{pn}. In addition, comparison of the extrapolated value $(d\sigma/d\Omega)_{q=0}$ with the Optical Point allowed to estimate the contribution of the spin-spin terms characterized by parameters β_{pp} and β_{pn}.

In a series of experiments carried out at Gatchina [6], the three parameters, $\rho_{pp}, b_{pp}, \beta_{pp}$, have been measured in the proton energy range from 0.6 GeV to 1 GeV. It was shown that the spin-spin contribution (parameter β_{pp}) becomes very small when the proton energy approaches 1 GeV (Fig.3). Also, it was demonstrated that the measured real part of the central amplitude (parameter ρ_{pp}) is in perfect agreement with Dispersion Relation calculations and that the previous experimental data on ρ_{pp} in this energy range were wrong (Fig.4).

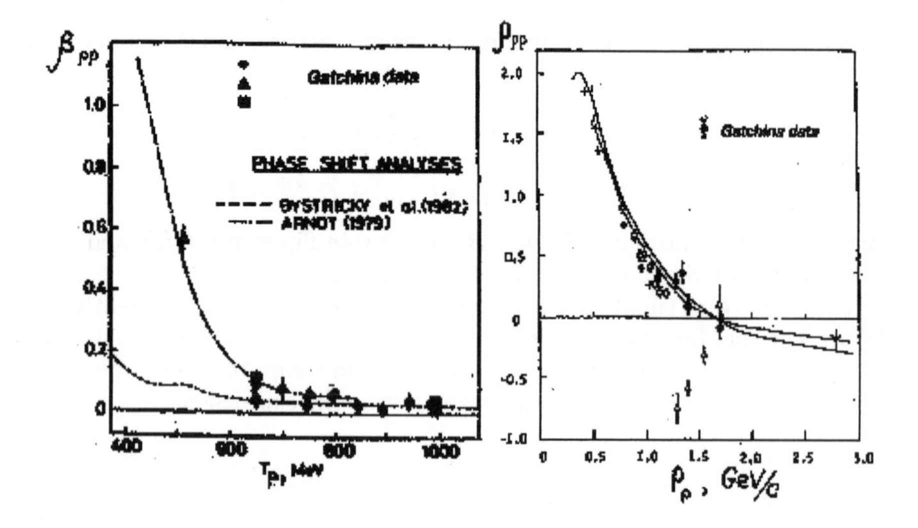

Figure 3: Spin-spin terms contribution to the pp-forward elastic scattering cross section. The parameter β_{pp} is evaluated from the excess of the forward cross section over Optical Point: $\left(\frac{d\sigma}{d\Omega}\right)_{q=0} = \frac{k^2\sigma_{pp}^2}{16\pi^2}(1+\rho_{pp}^2+\beta_{pp})$

Figure 4: Real to imaginary part ratio of the central amplitude for pp-scattering. The ratio $\rho_{pp} = ReA_p(0)/ImA_p(0)$ is determined from Coulomb interference region. Solid curves – calculations with Dispersion relations.

The next step was investigation of the *np* small angle scattering. This experiment was performed at Saclay in the neutron beam from Saturne-2 [7]. A modified IKAR detector was used in this experiment. The absolute differential cross section have been measured in the q^2-range from $1{\cdot}10^{-2}$ (GeV/c)^2 to $8{\cdot}10^{-2}$ (GeV/c)^2 at the neutron beam energies from 0.4 GeV to 1.1 GeV. The slope parameter b_{pn} was determined. Also, it was shown that the spin-spin terms are small and could be ignored at the energies around 1 GeV.

So conclusions from the small angle *pp* and *pn* experiments were as follows:
— The spin-spin terms could be neglected: $\beta_{pp} \simeq \beta_{pn} \simeq 0$;
— The pN-amplitudes could be described by expressions (3) and (4);
— All parameters entering the above expressions are fixed values:

σ_{pp}, σ_{pn} - the world data,
ρ_{pp} - Gatchina experiment and Dispersion relation calculations,
b_{pp} - Gatchina experiment,
b_{pn} - Saclay experiment,
ρ_{pn} - Dispersion relation calculations.

These amplitudes were used for analysis of the pA-scattering data.

4 Glauber fit to the pA differential cross sections and nuclear matter distributions

Glauber theory in its simplest form represents the pA-scattering amplitude as some integral function on proton-nucleon scattering amplitudes, $f_{pp}(q)$ and $f_{pn}(q)$, and on the one-body nucleon density distributions, $\rho_p(r)$ and $\rho_n(r)$. As we just discussed above, the amplitudes $f_{pp}(q)$ and $f_{pn}(q)$ were determined in the small angle scattering pN-experiments. Also, the proton density distributions $\rho_p(r)$ are known from the electron-nuclei scattering experiments. What remains is the neutron density distributions $\rho_n(r)$ which can be found from the fits to the measured pA-differential cross sections. In the fitting procedure, one can use some parametrization for $\rho_n(r)$, for example, the Fermi function with three free parameters

$$\rho_n(r) = \rho_o \frac{1 + W_n(r/R_n)^2}{1 + exp(r - R_n)/a_n}. \tag{5}$$

The parameters R_n and a_n are determined from the fit, while it is anticipated that $W_n = W_p$. As one can see from Fig.2, the quality of the fit is just perfect in the whole θ-range where the magnitude of the differential cross section is changed by four orders. And, what is also remarkable, such good description is reached for all nuclei, without any exception, from the lightest D, ^{3}He, ^{4}He to the heaviest ^{90}Zr, ^{208}Pb. At first sight, such high quality of the Glauber fit looks surprising, having in mind that the theory represents the nucleus in a simplified way as the gas of independent nucleons and ignores the nucleon-nucleon correlations, except the center-of-mass correlation. However, the detailed theoretical studies showed [8] that, at the proton energy of ~1 GeV, the calculated differential cross sections are not much sensitive to the NN-correlations. Moreover, some of the correlations cancel the effects from the others. That is why the independent-particle approximation of the Glauber theory proved to be so successful.

Nucleus	R_p fm	a_p fm	$\frac{W_n}{W_p}$	R_n fm	a_n fm	$\langle r^2\rangle_p^{1/2}$ fm	$\langle r^2\rangle_n^{1/2}$ fm
^{16}O	2.61	0.51	-0.05	2.48	0.55	2.73	2.74
^{28}Si	3.21	0.57	-0.12	3.13	0.61	3.14	3.15
^{32}S	3.44	0.62	-0.21	3.51	0.61	3.24	3.27
^{39}K	3.74	0.59	-0.20	3.75	0.58	3.41	3.41
^{40}Ca	3.71	0.59	-0.13	3.70	0.59	3.49	3.48
^{48}Ca	3.81	0.53	-0.08	4.06	0.52	3.48	3.64
^{90}Zr	4.86	0.57	-0.09	5.09	0.57	4.26	4.41
^{208}Pb	6.72	0.51	-0.06	6.69	0.57	5.50	5.56

Table 1: Fermi distribution parameters for proton and neutron densities.

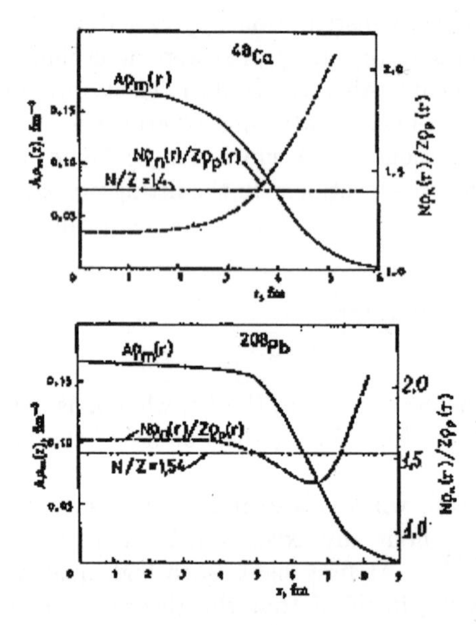

Figure 5: Radial dependence of the ratio of the neutron density $N\rho_n(r)$ to the proton density $Z\rho_p(r)$ in ^{48}Ca and ^{208}Pb.

So we believe that the Glauber fit of the pA-differential cross sections at $E_p \sim$1 GeV provides the information on the neutron distributions in nuclei with the highest precision, not reachable so far by any other experimental method. In terms of $\langle r^2\rangle_n^{1/2}$ of the neutron distribution, the precision of absolute measurements is estimated as $\sigma_r =$ 0.03 fm. Table 1 shows some of the obtained results (R_n, a_n, $\langle r^2\rangle_n^{1/2}$) in comparison with the corresponding values for proton distributions. Fig.5 illustrates the ratio $\rho_n(r)/\rho_p(r)$ in ^{48}Ca and ^{208}Pb. One can see from this figure that the information on $\rho_n(r)$ obtained from our experiments is much more rich than just the measurements of $\langle r^2\rangle_n^{1/2}$.

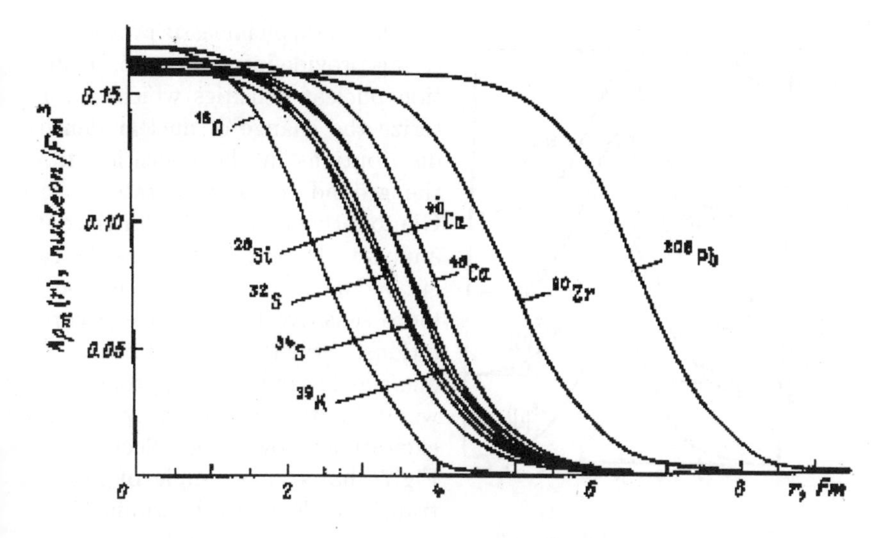

Figure 6: Nuclear matter distributions found from elastic pA-scattering data.

Having proton and neutron distributions, one can determine the nuclear matter distribution $\rho_m(r)$. Fig.6 demonstrates some of such distributions. One can clearly see the saturation of the nuclear matter central density. It is also interesting to note that the surface thickness parameters a_p, a_n proved to be similar for all nuclei with the magnitude around 0.55 fm.

The measured nucleon distributions in nuclei were widely used for comparison with various nuclear theories (Hartry-Fock theory, method of partially self-consistent field, quasi-particle Lagrange method) and stimulated further development of these theories. The result was that, with some tuning of its parameters, the self-consistent nuclear field theory was able to describe reasonably well the experimental nuclear matter distributions [9].

There is, however, one exception – ^{4}He nucleus. The slope $\beta_{p\mathrm{He}}$ of the p^{4}He differential cross section proved to be by 3 (GeV/c)^2 larger than the slope calculated with the assumption of equal proton and neutron distribution radii [10]. To achieve the agreement, one should introduce the difference $\Delta r = \langle r^2 \rangle_p^{1/2} - \langle r^2 \rangle_n^{1/2} \sim 0.2$ fm which is much larger that one could expect taking into account the Coulomb repulsion. This striking disagreement opened an area for various theoretical speculations. For example, one hypothesis assumed that ^{4}He ground state contains a 12% admixture of the 12-quark bag [11]. Unfortunately, at present we have no independent experimental prove for such conclusion.

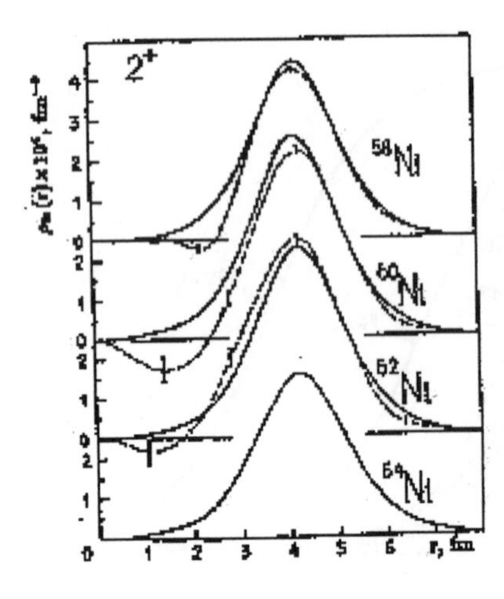

Figure 7: Comparison of the one-body isoscalar transition densities obtained from the data on inelastic scattering of 1 GeV protons (solid lines) with the charge densities obtained from the data on inelastic scattering of electrons (dashed lines) for $O^+ \to 2^+$ transitions in Ni isotopes.

The data on inelastic proton scattering provide information on transition nuclear densities which characterize the change of nuclear density distributions in the transition from the ground to excited states. The Glauber theory is applied for the data analysis, thought in this case the calculated inelastic cross-sections may be more sensitive to the nucleon correlations. Still the calculations in the one-elastic transition approximation were able to describe fairly well the excitation to low-lying collective states. Fig.7 shows an example of nucleon transition densities determined from the pA-scattering experiment in comparison with the charge transition densities measured in the eA-scattering experiment. The obtained results were used as a critical test for various nuclear models.

5 Summary

Started 25 years ago, the Saclay – Gatchina collaboration proved to be remarkably stable and fruitful. A large series of joint experiments was successfully carried out both at Saclay and at Gatchina. Many important results have been obtained. As an example, the reviewed in this report studies of the nuclear matter distribution became nowadays a classics of nuclear physics. As a member of the Gatchina group, I would like to point out the high professionalism and friendly atmosphere in the Saclay group which was so helpful for the success of the collaboration. This success was also due to the efficient support by the French and Russian authorities. Today, saying good bye to Saturne-2, I would like to wish to all our friends and colleagues many successes in their new activities and to express my hope that our friendly relations and scientific cooperation will continue in the future.

References

1. H.Palevsky et al., Phys.Rev.Lett. **18** (1967) 1200
2. R.H.Bassel and C.Wilkin, Phys.Rev. **174** (1968) 1179
3. J.Thirion, Int.Conf. on Nuclear Physics, Munich (1973) 781
4. G.Alkhazov et al., Nucl.Phys.**A274** (1976) 443
   G.Alkhazov et al., Phys.Lett. **67B** (1977) 402
   G.Alkhazov et al., Nucl.Phys. **A381** (1982) 430
   G.Alkhazov et al., Yad.Fiz. **41** (1985) 561
   G.Alkhazov et al., Yad.Fiz. **42** (1985) 8
5. A.Vorobyov et al., Nucl.Inst.Meth. **119** (1974) 509
6. A.Dobrovolsky et al., Nucl.Phys. **B214** (1983) 1
   A.Vorobyov et al., Phys.Lett. **41B** (1972) 639
   G.Velichko et al., PNPI report **N656** (1981)
7. A.Vorobyov et al., Nucl.Inst.Meth. **A270** (1988) 419
   B.Silverman et al., Nucl.Phys. **A499** (1989) 763
8. G.Alkhazov, S.Belostotsky, A.Vorobyov, Phys.Rep. **42C** (1978) 89
9. G.Alkhazov, B.Birbrair et al., Yad.Fiz. **27** (1978) 333
10. G.Grebenyuk et al., Nucl.Phys. **A500** (1989) 637
11. L.Dakhno and N.Nikolaev, Nucl.Phys. **A436** (1985) 653

PION AND DELTA DYNAMICS

M. ROY-STEPHAN
Institut de Physique Nucléaire,
F91406 Orsay cedex, France

Pion and Delta dynamics was an important subject of Laboratoire National Saturne research program. Results of the charge-exchange program will be presented, with a special emphasis on two aspects: the many body feature of the Δ excitation, and the evidence for a collective behavior of the pionic response of nuclei.

La dynamique du pion et du Delta a été l'un des thèmes essentiels du programme de recherche au Laboratoire National Saturne. Nous présenterons les résultats du programme d'échange de charge, en insistant plus particulièrement sur l'aspect problème à N corps de l'excitation du Δ et sur la mise en évidence du caractère collectif de la réponse pionique des noyaux.

1 Introduction

The effect of the nucleon structure appears in nuclei, this is well established to-day.[1] Of course, the relevant degrees of freedom depend on the resolution of the probe. At intermediate energies, the π plays a special role: to first order, it is the carrier of the nuclear interaction.

In this domain, the π-N system is dominated by the Δ resonance. The Δ_{1232} resonance (mass 1232 $\pm$ 3 MeV and width 116 $\pm$ 5 MeV) is the lowest π-N resonance. It is really a most typical resonant system: the phase shift actually crosses 90^0 at the mass 1232 MeV. Below 300 MeV π kinetic energy, the inelastic channels are absent and the other partial waves have much less importance, except very close to threshold where the s wave plays some role. With J=3/2 and T=3/2, the Δ resonance is the spin isospin excited state of the nucleon. In quark formalism it is obtained by flipping the spin and the isospin of one quark without any change in the spatial configuration. It may be considered as the Gamow-Teller excitation of the nucleon.

How do the π and the Δ resonance behave in nuclei?

In the nuclear medium, free pions cannot exist and the Δ becomes a quasi-particle. We have to deal with virtual pions which propagate as Δ-hole pairs.[2] Theory predicts that Δ-hole correlations can build collective pionic modes,[3] i.e. strong attractive correlations in the longitudinal spin-isospin channel are predicted, in contrast to the spin-transverse channel, where the correlations are expected to be small and slightly repulsive. The investigation of the longitudinal spin-isospin response of the nucleus is the main topic of the extensive charge-exchange program at SATURNE. In this prospect, the charge-exchange program developed more and more exclusive experiments. First, inclusive charge-exchange experiments with various compound projectiles were performed at SPES4,[4–14] then polarization observables were measured,[15–18] finally, in coïncidence experiments,[19–24] evidence for coherent pion production was found and this process was studied in details with SPES4π.[25–26] The first results of this last experiment show that the spin longitudinal response is actually shifted down in energy. The Δ-hole excitations play also a sizeable role in the SATURNE program on transmutation,[27–29] because at forward angle, the high energy neutrons are produced by direct charge-exchange (p,n) reaction. One should also refer to the extensive (p,n) program at LAMPF,[30–40] and to charge-exchange experiments at Dubna,[41–42] Gatchina,[43] and KEK.[44]

Another aspect of π and Δ dynamics was studied at SATURNE: the way this mixture of virtual pions and Δ-hole states, that G. E. Brown calls pisobars, propagate and decay. It must be understood, since pion production is commonly considered as a tool to determine the nuclear matter equation of state. It motivated exclusive experiments where the charge-exchange reaction was studied in coïncidence with the decay products of these Δ-hole states : pions, correlated π,p pairs, and correlated proton pairs.[19–24] The detector Diogène[45] was used in this part of the charge-exchange program. The quasi-free decay of Δ's emitting correlated π +p pairs was also studied in proton-nucleus and nucleus-nucleus collisions as a function of the centrality of the collision.[46–48] This was a part of the genuine Diogène program. A complete review on π and Δ dynamics should include a comparison with π and γ induced reactions, and a presentation of the Δ-hole model which was developed to interpret these data. On that matter, the reader is referred to reference 1 or to some specific papers.[2,14,49–50] Of course, beside pions and Δ in nuclei, the free Δ excitation was also studied at SATURNE,[51–57] for example, polarization observables were measured with Arcole,[54] Diogène[55–56] and SPES0.[57] The double Δ excitation was investigated with SPES3.[58] One should also mention older experiments of SATURNE I on possible Δ components in the ground state of nuclei.[59]

2 Inclusive charge exchange reactions

In the charge exchange program at SATURNE, (^{3}He,t),[4–9] ($\vec{d}$,2p[$^1S_0$]),[15–18] (^{6}L$\vec{i}$,^{6}He)[13] and heavy ion charge exchange[10–12] have been studied using the spectrometer SPES4.[60–61] This spectrometer is well suited for 0 degree and forward angle measurements, which is essential here, because the angular distributions are very forward peaked. The data were taken at several incident energies depending on the the main topic of investigation: lower incident ener-

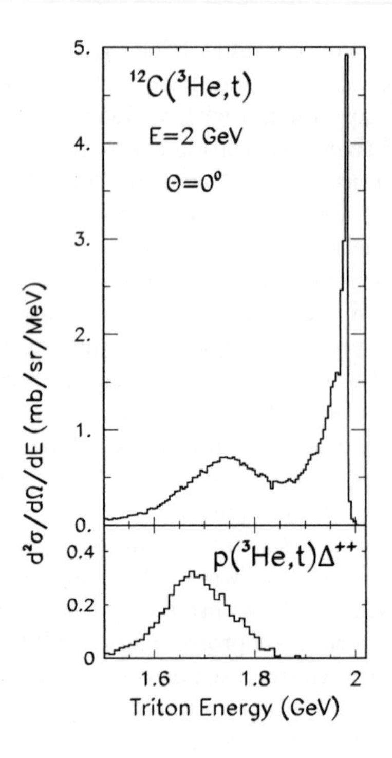

Figure 1: Triton energy spectrum at 2.GeV 0^0 in (^{3}He,t) on ^{12}C and the proton.

gies (from 200 MeV per nucleon) when nuclear structure was concerned, higher incident energies (up to 1.1 GeV per nucleon) when the spin-isospin response in the quasi-elastic or the Δ region was addressed.

2.1 (^{3}He,t) reaction

The upper part of figure 1 shows the triton energy spectrum for ^{12}C(^{3}He,t) at 0^0 and 2 GeV. This spectrum demonstrates that the spin-isospin response of nuclei is concentrated in two excitation energy domains with roughly equal strength:

- At low excitation energy, it corresponds to the excitation of particle-hole states, for example the Gamow-Teller giant resonance or higher multipolarities (spin-dipole, spin quadrupole,). At larger triton angles, the giant resonance excitations are damped and the quasi-elastic peak develops.
- At high excitation energy, around 300 MeV, it corresponds to the excitation

of Δ-hole states. This twofold spectrum is a common feature of all charge-exchange reactions around 1 GeV.

Before the charge-exchange program starts at SATURNE, a very successful program on (p.n) in Indiana gave evidence for the Gamow-Teller giant resonance in many nuclei, up to Uranium, and demonstrated in a model independent analysis that this resonance is quenched by 50 %.[62] The prospect of investigating the coupling of the Gamow-Teller resonance with the Δ resonance, which could be responsible for the missing strength,[63] motivated a Scandinavian team to come to SATURNE and join French physicists who had in mind an experiment on Δ excitation in heavy ion charge-exchange. It was the beginning of about 15 years fellowship in physics. The outcome of the giant resonance investigation is outside the scope of this talk, therefore I quote but one result: by comparing the intensity of well known Fermi transitions to the intensity of well known Gamow-Teller transitions, it was shown that (^{3}He,t) favors the spin-isospin transitions over non-spin transitions by a large factor (factor 4 in amplitude).[8] Therefore SATURNE delivers beams in the ideal energy range.

When comparing the upper and lower part of figure 1, one sees an effect which was pointed out by (^{3}He,t) : the peak corresponding to the Δ excitation in nuclei is shifted towards lower energy transfer than the peak of the free Δ created in the reaction on Hydrogen. New experiments, later on, and a close look to older data, show that this feature is common to all charge-exchange reactions around 1 GeV per nucleon. In (^{3}He,t) the peak position and its width are independent of the target, from ^{12}C to ^{208}Pb.[6] At 2 GeV and 0 degree this shift amounts to 70 MeV, it increases when the triton angle increases, i.e. when the momentum transfer increases.[8]

2.2 Theoretical interpretation: the spin-isospin response

This large shift motivated theoretical and experimental works. On the theory side, the basic idea was given by Magda Ericson and Guy Chanfray.[64] It relates the observed shift to the collective behavior of the longitudinal spin-isospin response. Then, other groups developed detailed calculations in the same spirit, they basically agree.[65–78]

Schematically, the spin-isospin response can be split into two components, depending on the coupling of $\vec{q}$, the momentum transfer, with the spin transition operator $\boldsymbol{\sigma}$ (in the N N^{-1} case), or $\boldsymbol{S}$ (in the Δ N^{-1} case), at the target nucleon vertex. The longitudinal component, corresponds to $\boldsymbol{\sigma}.\vec{q}$ or $\boldsymbol{S}.\vec{q}$, the transverse component, corresponds to $\boldsymbol{\sigma} \times \vec{q}$ or $\boldsymbol{S} \times \vec{q}$. Pure pion exchange excites the longitudinal component, photon absorption and pure rho exchange excite the transverse component. In reality, the ρ exchange is very phenomenological, and the transverse response is mostly triggered by the short range part of the interaction, for example the short range cut of pion exchange.

Theory predicts that in the longitudinal channel, the residual Δ-hole correlations due to pion exchange are strong and attractive. They give rise to a collective behavior of the longitudinal response. In the q, ω plane (ω is the energy transfer), most of the strength is removed from the Δ dispersion line and from the π dispersion line, and it is collected along another dispersion line which is often called the "pion branch".[3] Gerry Brown calls "pisobars" this mixing of virtual pions with Δ-hole states. On the contrary, the Δ-hole correlations are weak (slightly repulsive) in the transverse channel, therefore the transverse response looks like the bare response.

The charge-exchange reaction around 1 GeV per nucleon has several interesting features. Close to 0 degree its kinematics corresponds precisely to the pion branch dispersion line in the q, ω plane. It can trigger both components of the response, with a ratio in intensity determined by the driving force. In the case of compound particles, the projectile form factor is different in the longitudinal and transverse channels, which might emphasize either component. For example, microscopic calculations of the (^{3}He,t) form factor show that the sensitivity to correlations in the spin longitudinal response is enhanced by the interference of the ^{3}He and triton S and D states.[68] In Delorme and Guichon's calculations one half of the shift is due to attractive Δ-hole correlations in the spin longitudinal response.[65–68] The rest is due to mean field effects (Δ binding) and to Fermi broadening combined with the effect of the steep (^{3}He,t) form factor. In nuclear matter at normal density, the correlations would give a much larger shift: around 150 MeV. We see only part of the collective effect because (^{3}He,t) is a surface reaction. The model reproduces also the (^{3}He,t) shift increase when the triton angle increases. It is the result of two effects: the spin-isospin correlations become more and more attractive when the momentum transfer increases and the ratio of the longitudinal to transverse (^{3}He,t) projectile form factor is an increasing function of q.

2.3 Heavy ion charge-exchange

The Δ excitation plays a very important role in pion production and propagation in nucleus-nucleus collisions. This argument by itself justifies why heavy ion charge-exchange should be studied. Moreover, following the (^{3}He,t) interpretation, it was interesting to investigate the role of the density and of the projectile form factor with heavy ion charge-exchange. Even if this reaction is very peripheral, the density is not so small where the interaction takes place, because the target and the projectile overlap,[74–79] for example, with ^{12}C at 1 GeV per nucleon, it is around 0.6 ρ_0.

We studied the heavy ion charge-exchange with ^{12}C, ^{20}Ne and ^{40}Ar beams around 1 GeV per nucleon on ^{1}H and several nuclei.[10–12] We measured integrated cross-sections, using SPES4 at 0^0 with ±14 mrad horizontal and vertical aperture. In fact, the angular distribution is so forward peaked that more than 85% of the total cross-section was recorded in such an opening. This program took advantage of the intensity (6 10^8 ^{12}C ions per Saturne spill) and small emittance of the beam delivered by the source DIONE coupled to MIMAS and SATURNE, the high purity of the ^{12}C beam was especially appreciated.

These experiments gave evidence of the Δ excitation in nuclei by means of heavy ions. Like in (^{3}He,t), the strength is concentrated in two excitation energy domains : at low ω, the region of particle-hole states, and around ω=300 MeV, the region of Δ-hole states. At high enough incident energy, the Δ excitation may be very strong, depending on the nuclear structure of the projectile and the ejectile. The role of some specific spin-isospin transitions to ejectile bound states (the ejectile must be bound in order to be detected in the spectrometer focal plane) is discussed in references 10 and 11. In short, some transitions to ejectile bound states carry angular momentum transfer L=1 or L=2, therefore they are more efficient to excite the Δ, than Gamow-Teller transitions, because their form factor is maximum in the momentum transfer range of interest (around 1 fm^{-1}). For example, this is the case for ^{12}B, and it explains why the intensity of the Δ excitation is larger in (^{12}C,^{12}B) than in (^{12}C,^{12}N) which proceeds through a Gamow-Teller transition.

Whatever the projectile, the Δ peak in nuclei is shifted down in energy, from the free Δ peak. The shift is roughly equal to the shift observed with light ions, but it depends on the projectile-ejectile pair and in some cases it is larger (for example, in (^{12}C,^{12}B) and ^{40}Ar induced reactions). Let us compare the Δ peak energy in a (p,n) type reaction and in the corresponding (n,p) channel. In the (p,n) channel, (^{12}C,^{12}B) or (^{20}Ne,^{20}F), the energy of the Δ peak is independent of the target mass, like in (^{3}He,t). On the contrary, in the (n,p) channel, (^{12}C,^{12}N) or (^{20}Ne,^{20}Na), the shift of the Δ peak increases with the target mass.[11,14] This experimental fact is not explained up to now, it is probably related to the isospin dependence of the Δ N interaction in nuclei, or to Coulomb effect.

The dependence of the Δ excitation cross-section on the target mass follows from absorption. The number of effective nucleons is weak : for example, with a ^{12}C beam it is 2.3$\pm$0.4 for ^{12}C target, and 4.6$\pm$0.9 for ^{208}Pb.[11] In the nuclear sector the contribution of spin-isospin transitions with low multipolarity is large. When assuming the same absorption factor in the nuclear sector and in the Δ sector, one finds that the A dependence of the cross-section in the nuclear sector is explained by the neutron excess in the target.

Several models have been developed to predict heavy ion charge-exchange cross-sections.[79–81] The reaction (^{12}C,^{12}N) is the simplest, it proceeds via a single transition in the projectile-ejectile system, i.e. the Gamow-Teller transition to ^{12}N ground state, the only ^{12}N bound state. Guet et al. describe this reaction by a coherent mechanism where the projectile and the target undergo collective spin-isospin excitations.[79] π exchange with a short range cut is used for Δ-hole excitation. This model explains the experimental cross-sections in the nuclear and in the Δ sector on the proton and on ^{12}C, and their dependence on the incident energy. The (^{12}C,^{12}N) form factor plays a crucial role for the Δ excitation.

Angular distributions were measured by performing ray-tracing. They are dictated by the projectile form factor. As in (^{3}He,t), the form factor is different in the spin longitudinal and the spin transverse channel, for example, above 1 fm^{-1}, the (^{12}C,^{12}N) form factor is predicted to be mostly longitudinal.[14]

3 $(\vec{d}, 2p[^1S_0])$ reaction with tensor polarized deuterons

The theoretical interpretation of (^{3}He,t) shows that the shift is a delicate balance of several effects. In order to isolate the spin-isospin longitudinal response, a specific experiment with tensor polarized deuterons was performed. The $(\vec{d}, 2p[^1S_0])$ reaction was studied at 2 GeV and 1.6 GeV on ^{12}C, ^{40}Ca, ^{1}H, and ^{2}H (liquid cryogenic targets).[15–18] Both protons were detected in the SPES4 focal plane after a 33 meter long flight path in the spectrometer. Due to the SPES4 angular and momentum cuts, the 2 proton relative momentum was smaller than 0.2 fm^{-1}, which restricted the 2p system to the ^{1}S$_0$ state. This experiment took advantage of the 85% tensor polarized deuteron beam with intensity up to 10^{11} delivered by the ensemble MIMAS SATURNE.[82]

Using tensor polarized deuterons,[83–84] tensor analyzing powers T_{20} and T_{22} could be measured. Tensor analyzing powers depend on the square of the moduli of the spin dependent reaction amplitudes, therefore they are sensitive to the geometry of the interaction.[85–87] They are as informative as spin transfer coefficients that could be measured in $(\vec{n}, \vec{p})$ reaction. Such a very difficult experiment has not been performed up to now, but a $(\vec{p}, \vec{n})$ program was developed in the mean time at LAMPF.[39–40] In the case of Δ excitation at 0^0 and at small scattering angle, there is only one non vanishing tensor analyzing power, T_{20}^M which is a linear combination of T_{20} and T_{22}.

$T_{20}^M \simeq (T/L - 2.F(t))/(T/L + F(t))$

where T/L is the ratio of transverse to longitudinal cross-section in the underlying (n,p) transition, and F(t) is the ratio of the spin longitudinal to the spin transverse projectile form factor.[16,18] F(t) measures the emphasis that the probe puts on the longitudinal response.

3.1 $(\vec{d}, 2p[^1S_0])$ *reaction on the nucleon*

Concerning the Δ excitation, we refer to the 2 GeV data. In p$(\vec{d}, 2p[^1S_0])\Delta^0$ and d$(\vec{d}, 2p[^1S_0])\Delta$N at 0^0, we find a dominance of transverse versus longitudinal cross-section: T/L=1.93±0.22 on p and T/L=1.71±0.09 on d, essentially constant over the whole Δ region. When the scattering angle increases, the polarization response gets closer to the longitudinal response, but mainly as a consequence of the projectile form factor emphasizing the longitudinal channel. Why such a large transverse contribution in the elementary Δ production?

A large transverse contribution has previously been found by Wicklund et al. in p$(\vec{p}, \Delta^{++})$.[88–89] The data are described by π exchange with a short range cut. The large transverse component comes from the removal of the pion contribution at short distances. When the energy is higher than 800 MeV per nucleon, the polarization data are best described using Poor Man's Absorption (PMA), i.e. a "π cut"at small impact parameter.[18] At smaller incident energy, the data are better described by the π+g' model, which assumes a δ function for the short range absorption.[23] In any case, the cut off mass of the π form factor is small $\simeq$ 600 MeV.[90]

In p$(\vec{d}, 2p[^1S_0])\Delta^0$ at 0^0, one cannot reproduce the constant T/L ratio, i.e. no model is able to account for the short range absorption in this particular geometry where the momentum transfer $\vec{q}$ is parallel to the incident momentum (inelastic kinematics).[18,91] On the contrary, above 2^0, when the kinematics is closer to elastic, the polarization response and the cross-section are correctly reproduced by the PMA model.[18]

In connection with Wicklund's data, it is interesting to quote here the results on the Δ^{++} polarization in p(^{3}He,t)Δ^{++} that we obtained at 2 GeV with Diogène.[23] The angular correlation of the π and p decay products of the Δ^{++} resonance was measured and analyzed in term of the spin density matrix elements, like in Wicklund's work. The data are best described by a π+g' model where g'=2/5, like in p$(\vec{p}, \Delta^{++})$. The analysis confirms the enhancement of the longitudinal contribution by the (^{3}He,t) form factor.

Three other groups studied NN→NNπ around 1 GeV at LNS. Comptour et al. found a strong contribution (90%) of Δ excitation in their measurements of $p(\vec{p}, pp)\pi^0$ differential cross-section with Diogène.[55,56] Analyzing power measurements in $\vec{n}p \to pp\pi^-$ with Arcole also show Δ dominance and demonstrate the need for full three body treatment of the NNπ system.[54] Rappenecker et al. measured differential cross-section and asymmetries in $p(\vec{p}, \pi^0)pp$ with SPES0.[57] Comparing their results to Arcole data, they derived the T=0 n+p cross-section and found a dominance of $^3D_1$ in this channel.

3.2 *($\vec{d}, 2p[^1S_0]$) on nuclei*

Like in any charge-exchange reaction, the peak corresponding to the Δ excitation in nuclei is shifted towards lower energy transfer than the free Δ peak. For example, the shift between Δ in ^{12}C and Δ in deuterium is 65±5 MeV at 2 GeV, 0^0. But the polarization response shows that the shift is identical in the longitudinal response and in the transverse response, in contradiction to the theoretical predictions on the softening of the longitudinal response alone. Moreover, the spin transverse response is enhanced in nuclei, for example, T/L=2.99±0.23 in ^{12}C at 0^0. When the angle increases, T/L decreases because the projectile form factor emphasizes the longitudinal response, but it stays over the free value.

A subsequent ^{12}C $(\vec{p}, \vec{n})$ experiment at 800 MeV at LAMPF similarly observed an enhancement of the spin transverse response.[40] The measured spin dependent partial cross-sections were compared to DWIA predictions. The calculated spin longitudinal cross-section agrees with the experimental one, when the Δ-hole correlations are taken into account. As expected in this channel the Δ-hole correlations are attractive and they enhance the spin longitudinal response. But the calculated transverse cross-section is much too small on the low ω side of the Δ peak. Similarly ^{12}C $(\vec{p}, \vec{n})$ at 500 MeV in the quasi elastic region shows an excess of transverse cross-section on the high ω side of the quasi elastic peak.[39] And the $(\vec{d}, 2p[^1S_0])$ results, in the quasi elastic peak at SATURNE, show an excess of cross-section in the same ω domain.[17–18,92]

In summary, the existence of attractive Δ-hole correlations in the spin longitudinal channel is confirmed by the polarization data, but there is a large excess of cross-section in the transverse channel around the dip region. It means that charge-exchange reactions see strength beyond 1particle-1hole in the transverse response, presumably due to 2particle-2hole correlations and meson exchange currents.[93] Two consequences can be drawn. The transverse response has to be further investigated, this motivates experimental projects on ρ production. The detailed investigation of the spin longitudinal response requires very specific experimental conditions, because otherwise the effect of correlations in this channel is hidden behind larger effects in the spin transverse channel, which are not yet under control. Finally, the most precise tool to select and investigate the spin longitudinal response is a well defined decay channel of the Δ-hole states: the coherent pion production.

4 (^{3}He,t) in coïncidence with decay products of the Δ-hole states

The Δ-hole decay was investigated at SATURNE in an exclusive (^{3}He,t) experiment at 2 GeV, and triton angles up to 4 degrees.[19–24] The triton momentum and angle were measured by using the dipole magnet Chalut with a set of two drift chambers. The pions and/or protons from the Δ-hole decay were detected in the "4π" detector Diogène, originally built to study nucleus-nucleus collisions.[45] The reaction was performed on ^{1}H, ^{2}H, ^{4}He (cryogenic targets), ^{12}C and ^{208}Pb. Δ-hole decay channels were also studied in a ^{12}C(p,n) coïncidence experiment at 800 MeV at KEK.[44] The results of this experiment basically agree with Diogène data which are more complete and detailed.

The Δ-hole decay follows from the way, virtual pions and Δ-hole states propagate in medium. Let us focus on three decay types:

- $\pi^+ + p$ decay: the quasi free Δ^{++} decay.
- $2p$ decay: Δ^{++}+n$\rightarrow$pp or Δ^++p$\rightarrow$pp, the absorption channel which shows the coupling of the Δ-hole states to 2particle - 2hole states.
- Coherent pion production.

4.1 $\pi^+ + p$ decay channel

In the energy transfer spectrum (fig. 2), the Δ peak has the same position and width for all targets from ^{1}H to ^{208}Pb.[19,21–22] The $\pi^+ + p$ events do not contribute to the shift in the inclusive spectrum. The $\pi^+ + p$ invariant mass is 25 MeV lower in ^{12}C(^{3}He,t$\pi^+ p$) than in p(^{3}He,t)Δ^{++}, i.e. in the free Δ case, this is explained by binding effects. The excitation energy of the residual nucleus is at most some MeV. A cascade calculation shows that the Δ's which decay in $\pi + p$ are formed at the very surface of the nucleus,[94] they cannot be sensitive to correlations. The $\pi^+ + p$ decay selects pure quasi-free formation of the Δ resonance.

The Diogène heavy ion collaboration studied quasi-free Δ^{++} decay in proton-nucleus and nucleus-nucleus collisions as a function of the impact parameter.[46–48] The Δ signal over background ratio decreases with the impact parameter: the quasi-free decay selects peripheral collisions. Cascade calculations show that the rescattering and pion absorption destroy the information on primary Δ 's.[95] An experiment on quasi-free Δ^{++} decay was also performed at the Bevalac.[96]

4.2 2p decay channel

Figure 2 shows the energy transfer spectrum of the $2p$ and $3p$ decay.[21–22] In ^{2}H, the $2p$ decay cross-section decreases monotonically with ω. The sharp edge

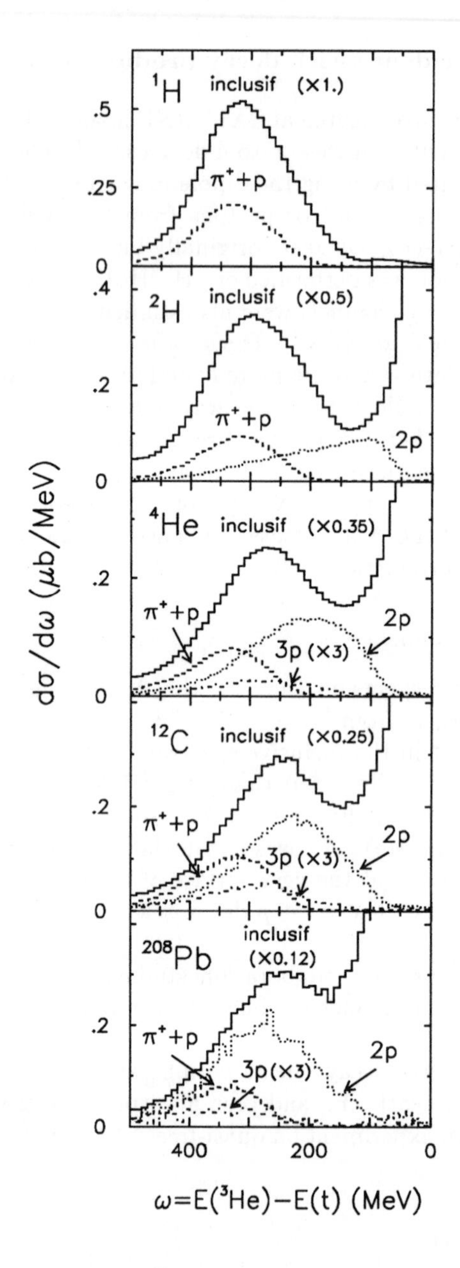

Figure 2: (^{3}He,t) on ^{1}H, ^{2}H, ^{4}He, ^{12}C and ^{208}Pb, at 2.GeV and $\theta_t < 4^0$. [21]

at low ω is an effect of the energy cuts. The difference between ^{2}H and the heavier targets illustrates the many body feature of the Δ excitation in nuclei. In ^{4}He, ^{12}C and ^{208}Pb, the $2p$ peak is considerably shifted down from the quasi-free Δ decay peak. This shift decreases with the target mass: 130 MeV in ^{4}He, 100 MeV in ^{12}C and 70 MeV in ^{208}Pb. In ^{12}C the $2p$ cross-section is twice the $\pi^+ + p$ cross-section, which tells us that the branching ratio is large (the acceptance cuts change this ratio, but not dramatically). The $2p$ events largely contribute to the shift observed in the inclusive spectrum.

For (^{3}He,t$2p$) on ^{4}He, ^{12}C and ^{208}Pb, the excitation energy spectrum of the residual nucleus peaks at 0 MeV, 10 MeV and 40 MeV respectively, and the width of the distribution increases with A. At the same time, the branching ratio of $1p$ events increases. The missing cross-section at low ω in ^{208}Pb may be a consequence of proton rescattering inside the target nucleus and of experimental cuts. In the reaction on ^{12}C, 20% of $2p$ are emitted back to back in the quasi-deuteron frame defined by the four momentum transfer $(\vec{q}, \omega)$ and an initial pn pair at rest $(\vec{\beta}_{qd} = \vec{q}/(\omega + 2m_p))$. The relative energy of back to back protons is maximum and the excitation energy of the residual nucleus is minimum. But the energy transfer distribution is the same for all $2p$ events, which means that in ^{12}C, most $2p$ events are genuine $2p$ with possible rescattering.[22,50]

In the reaction on ^{2}H, no significant Δ peak shows up. It might look surprising when compared to pion absorption on deuterium where the Δ dominates. In fact, $\pi^+ d \to pp$ represents only 5% of the $\pi + d$ total cross-section.[1] 5% of the inclusive (^{3}He,t) cross-section would be hardly visible in the $2p$ spectrum, where the tail of the quasi-elastic peak dominates even above pion production threshold. In π absorption (a typical longitudinal probe), the proton angular distribution in their center of mass has the characteristic shape $1 + 3cos^2\theta$, in γ absorption (typical transverse probe), the angular distribution is bell shaped. In (^{3}He,t) it evolves from a bell shape at low ω, to $1 + 3cos^2\theta$ at high ω.[21–22] This may indicate longitudinal dominance at high ω and transverse dominance at small ω. At high ω, the projectile form factor emphasizes the longitudinal response. At small ω, the quasi-elastic peak dominates, we know from polarization data, that it is mainly transverse. In ^{4}He we observe the same evolution of the angular distribution, but in nuclei, the quasi-elastic peak does not pollute the $2p$ spectrum, it contributes in $1p$ decay. Here, a large transverse component is related to the transverse extra strength which was observed in $(\vec{p}, \vec{n})$ on the low ω side of the Δ peak, and was interpreted as 2particle-2hole. To me, this demonstrate how interesting would be a measure of the polarization response in coïncidence with the Δ decay, but where could it be performed now, when SATURNE and LAMPF are closed?

The cascade calculation shows that the ΔN$\rightarrow$NN process takes place further inside the nucleus than the quasi-free decay $\Delta \rightarrow \pi^+ + p$.[94] Therefore $2p$ decay is sensitive to correlations: Δ-hole correlations in the longitudinal channel and/or 2particle-2hole correlations in the transverse?

The inclusive and partial decay cross-sections were calculated by Osterfeld et al., in DWIA using the Δ-hole model.[71] The predictions for ^{12}C(p,n) are the following. The quasi-free decay branching ratio is 60% and the T/L ratio is 2, exactly T/L in the driving force. This decay actually selects peripheral interactions, the effect of longitudinal correlations is small. The absorption channel corresponds to the inner region of the nucleus. The effect of longitudinal correlations is enhanced and T/L is around 1.2. The peak is largely shifted (100 MeV). The branching ratio is 30%. In (^{3}He,t), the effect of the projectile form factor increases the 2p branching ratio, decreases the $\pi^+ + p$ branching ratio and enhances the effect of the longitudinal correlations. Oset et al. performed a calculation of the inclusive (^{3}He,t) cross-section as a superposition of several initial mechanisms.[78] Knock-out of a target nucleon with possible multiple scattering in the projectile explains the low ω peak. The incoherent pion production represents $\simeq$75% of the Δ peak, its contribution is not shifted. The genuine $2p$ channel is weak, $<$15%, but the subsequent π absorption would redistribute the strength of incoherent pion production on other channels, for example it would feed the $2p$ decay. In both cases, one should take into account the experimental cuts when comparing to data, we anticipate a decrease of the ratio of $\pi^+ + p$ over $2p$ predicted counting rates. Besides $\pi^+ + p$ and $2p$ decay channels, coherent pion production was addressed in both these theoretical works and publications of other groups.[71,75,78,97]

4.3 Coherent pion production

In coherent pion production one pion is emitted by the target nucleus which stays in its ground state, for example, ^{3}He+^{12}C$\rightarrow$t+^{12}C$_{gs}$+π^+. This process qualifies as production of virtual pions followed by elastic scattering.[78] In the energy transfer spectrum, the coherent pion production peak is predicted to be precisely in line with the maximum of the longitudinal cross-section.[71] The cross-section and the energy of the peak of coherent pion production provide a direct measure of the strength of correlations in the longitudinal channel.

The specific selectivity of coherent pion production for the longitudinal response follows from the spin structure of the Δ-hole excitation and decay vertices. In charge-exchange the coupling at the creation vertex might be transverse or longitudinal. The decay vertex is pure longitudinal, since a real pion is created. Therefore, when the probe is longitudinal, the coherent pion production amplitude is proportional to $(S.\vec{k}_\pi)(S^\dagger.\vec{q})$ where $\vec{k}_\pi$ is the pion momentum and $\vec{q}$ the momentum transfer in the (^{3}He,t) reaction. Therefore, in a spin saturated target nucleus, the cross-section is proportional to $(\vec{k}_\pi.\vec{q})^2 = cos^2\theta_{q\pi}$, where $\theta_{q\pi}$ is the angle of $\vec{k}_\pi$ with $\vec{q}$. On the contrary, when the probe is transverse (coupling $S^\dagger \times \vec{q}$), the cross-section is proportional to $sin^2\theta_{q\pi}$. Moreover, the amplitude is multiplied by the target form factor, which accounts for coherence in the target. This form factor decreases rapidly with the four-momentum transfer to the target, in particular, it decreases rapidly with $\theta_{q\pi}$. Therefore the angular distribution of the coherent pion production in the longitudinal channel is peaked in $\vec{q}$ direction. Its width follows from the target size. The transverse cross-section is considerably reduced by the contrary effect of $sin^2\theta_{q\pi}$ and of the target form factor. In Osterfeld's calculation of ^{12}C(p,n), the ratio of the integrated cross-sections which is T/L=2 in the probe and in the quasi-free decay, becomes 1/3 in coherent pion production.[71] One can still improve this impressive selectivity by restricting the π angular range around $\vec{q}$ direction, which was done with the SPES4π detector.[25]

Previously, a clear indication of coherent pion production was found in Diogène data.[20–22] The target excitation energy spectra were recorded in the reaction (^{3}He,tπ^+) on ^{4}He, ^{12}C and ^{208}Pb, for events where one pion alone was detected in Diogène. These spectra show a maximum at ground state in ^{4}He and ^{12}C. This indicates a significant contribution of coherent pion production, but the energy resolution was not sufficient to separate the ground state contribution. The pion angular distribution added an extra evidence for coherent pion production. When selecting the ground state region, the pions are found to be emitted at small angle around $\vec{q}$, the angular distribution width (20^0 in ^{12}C) agrees with the theoretical predictions. On the contrary, when selecting the region around 50 MeV excitation energy, the angular distribution is almost flat. In ^{208}Pb, the excitation energy spectrum shows no significant contribution of coherent pion production. For ^{4}He and ^{12}C, the peak of "ground state region pions" is shifted to lower ω than the inclusive (^{3}He,t) peak. This is in rather good agreement with theoretical calculations, but one needed better energy resolution to really identify coherent pions in order to check theory versus experiment. Moreover the Diogène Chalut set up was not fully efficient for coherent pion production below 2.5^0 triton angle, where the cross-section is maximum. Both reasons motivated the construction of SPES4π.

SPES4π consists of the spectrometer SPES4 to analyze the ejectile momentum and angle and a dipole magnet TETHYS surrounding the target to analyze the pion momentum and angle. Pions were ray-traced in TETHYS field, through position measurement in a set of multiwire chambers sitting outside, they were identified through momentum and time of flight informations.[25]

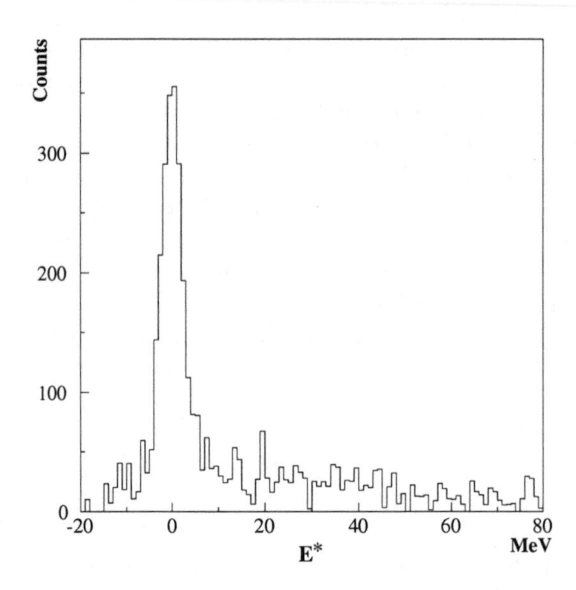

Figure 3: ^{12}C excitation energy obtained with SPES4π in ^{12}C(^{3}He,tπ^{+}) at 1^{0}, 2.GeV.[26]

The acceptance cut corrections were evaluated through a Monte-Carlo calculation using GEANT CERN package. The coherent pion production on ^{12}C and ^{40}Ca was studied, in (^{3}He,tπ^{+}) at 2 GeV around θ_t=1^{0} and (^{12}C,^{12}Nπ^{-}) at 1.1xA GeV (integrated cross-section around 0^{0} ^{12}N scattering angle). The analysis is on progress.

Presently, in (^{3}He,tπ^{+}), we have the following results.[25–26] The target excitation energy resolution is 6.5 MeV (FWHM). In the excitation energy spectrum of both targets there is almost nothing but the ground state peak (fig 3.). The coherent pion production overwhelms the incoherent contribution. This agrees with calculations showing that pion production leading to ^{12}C excited states is weak.[97] The coherent pions are emitted at small angle around $\vec{q}$, the width of the angular distribution agrees with the Diogène data. The detailed width dependence on ω (which follows from the target form factor) is reproduced by theoretical calculations.[97] In ^{12}C, the coherent pion peak is shifted 20 MeV lower in ω than the inclusive (^{3}He,t) peak. The branching ratio is of the order of 10%. This number will be precisely determined in the final analysis. In (^{12}C,^{12}Nπ^{-}) the experimental resolution (15 to 20 MeV) is not as good, but we see the same predominance of the ground-state peak in both target excitation energy spectra. The pion angular distribution follows the target form factor, in particular, the width is determined by the target size.

The branching ratio and the energy transfer spectrum are presently analyzed.

5 Conclusion

SATURNE experimental programs brought major contributions to the understanding of the many-body feature of π and Δ dynamics. The charge-exchange program showed that Δ excitation is strong at intermediate energies, it represents a large fraction of the nucleus spin-isospin response. Whatever the projectile, from proton to ^{40}Ar, the peak corresponding to the Δ excitation in nuclei is shifted towards lower energy transfer than the peak of the free Δ created in the reaction on Hydrogen. In ^{12}C(^{3}He,t) for example, theoretical calculations attribute half of this shift to attractive correlations in the spin longitudinal response, i.e. to a collective behavior of the pionic response.

The spin structure of the Δ excitation was investigated with tensor polarized deuterons. An enhancement of the transverse response was observed in nuclei, and was confirmed by $(\vec{p},\vec{n})$ at LAMPF, which demonstrates that 2particle-2hole correlations and meson exchange currents in the transverse response have to be taken into account in charge-exchange.

The Δ-hole decay was investigated in coïncidence experiments with special attention to three channels. The quasi-free decay selects quasi-free Δ formation at the nuclear surface, it does not contribute to the shift observed in the inclusive charge-exchange spectrum. The absorption channel largely contributes to the shift, this channel selects Δ produced deeper in the nucleus which are sensitive to correlations.

In the end, the coherent pion production is the most precise tool to select and investigate the spin longitudinal response. In the energy transfer spectrum, the coherent pion production peak is predicted to be precisely in line with the maximum of the longitudinal cross-section. The cross-section and the energy of the peak provide a direct measure of the strength of correlations in the longitudinal channel. Striking evidence was found for coherent pion production. The branching ratio is significant and the spin longitudinal response is actually shifted down in energy.

Acknowledgments

The SATURNE charge-exchange program was performed by a franco-scandinavian collaboration which involved Laboratoire National Saturne, N.B.I. (Copenhagen) and I.P.N. (Orsay), and at various stages, I.P.N. (Lyon), Lund University, DAPNIA/SPhN (Saclay) and Soltan Institute for Nuclear Studies (Warsaw). I thank all the collaboration members and I would like to mention especially the participants of the coherent pion production experiment with SPES4π: L. Farhi, R. Dahl, W. Augustiniak, J.L. Boyard, M. Drews, C. Ellegaard, C. Gaarde, T. Hennino, J. Jensen, J.C. Jourdain, R. Kunne, J. Larsen, P. Morsch, P. Radvanyi, B. Ramstein, M. Skousen and P. Zupranski. We warmly thank Madeleine Soyeur for so fruitful discussions and collaboration.

We gratefully acknowledge the technical staff of Laboratoire National Saturne. They developed efficient investigations to reach the highest technical performances and they always operated SATURNE accelerators and equipments at the highest level.

This paper is dedicated to the memory of Carl Gaarde.

References

1. T. Ericson et W.Weise, Pions and nuclei. Clarendon, Oxford (1988).
2. E. Oset, H. Toki and W. Weise, *Phys. Rep.* **83**, 281 (1982).
3. R. F. Sawyer, *Nucl. Phys.* A **335**, 315 (1980).
4. C.Ellegaard *et al.*, *Phys. Rev. Lett.* **50**, 1745 (1983).
5. C. Ellegaard *et al.*, *Phys. Lett.* B **154**, 110 (1985).
6. D. Contardo *et al.*, *Phys. Lett.* B **168**, 331 (1986).
7. D. Contardo, Thèse de troisième cycle, Université Lyon I, 1984.
8. I. Bergqvist *et al.*, *Nucl. Phys.* A **469**, 648 (1987).
9. A. Brockstedt *et al.*, *Nucl. Phys.* A **530**, 571 (1991).
10. D. Bachelier *et al.*, *Phys. Lett.* B **172**, 23 (1986).
11. M. Roy-Stéphan, *Nucl. Phys.* A **488**, 187c (1988).
12. T. Hennino, Cinquièmes Journées d'études SATURNE, (1989) p. 95.
13. C. Gaarde, Spin and Isospin in Nuclear Interactions, ed. S.W. Wissink, C.D. Goodman, G.E. Walker, (Plenum Press, New-York, 1991) p. 295.
14. C. Gaarde, *Ann. Rev. Nucl. Part. Sci.* **41**, 187 (1991).
15. C. Ellegaard *et al.*, *Phys. Rev. Lett.* **59**, 974 (1987).
16. C. Ellegaard *et al.*, *Phys. Lett.* B **231**, 365 (1989).
17. T. Sams *et al.*, *Phys. Rev.* C **51**, 1945 (1995).
18. T. Sams, Phd thesis, N.B.I., Kobenhavns Universitet, 1990.
19. T. Hennino *et al.*,*Phys. Lett.* B **283**, 42 (1992).
20. T. Hennino *et al.*,*Phys. Lett.* B **303**, 236 (1993).
21. B. Ramstein, *et al., Acta Phys. Pol.* B **27**, 3229 (1996).
22. S. Tarlé-Rousteau, Thèse, Université Grenoble I, 1995.
23. R. Dahl, Master thesis, N.B.I., Kobenhavns Universitet, 1995.
24. M. Stampe, Master thesis, N.B.I., Kobenhavns Universitet, 1996.
25. L. Farhi, Thèse, Université Paris VII, 1997.
26. R. Dahl, *et al., Acta Phys. Pol.* B to be published.

27. E. Martinez, Thèse, Université de Caen, 1997.
28. S. Leray, talk at this conference.
29. F. Borne, Thèse, in preparation.
30. C. W. Bjork *et al.*, *Phys. Lett.* B **63**, 31 (1976).
31. C. G. Cassapakis *et al.*, *Phys. Lett.* B **63**, 35 (1976).
32. G. Glass *et al.*, *Phys. Rev.* D **15**, 36 (1977).
33. P. J. Riley *et al.*, *Phys. Lett.* B **68**, 217 (1977).
34. B. E. Bonner *et al.*, *Phys. Rev. Lett.* **41**, 1200 (1978).
35. B. E. Bonner *et al.*, *Phys. Rev.* C **18**, 1418 (1978).
36. G. Glass *et al.*, *Phys. Lett.* B **129**, 27 (1983).
37. A. D. Hancock *et al.*, *Phys. Rev.* C **27**, 2742 (1983).
38. C. L. Hollas *et al.*, *Phys. Rev. Lett.* **55**, 29 (1985).
39. T. N. Taddeucci *et al.*, *Phys. Rev. Lett.* **73**, 3516 (1994).
40. D. L. Prout *et al.*, *Phys. Rev. Lett.* **76**, 4488 (1996).
41. V. G. Ableev *et al.*, *Sov. Phys. JETP. Lett.* **40**, 763 (1984).
42. S. A. Avramenko *et al.*, *Nucl. Phys.* A **596**, 355 (1996).
43. V. N. Baturin *et al.*, *Yad. Fiz.* **31**, 396 (1980).
44. J. Chiba *et al.*, *Phys. Rev. Lett.* **67**, 1982 (1991).
45. J. P. Alard *et al.*, *Nucl. Instrum. Methods* A **261**, 379 (1987).
46. M. C. Lemaire *et al.*, *Phys. Rev.* C **43**, 2711 (1991).
47. M. Trzaska *et al.*, *Z. Phys.* A **340**, 325 (1991).
48. C. Cavata, Thèse, Université Paris Sud, 1989.
49. M. Roy-Stéphan, *Ann. Phys. Fr.* **15**, 393 (1990).
50. M. Roy-Stéphan, *Nucl. Phys.* A **553**, 209c (1993).
51. G. Bizard *et al.*, *Nucl. Phys.* B **108**, 189 (1976).
52. J. L. Laville, Thèse, Université de Caen (1976).
53. T. Hennino *et al.*,*Phys. Lett.* B **48**, 997 (1982).
54. Y. Terrien *et al.*, *Phys. Lett.* B **294**, 40 (1992).
55. C. Comptour *et al.*, *Nucl. Phys.* A **579**, 369 (1994).
56. C. Comptour, Thèse, Université Clermont II, 1993.
57. G. Rappenecker *et al.*, *Nucl. Phys.* A **590**, 763 (1995).
58. J. Yonnet *et al.*, *Nucl.Phys.* A to be published.
59. B. Tatischeff *et al.*, *Phys. Lett.* B **77**, 254 (1978).
60. E. Grorud *et al.*, *Nucl. Instrum. Methods* A **188**, 549 (1981).
61. M. Bedjidian *et al.*, *Nucl. Instrum. Methods* A **257**, 132 (1987).
62. C. Gaarde *et al.*, *Nucl. Phys.* A **369**, 258 (1981).
63. G. E. Brown and M. Rho, *Nucl. Phys.* A **372**, 397 (1981).
64. G. Chanfray and M. Ericson, *Phys. Lett.* B **141**, 163 (1984).
65. J. Delorme et P. A. M. Guichon, 10eme Session d'Etudes Biennale d'Aussois, Mars 1989, Rapport LYCEN 89-02, page C4.1.

66. J. Delorme, 7th Int. Conf. Polarization Phenomena in Nuclear Physics. Paris 1990, Colloque de physique 51, C6-125 (1990).
67. J. Delorme and P. A. M. Guichon, *Phys. Lett.* B **263**, 157 (1991).
68. P.Desgrolard, J.Delorme and C.Gignoux, *Nucl. Phys.* A **544**, 811 (1992).
69. T. Udagawa, S.W. Hong and F. Osterfeld, *Phys. Lett.* B **245**, 1 (1990).
70. P.Oltmanns, F.Osterfeld and T.Udagawa, *Phys. Lett.* B **299**, 194 (1993).
71. T.Udagawa, P.Oltmanns, F.Osterfeld and S.W.Hong, *Phys. Rev.* C **49**, 3162 (1994).
72. B. Körfgen, F. Osterfeld and T.Udagawa, *Phys. Rev.* C **50**, 1637 (1994).
73. V.F. Dmitriev and T. Suzuki, *Nucl. Phys.* A **438**, 697 (1985).
74. V.F. Dmitriev, *Phys. Lett.* B **226**, 219 (1989).
75. V.F. Dmitriev, *Phys. Rev.* C **48**, 357 (1993).
76. E. Oset, E. Shiino and H. Toki, *Phys. Lett.* B **224**, 249 (1989).
77. P. Fernandez de Cordoba, and E. Oset, *Nucl. Phys.* A **544**, 793 (1992).
78. P. Fernandez de Cordoba, J. Nieves, E. Oset and M. J. Vicente-Vacas, *Phys. Lett.* B **319**, 416 (1993).
79. C. Guet, M. Soyeur, J. Bowlin and G. E. Brown, *Nucl. Phys.* A **494**, 558 (1989).
80. L. S. Celenza, J. Hüfner and C. Sander, *Nucl. Phys.* A **276**, 509 (1977).
81. P. A. Deutchman, *et al.*, *Phys. Rev.* C **34**, 2377 (1986) and references therein.
82. P.A. Chamouard *et al.*, 7th Int. Conf. Polarization Phenomena in Nuclear Physics. Paris 1990, Colloque de physique 51, C6-569 (1990).
83. J. Arvieux *et al.*, *Nucl. Instrum. Methods* A **273**, 48 (1988).
84. A. Boudard, Techniques de polarisation et polarimetrie, note CEA-N2584, 1988.
85. D. V. Bugg and C. Wilkin, *Phys. Lett.* B **152**, 57 (1985).
86. C. Wilkin and D. V. Bugg, *Phys. Lett.* B **154**, 243 (1985).
87. D. V. Bugg and C. Wilkin, *Nucl. Phys.* A **467**, 575 (1987).
88. A. B. Wicklund *et al.*, *Phys. Rev.* D **34**, 19 (1986).
89. A. B. Wicklund *et al.*, *Phys. Rev.* D **35**, 2670 (1987).
90. V.F.Dmitriev, O.Sushkov and C.Gaarde, *Nucl. Phys.* A **459**, 503 (1986).
91. S. Mundigl and W. Weise, *Phys. Rev.* C **39**, 710 (1989).
92. T. Sams and V. F. Dmitriev, *Phys. Rev.* C **45**, R2555 (1992).
93. C. Gaarde, *Nucl. Phys.* A **606**, 227 (1996).
94. K. Sneppen and C. Gaarde, *Phys. Rev.* C **50**, 338 (1994).
95. J. Cugnon and M. C. Lemaire, *Nucl. Phys.* A **489**, 781 (1988).
96. E. L. Hjort *et al.*, *Phys. Rev. Lett.* **79**, 4345 (1997).
97. M. A. Kagarlis and V. F. Dmitriev, *Phys. Lett.* B **408**, 12 (1997), and private communication.

ETA MESON AND BARYONIC RESONANCE PHYSICS

G. DELLACASA
Dipartimento di Scienze e Tecnologie Avanzate,
II Facolta' di Scienze dell'Universita di Torino, sede di Alessandria
and
INFN Torino

In its 20 year of life, SATURNE-2 has contributed greatly to providing systematic and comprehensive experimental information in the field of intermediate energy hadron physics. High quality beams and extremely suitable detectors, together with a continuous and heart-felt enthusiasm of researchers, machine physicists, engineers and technicians, have made possible the achievement of excellent physics results. In the following some of them, namely results on the hadroproduction of the Roper resonance, the production of the η meson on nucleons and nuclei, the measurement of the mass and selected branching ratios of the η meson and the subthreshold production of the K^+ meson on nuclei, will be reviewed.

Dans ses 20 ans de vie, SATURNE-2 a donné une contribution enorme du point de vue experimental dans le domain de la physique hadronique à moyenne energie. La qualité des ces faiseaux et an enthousiasme sans cesse des chercheurs, des ingénieurs de la machine et des techniciens a permis une vaste production d'excellents resultats de physique. Dans la suite quelques-uns de ces resultats, notamment la hadroproduction de la resonance de Roper, la physique du meson η, en particulier la production dans les reactions à petit nombre de nucleons, la production inclusive sur noyaux, la mesure de la masse et de quelques rapports de branchement, et sur la production sous seuil du meson K^+ par des proton sur des noyaux, seron rappeler et discuter.

1 The Roper Resonance

1.1 Introduction

The Roper resonance, $P_{11}(1440)$, is a baryonic resonance with the same quantum numbers as the nucleon, $S=\frac{1}{2}$ and $T=\frac{1}{2}$and positive parity, which was discovered [1] in the phase shift analysis of πN elastic scattering in 1964. It is interesting to note that it is the only resonance with the name of its discoverer, probably because of its highly elusive character. There is a relative ignorance of some of its properties. The Particle Data Group quote a width $\Gamma = 250-450$ MeV and the branching ratios still have uncertainties of $20-30$ %. In general the study of the baryon properties is related to the understanding of the structure of the QCD in the non-perturbative regime. The Roper resonance, in particular, is directly connected to the dynamical properties of the size degree of freedom of the nucleon: the Roper resonance is thought in fact to be a good candidate for the lowest energy radial excitation mode of the nucleon

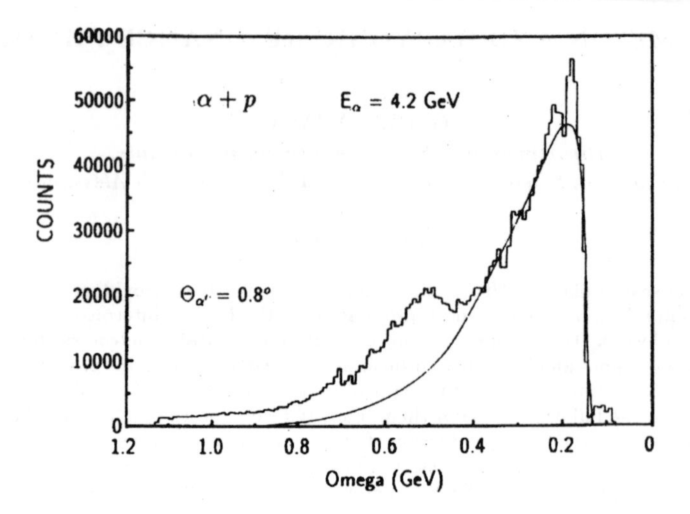

Figure 1: Missing-energy spectrum of inelastically scattered α-particles from a hydrogen target [2]. The solid line shows the spectral shape calculated for projectile excitation.

and therefore its study provides information about the nucleon compressibility.

1.2 The Saturne experiment

The Roper resonance can be excited either electromagnetically or by strong interaction, though this lowest compressional mode of the nucleon is weakly excited in electromagnetic interactions due to its particular structure. From such reactions, it is possible to obtain information about the N* wave function and about the mechanism of the interaction of virtual photons with quarks and gluons in a compound system such as the N*, a field which will be covered by CEBAF. In proton-nucleon scattering experiments, the study of the radial excitations is difficult since the excitation spectra are strongly dominated by spin-isospin modes. Choosing selective probes that may enhance the cross section is then extremely important and forward inelastic scattering of α-particles on protons seems to be a favorable reaction. The structure of the α particle (S=T=0) is particularly suited to induce scalar excitation (ΔS=0, ΔT=0). Such an experiment [2] was carried out at Saturne by a LNS-Orsay-CENS-Stockholm collaboration using α particles of 7 GeV/c, quite close to the maximum momentum available at Saturne, and a 4 cm liquid hydrogen target, with the scattered α particles being detected and momentum analysed in the SPES-4 magnetic spectrometer. Missing-energy spectra ($\omega = E_i - E_f$)

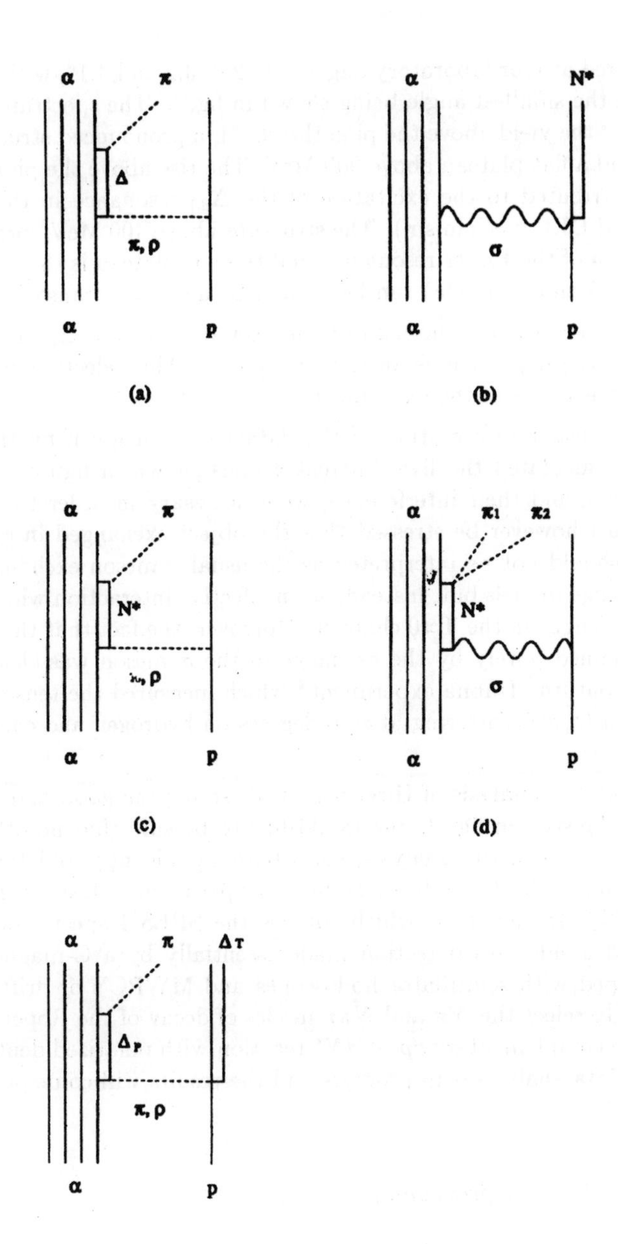

Figure 2: Diagrams for the (α, α') reaction [3].

were measured at four laboratory angles 0.8, 2.0, 3.2 and 4.1^0, with the results obtained at the smallest angle being shown in fig. 1. The spectrum exhibits a strong rise of the yield above the pion threshold, a pronounced structure above 400 MeV and a flat plateau above 900 MeV. The rise above the pion threshold has been attributed to the excitation of the Δ_{33} resonance in the projectile (the so called DEP mechanism). The structure above 400 MeV corresponds to the excitation of the Roper resonance, and the flat plateau is the region where the $D_{13}(1520)$ and $S_{11}(1535)$ can be excited by an $L = 1$ transfer.

As expected, there is no sign of the excitation of the Δ_{33} isobar in the target, since the α particle is an isoscalar probe. This selection rule was one of the prime features of the experiment.

A theoretical interpretation of the data has been given by Hirenzaki et al.[3]. They calculated the five different graphs shown in figure 2 and found that only two, and their interference, were necessary in order to explain the data. It must however be stressed that the object exchanged in graph *b* and called "σ" should not be interpreted as the usual σ-meson exchanged in one-boson-exchange models but, instead, as an effective interaction which accounts for all the strength in the T=0 channel. Moreover, the fact that the data could not be explained purely by the exchange of the σ meson was clear from the data of the Saturne-Dubna experiment[4] which measured the tensor analysing power T_{20} in $(\vec{d}, d')$ scattering at zero degrees on hydrogen and carbon at 4, 5 and 5.5 GeV/c.

Although the analysis of Hirenzaki et al. shows the good agreement with the data to be seen in fig. 3, one would like to be sure that no other graphs, except those used in the analysis, contribute significantly and the only way to do that would be to perform exclusive experiments. Data were taken at the new SPES-4π facility[5], which couples the SPES-4 spectrometer with a forward and a sideways detection made essentially by a C-magnet and two arms equipped with scintillator hodoscopes and MWPC's or drift chambers. The idea is to select the $N\pi$ and $N\pi\pi$ modes of decay of the Roper resonance. Data on the complementary $\vec{d}p \to dN^*$ reaction with polarized deuterons were taken; the data analysis is in progress and the results will come out soon.

1.3 The nucleon compressibility

A very important topic related to the Roper resonance is the question of the nucleon compressibility. For the first time, thanks to the Saturne experiment, it was possible to deduce a value of K_N, the nucleon compressibility defined

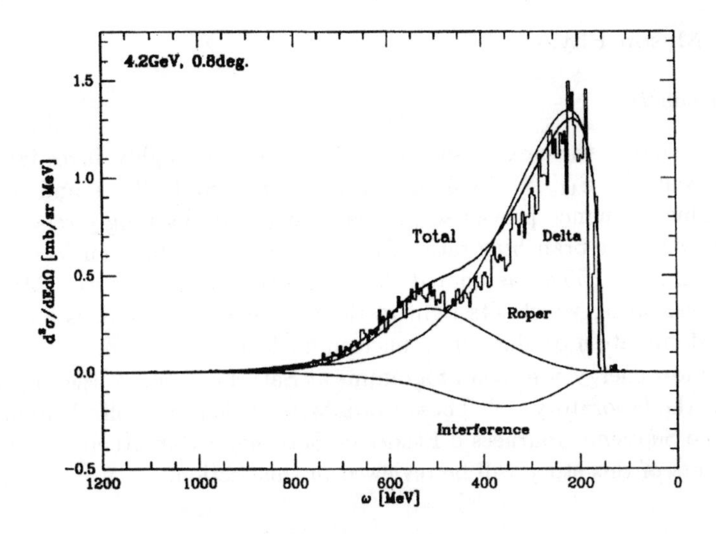

Figure 3: Calculated cross sections of the target Roper process and the projectile Δ process at $E_\alpha = 4.2$ GeV and $\theta = 0.8^o$ compared with experimental data[3].

for a spherical system of dimension r as

$$K_N = r^2 \left(\frac{d^2E}{dr^2}\right)_{E=E_0} \tag{1}$$

corresponding to the curvature in the equation of state at the ground- state minimum. K_N is also related to the characteristics of the resonance by the relation

$$K_N = \left(\frac{m_N}{3\hbar^2}\right) E_x^2 < r_N^2 > , \tag{2}$$

where E_x is the excitation energy (roughly 500 MeV), m_N is the nucleon mass (939 MeV) and $< r_N^2 >$ is the mean square radius of the nucleon as seen by the α-particle. From α-proton elastic scattering one gets a $< r_N^2 >$ value of (0.6 ± 0.1) fm^2, which yields

$$K_N = (1.4 \pm 0.3) \text{ GeV} . \tag{3}$$

This means that the surface of the nucleon is not very hard. Calculations of K_N performed in the framework of existing nucleon models give values between 800 and 1000 MeV (the Skyrmion model), between 900 and 1200 MeV (the MIT bag model), and 3 GeV (the constituent quark model using an harmonic oscillator with $\hbar\omega = 500$ MeV). The experimental value thus lies between the values obtained in different models.

2 Eta Meson Physics

2.1 *Introduction*

Rising in energy, the next resonance which was thoroughly investigated at Saturne, was the S_{11} $N^*(1535)$ isobar which has S=$\frac{1}{2}$, T=$\frac{1}{2}$ and negative parity. This resonance proved so popular because it is strongly related to η-meson physics; the branching ratio of the $N^*(1535)$ into the η-nucleon channel is very high, $30-55\%$ (as quoted by the Particle Data Group). Moreover, this S_{11} resonance, which sits right at the η-nucleon threshold, is believed to strongly dominate η production in nucleon-nucleon interaction.

"The low energy η-meson programme at Saturne is one of the big success stories of the laboratory ⋯" These words were spoken by Colin Wilkin in his talk at the Septième Journées d'Études de Saturne in Ramatuelle in 1996 and some events of this story will be reviewed in this section.

2.2 *The $dp \to {}^3He\,\eta$ and $pd \to {}^3He\,\eta$ reactions*

After earlier measurements performed also at Saturne [6], the $dp \to {}^3\text{He}\,\eta$ reaction was measured [7] by detecting the ^{3}He particles in the SPES-4 spectrometer with the aim of comparing its tensor analysing power, T_{20}, with that of the analogous reaction $dp \to {}^3\text{He}\,\pi^0$, which had earlier been shown to be large and negative near threshold [8]. This was consistent with a reaction mechanism consisting of the emission of the pion by one nucleon and its subsequent scattering on a second nucleon before emerging. Since the two mesons have the same spin-parity, despite the difference in their isospin, if the mechanism is the same, apart from the change of the π into an η in the second step, the tensor analysing power for the dp$\to^3$He η reaction should also be large and negative near threshold.

The experiment was done at five energies very near threshold, *viz.* $\Delta T_d=$ 0.2, 3.0, 7.9, 16.6 and 71.7 MeV [T_d(threshold)= 1786.7 MeV] and the results showed that the tensor analysing power was almost a factor of ten less than for threshold π^0 production, in complete disagreement with the foreseen mechanism. The discrepancy was thought to be due to the large momentum transfer involved and it pushed theoreticians to invoke genuine three-nucleon processes and strong final state interactions.

It is interesting to note that in this energy domain, supposing that the η production proceeds *via* a $pp \to d\pi^+$ reaction followed by a $\pi^+ n \to p\eta$ reaction, the intermediate pion is almost "real"; this observation was dubbed the "kinematic miracle" [9] and it means that the second process does not require a high Fermi motion of the nucleon. As is clear from the experimental spectra

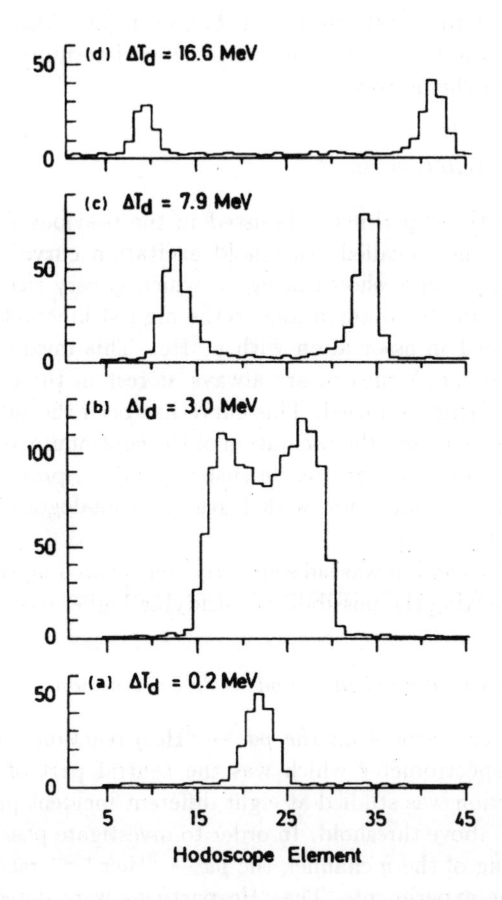

Figure 4: Number of events per machine burst (10^{10} d) for individual hodoscope elements of the SPES-4 focal plane of about 5 MeV/c, the center of 25th element, corresponding to p/Z=1.317 GeV/c[7].

of fig. 4, this experiment showed the possibility of building a tagged η-meson facility profiting from the high intensity of Saturne. In fact in the article the authors declare that: "While investigating this, it became apparent that the high efficiency for selecting events near threshold, based upon ^{3}He detection, was a matter of great practical interest in its own right. Also of unforeseen importance is the method of intrinsic incident energy determination which was hinted at by the data themselves."

2.3 The threshold excitation curve

The continuation of the experiment discussed in the previous subsection was the measurement of the so-called "threshold excitation curve" of the $pd \rightarrow {}^3\mathrm{He}X$ reaction [10]. The curve shown in fig. 5, which is very rich in structure, was obtained by limiting the observations to the highest kinematically allowed missing mass produced in association with a ^{3}He. This means that the two objects, the ^{3}He and the X-meson, are always at rest in the c.m. system as the incident proton energy is varied. This curve has been the subject of many studies. The sharp η peak and the oscillation of the continuum at the η threshold, called after its interpretation the "η-cusp region" [11], proved particularly interesting especially in connection with features of analogous reactions, for example $pd \rightarrow {}^3\mathrm{He}\,\pi^0$.

The $pd \rightarrow {}^3\mathrm{He}X$ reaction was subsequently remeasured up to a proton energy of 1.86 GeV, showing the possibility of studying higher mass X-mesons [12].

2.4 The SPES-2 measurement of the $pd \rightarrow {}^3He\,\eta$ reaction

The last Saturne measurement on the $pd \rightarrow {}^3\mathrm{He}\,\eta$ reaction was carried out using the SPES-2 spectrometer which was the central part of the tagged η facility [13]. The reaction was studied at eight different incident proton energies from 0.2 to 11 MeV above threshold. In order to investigate possible threshold effects at the opening of the η channel, the $pd \rightarrow {}^3\mathrm{He}\,\pi^+\pi^-$ reaction was also studied in the same experiment. The ^{3}He particles were detected by using MWPC's and scintillator counters and selected by applying cuts on the energy loss and on the time of flight. Particular care was taken in the measurements at low proton energy since the energy loss in the target was such that only part of the target thickness was then active. The most important results of the experiment were:

1. Confirmation of the forward-backward asymmetry being compatible with zero, indicating that the production is in an S wave;

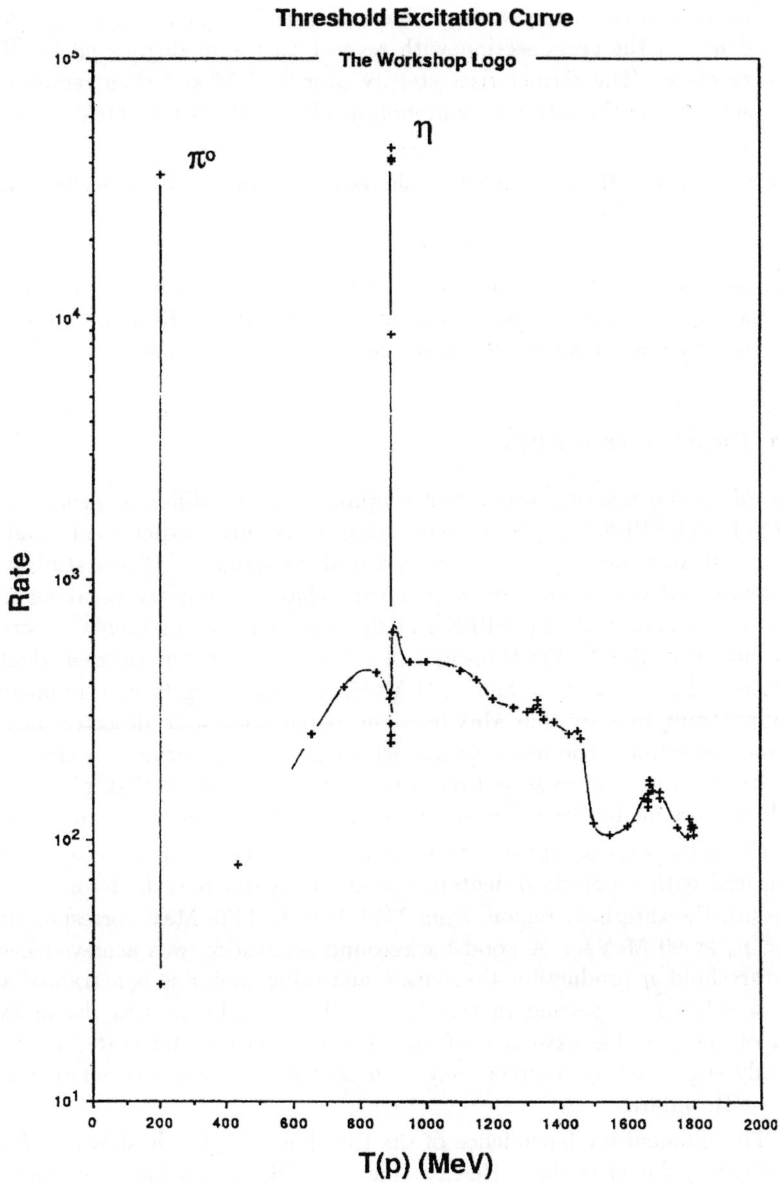

Threshold Excitation Curve

The Workshop Logo

Figure 5: The threshold excitation curve [10]

2. Confirmation of the strong difference in the angular and energy dependence of the cross section with respect to the analogous $pd \to {}^3\mathrm{He}\,\pi^0$ reaction. The former rises steeply over 2 MeV and then remains constant while the latter rises monotonically to at least 30 MeV;

3. The $pd \to {}^3\mathrm{He}\,\pi^+\pi^-$ reaction shows no evidence of strange effects at the threshold of η production;

4. At ΔT_p=1.5 MeV, which is the energy chosen for the use of SPES-2 as a tagged η facility, the background due to $pd \to {}^3\mathrm{He}\,\pi^+\pi^-$ production is very low, about $1-2\%$ if the reaction were isotropic.

2.5 The dd $\to \alpha\,\eta$ reaction

The $dd \to \alpha\,\eta$ reaction was recently studied by two different groups at the SPES-4 and SPES-3 spectrometers. [Earlier Saturne experiments had provided first an upper limit[14] and then a positive signal[15].] These studies were motivated mainly by the connection with charge-symmetry breaking in the $dd \to \alpha\pi^0$ reaction. At the SPES-4 facility the new measurements[16] were carried out very close to the threshold of 1120.3 MeV at four incident deuteron energies, 1121, 1122, 1123 and 1124 MeV, corresponding to c.m. momenta p_η ranging from 14.4 to 38.0 MeV/c. The α-particles were detected near the forward direction. The main results achieved in the experiment were a determination of the total cross sections and a strong suggestion that $\pi^0-\eta$ mixing could be responsible for the positive signal reported in the reaction $dd \to \alpha\pi^0$.

Using the wide acceptance SPES-3 spectrometer[17], the same reaction was measured with a polarized deuteron beam at six different deuteron beam energies in the threshold region, from 1121 Mev to 1139 MeV corresponding to $10 < p_\eta < 90$ MeV/c. A good background separation was achieved because for threshold η production the tensor analysing power is constrained to be $T_{20} = +1/\sqrt{2}$. A second motivation for this experiment was the investigation of the possible existence of an η-helium quasi-bound state, which was already suggested by the relatively large scattering length extracted from the $pd \to {}^3\mathrm{He}\,\eta$ data.

The momentum dependence of the threshold amplitude shown in fig. 6 is significantly flatter in the η^4He case than for η^3He, suggesting that the former system is much more "bound" than the latter. With an unconcealed sensation of melancholy the authors conclude the article by stating that "this was the last experiment carried out before the closure of the SPES-3 spectrometer".

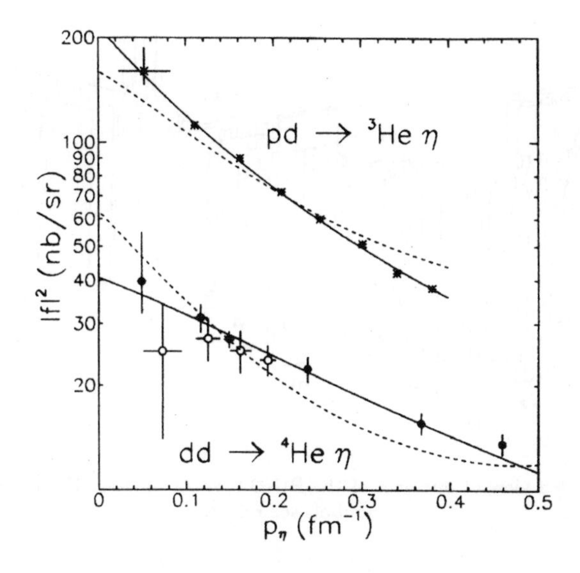

Figure 6: Averaged squared amplitudes of the $dd \to {}^4\mathrm{He}\,\eta$ and $pd \to {}^3\mathrm{He}\,\eta$ reactions as functions of the η c.m. momentum [17].

2.6 η-meson inclusive production

I have so far dealt with η production in exclusive experiments on light nuclear system, in which the heavy particle associated with the meson was detected by a magnetic spectrometer. For exclusive experiments on heavier nuclear or inclusive production studies, this is not possible and one must detect the produced meson directly. The almost 40% decay branching ratio of the η-meson into the $\gamma\gamma$ channel suggests the possibility of detecting it through the measurement of the invariant mass of the $\gamma\gamma$ pair. This was achieved at Saturne using the PINOT spectrometer. Fig. 7 shows one of the two arms of which the spectrometer consisted. Each arm was designed to detect the energy and direction of a γ-ray.

In addition to measurements of NN total production cross section, from which the strong difference in η production between pn and pp was demonstrated for the first time [18,19], a series of measurements of η production in proton-nucleus reactions was made, varying the proton incident energy, the laboratory kinetic energy of the detected η's, and the target mass A, in order to obtain as complete a set of data as possible [20,21,22].

A typical spectrum, obtained with the Pinot spectrometer, is displayed in fig. 8 and it clearly shows two peaks ascribed to π^0 and η production with a

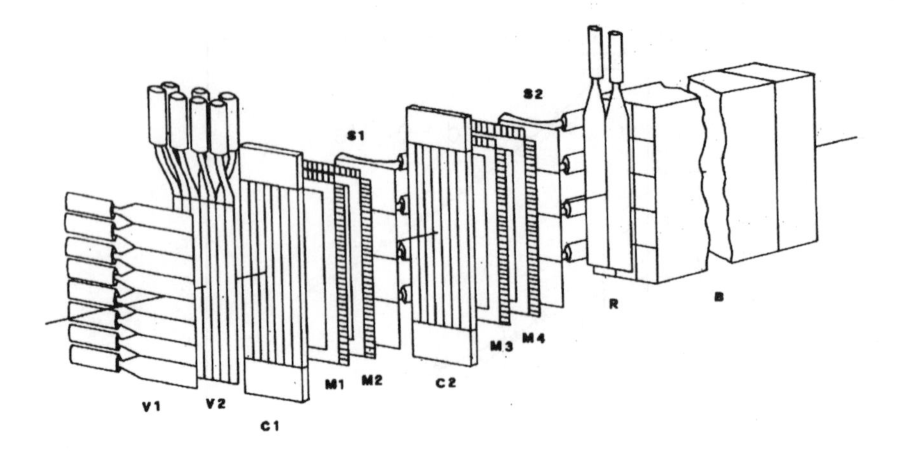

Figure 7: Exploded view of one arm of the PINOT Spectrometer. For the meaning of the symbols see ref. [20].

shoulder between due to multiple π^0 production. The interpretation of the data, presented in the form of doubly differential cross sections as a function of meson energy and production angle, was normally done using the so-called folding model [23]. One important ingredient of the model is the re-absorption of the meson in the nuclear medium which in turn leads to the determination of the η-nucleon cross section, which is impossible to measure directly due to the very short (10^{-19} s) lifetime of the η.

From the above-threshold data, where the production mechanism *via* a single NN collision and a two-step $NN \to NN\pi$ followed by a $\pi N \to N\eta$ collision is well treated, an ηN cross section of about 30 mb was deduced [24], in agreement with that measured [25] in $\gamma N \to N\eta$. Regarding the subthreshold production, there are still big discrepancies, as shown for example in fig. 9, between data and calculations, leaving room for speculations such as the variation of the meson mass in nuclear medium or unknown production mechanisms.

Though the PINOT data analysis is finished, not all of the many results have yet been completely written up and published.

2.7 The determination of the η-meson mass

The fundamental observation, which showed the feasibility of this experiment, was that all the objects apart from the η involved in the $dp \to {}^3\mathrm{He}\,\eta$ are charged, so that their momenta can therefore be analysed to high precision

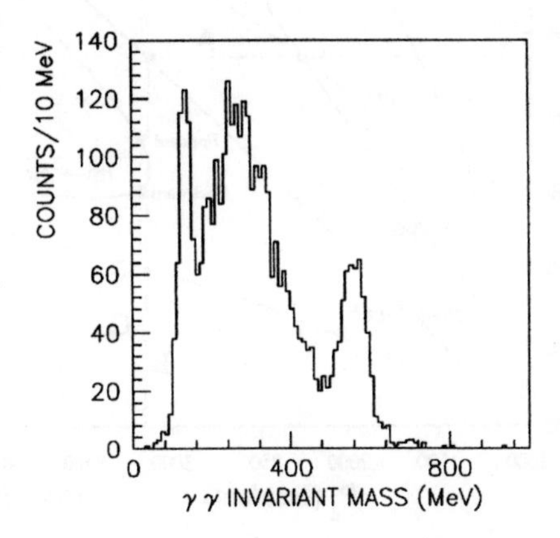

Figure 8: Typical $M_{\gamma\gamma}$ invariant mass spectrum[24].

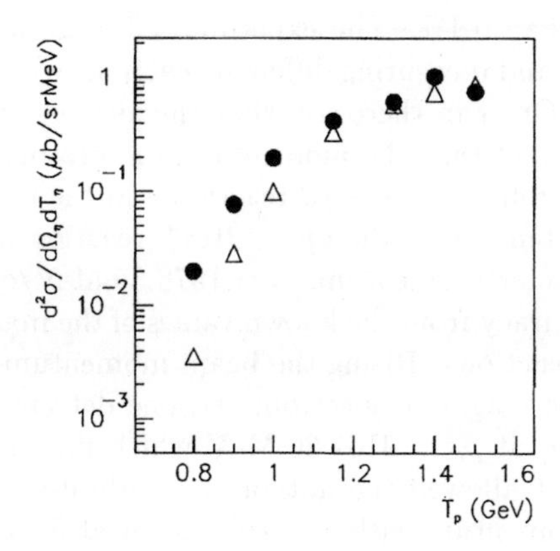

Figure 9: Doubly differential η production cross section as a function of the incident proton energy for the reaction $p\,^{12}\mathrm{C}\rightarrow \eta X$ [24].

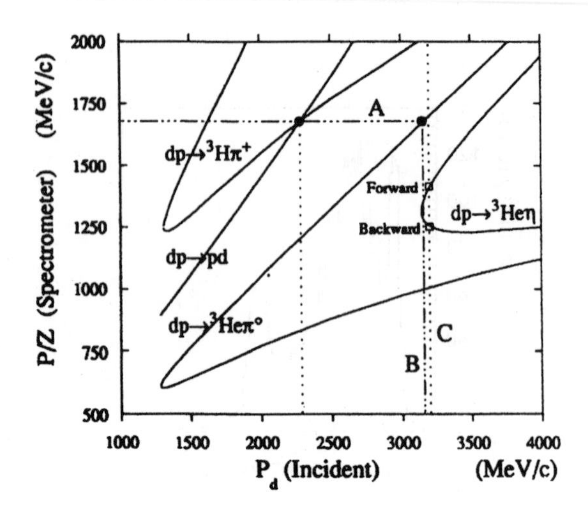

Figure 10: The secondary momentum per unit charge ($p/Z = p_s$) detected at 0° (lab.) in SPES-4, plotted against the incident deuteron beam momentum for various kinematic curves[26].

in a magnetic spectrometer like those operating at the Laboratoire National Saturne. The η-meson mass was rather poorly known at that time, with an incertitude of more than 0.1%. The experiment [26] was carried out using the SPES-4 spectrometer and measuring different reactions whose kinematic curves are shown in fig. 10. One can there see what the authors call "our sheer good fortune", namely the fact that the momenta of the protons produced forward (in the c.m. system) from the $dp \to pd$ reaction and the tritons emitted backward (in the c.m. system) from the $dp \to {}^3\text{H}\pi^+$ reaction intersect at a single value of the spectrometer momentum, p_s = 1678.35 MeV/c, which can be calculated with high accuracy from the known values of the masses of the particles involved in the two reactions. Rising the beam momentum then, for the same settled value p_s of the magnetic spectrometer, one detects the ^{3}He's from the $dp \to {}^3\text{He}\pi^o$ reaction, at p_d = 3161.26 MeV which lies in the vicinity of the η threshold. This fact allowed the authors to calibrate both the spectrometer and the beam momentum without using classical field measurements. In addition, thanks to the high sensitivity of the $dp \to {}^3\text{He}\,\eta$ reaction to the incident deuteron energy, it was possible to observe perturbations arising from the details of the beam extraction from Saturne. In the plot of fig. 11 this dependence is shown for the case of the $dp \to {}^3\text{He}\pi^o$. As underlined by the authors, not correcting the data for this effect would have destroyed the precision of

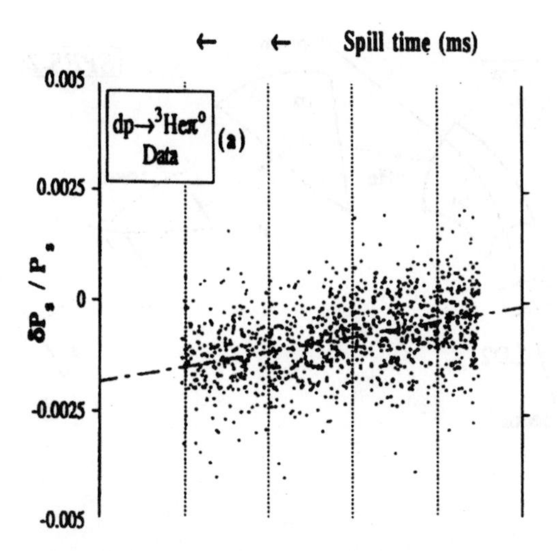

Figure 11: Variation of the fractional deviation from the central value of the ^{3}He momentum against the beam spill time for the $dp \to {}^3\text{He}\pi^o$ far from its threshold[26].

the measurement. After all corrections, and averaging over several runs, they found the value:

$$m_\eta = (547.30 \pm 0.15)\ \text{MeV/c}^2\ . \tag{4}$$

A calibration done using information from surveying the accelerator would have given:

$$m_\eta = (547.20 \pm 0.30)\ \text{MeV/c}^2\ , \tag{5}$$

which is completely consistent with the previous answer but with a much larger error bar.

It is instructive to compare the new measurement with earlier ones. The value quoted by the Particle Data Group was:

$$m_\eta(PDG) = (548.8 \pm 0.60)\ \text{MeV/c}^2\ , \tag{6}$$

which differs from the Saturne value by more than two of their standard deviations. The new value is actually in very good agreement with the one obtained in the only existing electronic experiment, *viz.*:

$$m_\eta(\text{Duane}) = (547.45 \pm 0.25)\ \text{MeV/c}^2\ . \tag{7}$$

By chance this value had not been included in the PDG average, though it was the most precise result. Since the aim of the Duane's experiment was

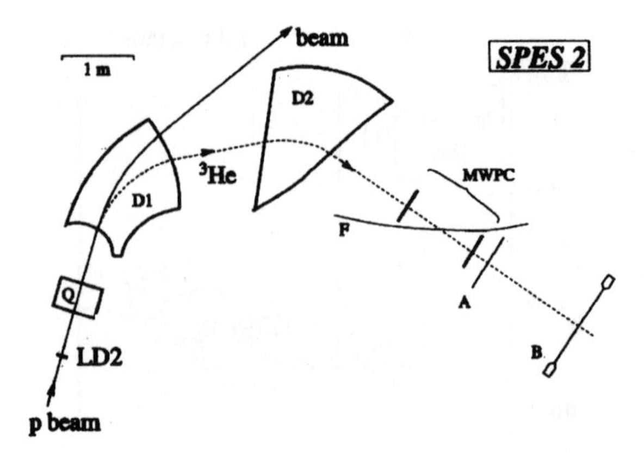

Figure 12: Top view of the SPES-2 tagged η meson facility. From ref. [13].

primarily the determination of the η'-meson mass, the Saturne result reinforces the confidence in that also.

2.8 The η-meson facility

That the high production rate of η-mesons *via* the $dp(pd) \rightarrow {}^3\text{He}\,\eta$ reactions could be a valuable source of tagged η's was already noted in 1985 by L. Satta et al. [27] on the basis of data taken away from threshold [6].

The η-meson facility at Saturne was built at the beginning of nineties using the even higher rates measured near threshold [7]. This was achieved by coupling the SPES-2 magnetic spectrometer with a detection designed for providing a careful identification of the recoiling ^{3}He particles. The facility, sketched in fig. 12, was able to tag a maximum of 800 quasi-monochromatic η-mesons per second ($p_\eta \approx 257$ MeV/c) using a proton beam with an average intensity of 10^{11} particles per spill of 0.7 s duration every 1.5 s. The facility was used to perform a series of experiments on η-meson decay branching ratios with very high precision.

2.9 The measurement of the $\eta \rightarrow \mu^+\mu^-$ branching ratio

$\eta \rightarrow \mu^+\mu^-$ is an example of a transition between a pseudoscalar meson and a pair of charged leptons. In the Standard Model the process is described by a graph which contains two photons in the intermediate state. The contribution

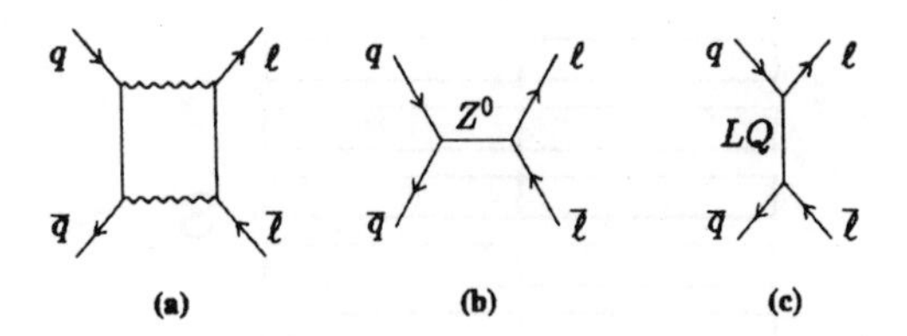

Figure 13: Diagrams for the decay $P^0 \to l^+l^-$: (a) QED contribution,(b) weak interaction contribution, and (c) hypothetical leptoquark contribution[28].

due to a weak interaction *via* an intermediate Z_o state is very low and should modify the transition amplitude by less than 1% and so can be neglected. Other processes may be involved involving physics beyond the Standard Model and it is this possibility of new physics that makes this decay so interesting. There could for example be leptoquark bosons carrying both quark and lepton flavours. The three processes discussed before are shown in fig. 13. For this experiment [28] the proton beam energy was kept 1.6 MeV above the threshold of the $pd \to {}^3\mathrm{He}\,\eta$ reaction and the ^{3}He's, emitted forward in a cone of semi-aperture $< 1.2^0$, were detected by the SPES-2 spectrometer. The corresponding tagged η-mesons were then emitted, also forward in the laboratory system, in a cone with $\theta < 6^0$. In this way the minimum $\mu^+\mu^-$ opening angle was 126^0. The muon detector, shown in fig. 14, was made from an ensemble of hodoscopes, for triggering and positioning purposes, stop counters and degraders suitably positioned and shaped. Prior to this experiment the situation on this decay was not clear, due essentially to the limited statistics and to normalisation problems. An upper limit was given in 1968 by an experiment performed at Brookhaven:

$$B(\eta \to \mu^+\mu^-) < 20 \times 10^{-6}\,. \tag{8}$$

In 1969 the decay was positively identified at CERN on the basis of 18 events with the result:

$$B(\eta \to \mu^+\mu^-) = (23 \pm 9) \times 10^{-6}\,. \tag{9}$$

The most recent result, obtained in Serphukov, was based on 27 events and gave:

$$B(\eta \to \mu^+\mu^-) = (6.5 \pm 2.1) \times 10^{-6}\,. \tag{10}$$

Both the measurements had difficulties with the normalisation. In the CERN experiment this was established by means of a calculated value of $\sigma(\pi^- p \to \eta X)$, and in the Serphukov experiment by measuring simultaneously the Dalitz decay $\eta \to \mu^+\mu^-\gamma$. The normalisation problems were automatically overcome

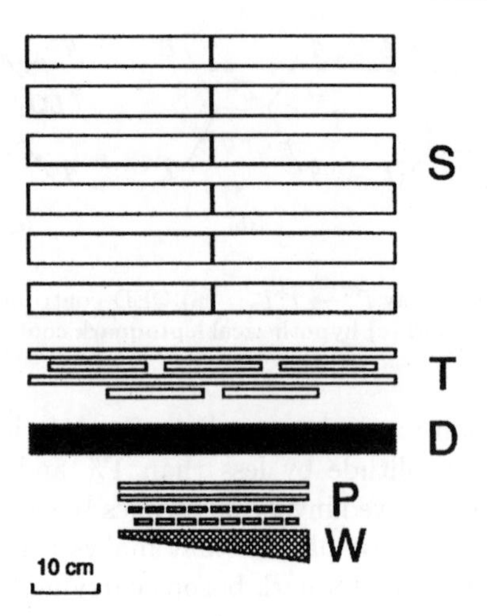

Figure 14: Top view of the left muon arm. For the meaning of the symbols see ref. [28].

by the Saturne tagging technique. The result of the experiment, which tagged more than 10^9 η's, leading to 114 good events was:

$$B(\eta \to \mu^+\mu^-) = [5.7 \pm 0.7(\mathrm{stat}) \pm 0.5(\mathrm{syst})] \times 10^{-6}\,. \tag{11}$$

This is $(1.3 \pm 0.2)\times$ (unitary limit) and is consistent with the Serphukov value, leaving no room for significant new physics since it agrees with the predictions for the electromagnetic contribution. In this field the Saturne experiment played a very crucial role as is demonstrated in fig. 15.

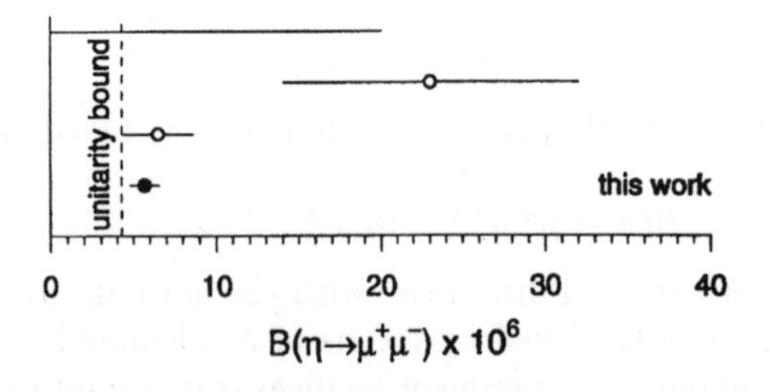

Figure 15: Results of measurements of the branching ratio $B(\eta \to \mu^+\mu^-)$ [28].

2.10 The precise measurement of the branching ratio $\eta \to \gamma\gamma$

The second branching ratio measured[29] at the η-meson facility with high precision was $B(\eta \to \gamma\gamma)$. A precise knowledge of this decay is important for several reasons:

1. It contributes to the determination of the η-meson total width, $\Gamma_{tot(\eta)}$, a quantity that cannot be directly measured with the present available experimental techniques;
2. It is needed for calculations of branching ratios for which the decay into two γ's is the intermediate state;
3. It is necessary for the absolute determination of many η production cross-section measurements which use the two-γ decay as signature for the meson.

The value reported by the PDG was:

$$B(\eta \to \gamma\gamma) = 0.388 \pm 0.005 . \tag{12}$$

This was deduced primarily from two other experiments. The ratio

$$\frac{\Gamma_{\eta \to neutrals}}{\Gamma_{tot(\eta)}} = 0.705 \pm 0.008 \tag{13}$$

was measured in a CERN experiment where the η-mesons were produced by the $\pi^- p \to \eta n$ reaction and tagged by detecting the neutron.

At IHEP was measured the ratio

$$\frac{\Gamma_{\eta \to \gamma\gamma}}{\Gamma_{\eta \to neutrals}} = 0.549 \pm 0.004 \tag{14}$$

where neutrals are in this case the sum of the 2γ, $3\pi^o$ and $\pi^o\gamma\gamma$ channels.

In the Saturne experiment, for the first time, the decay was directly measured with the advantage of reducing the systematic uncertainties. The photons were detected by calorimeters made of 61 BGO counters of hexagonal cross-section (with sides of 31.5 mm) 200 mm long (17.8 radiation lengths) arranged in a left and right position with respect to the incident beam in the horizontal plane, as shown in fig. 16. The beam intensity used in the experiment was 3×10^9 protons per spill, which in a 6 mm thick liquid deuterium target produced about 30 η's.

A careful data analysis, based on a sample of $(6.47 \pm 0.03) \times 10^4$ η's led to the result of:

$$B(\eta \to \gamma\gamma) = 0.3949 \pm 0.0017(\text{stat}) \pm 0.0030(\text{syst}). \tag{15}$$

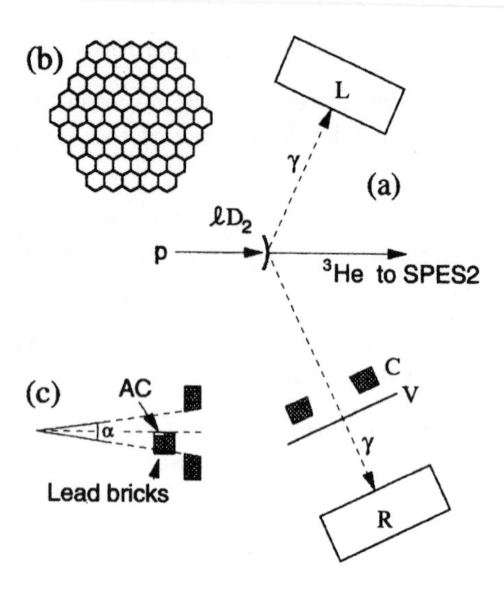

Figure 16: The photon detector : (a) top view, (b) front view of the L and R calorimeters, and (c) cross sectional view of the collimator. For further details see ref. [29].

This result, combined with those of other experiments, sets an upper limit on all neutral η decay modes not observed so far:

$$B(X^o) < 2.8\% \,. \tag{16}$$

There is therefore an upper limit of 2.8% on the sum of the branching ratios for the P- and CP-violating decay $\eta \to \pi^o\pi^o$, for the the C-violating decay $\eta \to \pi^o\pi^o\gamma$, and for the branching ratios of other exotic neutral decays.

2.11 The $\eta \to \mu e$ and $\eta \to e^+e^-$ decays

For the sake of completeness, the search [30] on two other decays, the lepton-family-violating decay $\eta \to \mu e$ and the rare decay $\eta \to e^+e^-$, done with almost the same detectors, has also to be mentioned.

No events were seen, and the result of the experiment was the determination of two new 90% confidence level upper limits: 6×10^{-6} for the former and 2×10^{-4} for the latter.

3 Subthreshold K^+ production by protons on nuclei

An experiment[31] aiming to study subthreshold K^+ production by protons on nuclei was primarily motivated by the suggestion that, on the basis of general features of the strong interaction, the properties of a kaon might change when it is surrounded by nuclear medium rather than vacuum. The subthreshold production of such a meson on nuclei must, for example, be very sensitive to a change of the meson mass. In addition, since the K^+ contains the antistrange quark, it has a reduced interaction with baryons and so the K^+ is regarded as a good messenger for its production mechanism. As in other subthreshold particle production, various nuclear medium phenomena can play a role: the internal Fermi motion, the baryon density, the abundance of baryonic resonances, the nuclear equation of state and possibly the in-medium modifications of hadron masses. The production in proton-nucleus collisions can be used in order to disentangle these different phenomena, since in this process the nuclear medium remains rather cold and there is no compression. The data in this field were very scarce since such a measurement requires a large momentum acceptance band, extremely good particle identification and momentum resolution, a selective trigger, high background suppression and a relatively good detection efficiency for kaons which, in turn, means a short flight path.

The SPES-3 spectrometer at Saturne was able to fulfill all the requirements except the short flight path, which was of the order of 10 m. However, since the flight paths could be determined with an accuracy of better than 1 cm, the decay losses were deduced with a corresponding accuracy of 9%.

The standard SPES-3 trigger was based mainly on TOF measurement over a flight path of 3 m. For this particular experiment, in order to increase the trigger selectivity with respect to the particle identification, a small counter called the *start detector*, see fig. 17, was added close to the production target for two reasons, firstly to make the time of flight longer,in this case up to 10 m, and secondly to provide a dE/dx signal to be used in the off-line analysis. Excellent particle identification was utterly crucial in the experiment because of the different particles production cross-sections in this energy regime with ratios $\sigma_p : \sigma_\pi : \sigma_K \approx 20000 : 2000 : 1$. Fig. 18 shows the improvement achieved by adding the start detector to the SPES-3 standard detection. Three drift chambers provided the coordinate measurements for the momentum reconstruction. A beam intensity of 10^9 particles/spill was used together with three different target thicknesses for studying the five combinations:

$p+^{12}$C at 2.5, 1.5 and 1.2 GeV incident proton energies

p + natPb at 1.5 and 1.2 GeV incident proton energies

The production angle in the laboratory system was chosen as 40^0 and two

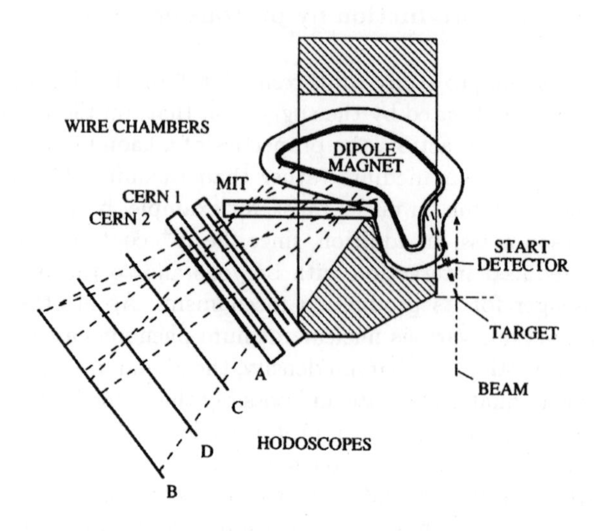

Figure 17: The *start detector* integrated in the detection of the SPES-3 spectrometer[31].

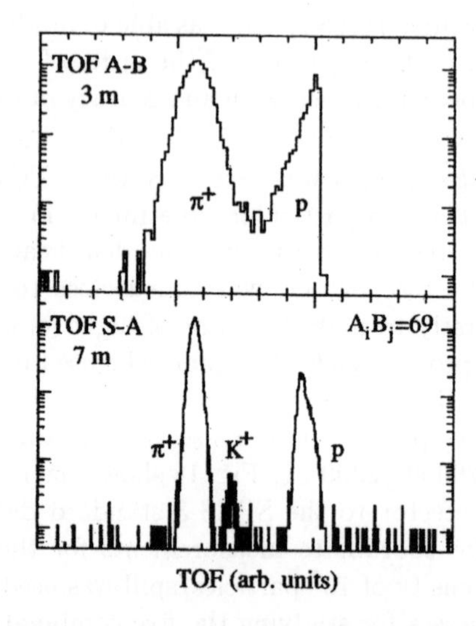

Figure 18: Particle identification using the standard detection (TOF A-B) compared to the standard detection plus the start detector (TOF S-A)[31].

momentum ranges were selected by appropriately setting the SPES-3 magnetic field : 0.35 – 0.8 GeV/c and 0.5 – 1.15 GeV/c. Protons and pions were also recorded in addition to the kaons. The data transformed to the c.m. system and summed over all momenta provide the integrated cross section. The data have also been analysed in term of their target mass dependence and have been compared to the elementary process, as shown in fig. 19. For pions the ratios of the experimental production yields between lead and carbon was found to be lower than the ratio of their reaction cross sections, as given in the geometrical model $(208/12)^{2/3} = 6.7$. The authors concluded that the pion ratios are probably a reflection of the high absorption of pions in the Pb target while the fact that the ratios for kaons are above the geometrical limit may indicate that they are mainly produced in multi-step processes.

The comparison with the elementary process suggests that, in contrast to what happens for the η-meson,

$$\sigma(pp \to K^+) \approx \sigma(pn \to K^+) \,, \tag{17}$$

and that nuclear medium effects extend the kaon spectrum to high momentum values.

An interpretation of the data has been given in the framework of the folding model in which, as discussed for the η case, one takes properly into account elementary processes, internal nucleon energy distributions and, eventually, absorption cross- section. The cross section is then calculated from the integral:

$$\sigma_{pA}^{K^+} = N \int d^3p \, n(p) \, \sigma_{pN}^{K^+} \tag{18}$$

In this particular analysis, the internal Fermi momentum distribution was replaced by the nuclear spectral function, the difference being that while the nucleon in the former case is treated on-mass shell, in the latter the nucleon is considered off-mass shell by taking into account an in-medium energy E_i. In this approach, called the Advanced Binary Collision Model, the integral becomes:

$$\sigma_{pA}^{K^+} = N \int d^3p_i \int dE_i \, S(p_i, E_i) \, \sigma_{pN}^{K^+} \tag{19}$$

The result of the calculations, shown in fig. 20 for the subthreshold cases, underestimates the data by a large factor, suggesting that other mechanisms must be considered. A two-step production mechanism was then considered, in which a kaon can be either produced by a $\pi N \to K^+\Lambda$ interaction or by a $\Delta N \to K^+\Lambda N$ interaction, where the π or Δ are produced in a primary NN collision. A great improvement is evident in fig. 21. The model reproduces

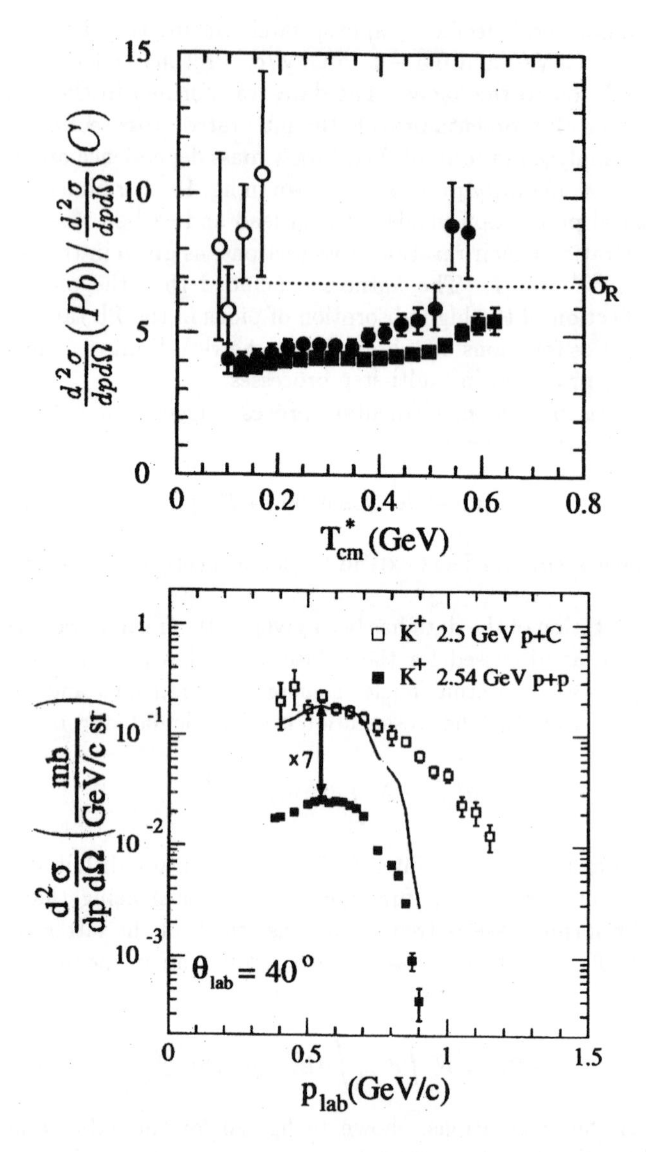

Figure 19: Ratio of measured cross sections for pions (full squares- π^+ and circles-π^-) and kaons (open circles-K^+) in p+Pb and p+C collisions as functions of particle c.m. system kinetic energy (upper) and double differential cross section for K^+ production in p+C collision and in elementary p+p collision as functions of kaon c.m. momentum (lower) [31]

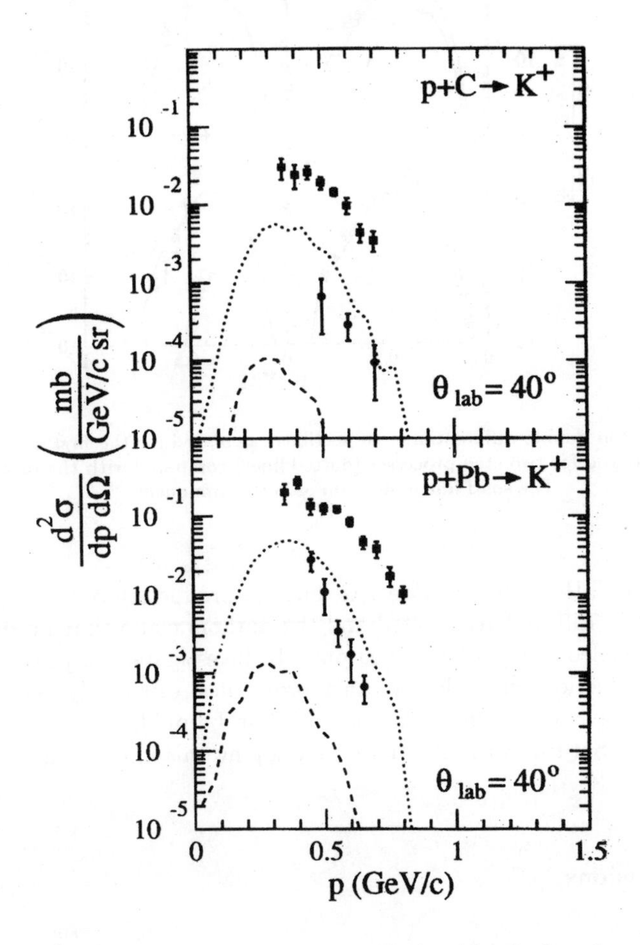

Figure 20: Kaon double differential cross sections calculated for the first chance collision for different beam energies compared to the measured data[31].

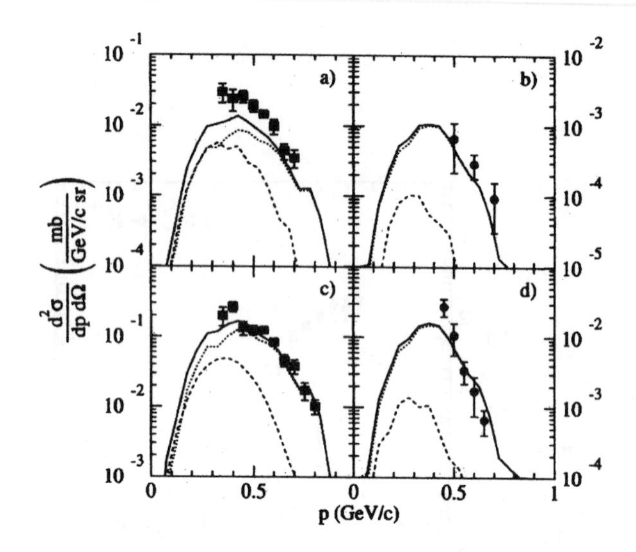

Figure 21: Kaon double differential cross sections calculated for the first chance collision (dashed lines) and for two-step processes (dotted lines) compared with the measured data. The solid line is the sum of both components[31].

reasonably well the different data sets, but the production of kaons on carbon at 1.5 GeV is still underestimated and the authors claim that production on clusters of nucleons cannot be ruled out. To investigate this possibility, data on kaon production on nuclei by other projectiles, essentially deuterons and α particles, have also been taken and will soon be published. Once more the work done at Saturne was essential in developing this kind of physics.

4 Conclusions

Despite the enormous work and the beautiful results obtained at SATURNE, most of the Physics discussed here will still need refinement. In the future this will have be accomplished in other laboratories such as COSY, GSI, CELSIUS, CEBAF, KEK, TRIUMF and JINR.

The role played by SATURNE however was such that I personally have no doubt that Intermediate Energy Physics and the associated community will long miss SATURNE.

References

1. L.D. Roper, *Phys. Rev. Lett.* **12**, 340 (1964).
2. H.P. Morsch *et al.*, *Phys. Rev. Lett.* **69**, 1336 (1992).
3. S. Hirenzaki *et al.*, *Phys. Rev.* C **53**, 277 (1996).
4. L.S. Azhgirey *et al.*, *Phys. Lett.* B **361**, 21 (1995).
5. The SPES-4 π Collaboration in *Nouvelles de Saturne 21*, November 1997, p69.
6. J. Banaigs *et al.*, *Nucl. Phys.* B **67**, 1 (1973).
7. J. Berger *et al.*, *Phys. Rev. Lett.* **61**, 922 (1988).
8. C. Kerboul *et al.*, *Phys. Lett.* B **181**, 28 (1986).
9. K. Kilian and H. Nann in *Meson Production near Threshold* , ed. H. Nann and E.J. Stephenson, AIP Conf. Proc. No. 221 (AIP, New York, 1990), 185.
10. F. Plouin in *Production and Decay of Light Mesons*, ed P. Fleury (World Scientific, Singapore, 1988) 114.
11. C. Wilkin in *Production and Decay of Light Mesons*, ed P. Fleury (World Scientific, Singapore, 1988) 187.
12. R. Wurzinger *et al.*, *Phys. Rev.* C **51**, R443 (1995).
13. B. Meyer *et al.*, *Phys. Rev.* C **53**, 2068 (1996).
14. J. Banaigs *et al.*, *Nucl. Phys.* B **105**, 52 (1976).
15. J. Banaigs *et al.*, *Phys. Rev.* C **32**, 1448 (1985).
16. R. Frascaria *et al.*, *Phys. Rev.* C **50**, R537 (1994).
17. N. Willis *et al.*, *Phys. Lett.* B **406**, 14 (1997).
18. E. Chiavassa *et al.*, *Phys. Lett.* B **322**, 270 (1994).
19. E. Chiavassa *et al.*, *Phys. Lett.* B **337**, 192 (1994).
20. E. Chiavassa *et al.*, *Z. Phys.* A **342**, 107 (1992).
21. E. Chiavassa *et al.*, *Z. Phys.* A **344**, 345 (1993).
22. E. Chiavassa *et al.*, *Nuovo Cimento* A **107**, 1195 (1994).
23. W. Cassing *et al.*, *Z. Phys.* A **340**, 51 (1991).
24. E. Chiavassa *et al.*, *Europhys. Lett.* **41**, 365 (1998).
25. M. Robig-Landau *et al.*, *Phys. Lett.* B **373**, 45 (1996).
26. F. Plouin *et al.*, *Phys. Lett.* B **276**, 526 (1992).
27. L. Satta *et al.*, presented at the *Workshop on Nuclear Physics at Intermediate Energies with Hadrons*, held at Miramare, Trieste, Italy, 1-3 April 1985, materials quoted by R. Bertini in his rapporteur talk.
28. R. Abegg *et al.*, *Phys. Rev.* D **50**, 92 (1994).
29. R. Abegg *et al.*, *Phys. Rev.* D **53**, 11 (1996).
30. D. B. White *et al.*, *Phys. Rev.* D **53**, 6658 (1996).
31. M. Dębowski *et al.*, *Z. Phys.* A **356**, 313 (1996).

THEORY AND DESIGN OF PARTICLE ACCELERATORS

André Tkatchenko
Institut de Physique Nucléaire, 91406 ORSAY Cedex, France

L'activité en théorie et conception d' accélérateurs au LNS est illustrée par quelques exemples marquants. Certains aspects des études de conception de Saturne 2 sont d'abord évoqués.Sont ensuite résumées les contributions à la définition ou l'amélioration des performances d'autres machines construites ou envisagées en France.Enfin, des travaux entrepris pour répondre à des questions générales en Physique des accélérateurs ou développer des codes de simulations sont mis en relief.

Activities in theory and design of particle accelerators at LNS are illustrated by a few examples. First, some aspects of the design studies of Saturne 2 are evoked. Then, contributions of LNS to the design or the improvement of other facilities built or contemplated in France are listed. Finally, works started in the spirit of answering general questions in accelerator physics or to develop multi-purpose computer codes are emphasized.

1 Introduction

Over the past two decades, LNS has designed,constructed and operated two synchrotrons, one RFQ and several other equipments such as ion sources or magnetic spectrometers for the needs of Nuclear Physics research at Saclay. In addition, LNS has played an important role on the national level in contributing to the construction or the improvement of many other french projects and in helping European efforts directed towards the design of large international facilities.

There are of course many elements that made possible the success of these accomplishments, but certainly one of them was the fact that it was recognized at LNS like in many other laboratories in the world that modern accelerators are not simply a collection of technologies and therefore must be studied from a theoretical point of view in order to have a clear understanding of the phenomena which may limit the performances of the high quality beams required by more and more demanding users.

This is evidently the reason why a group specially assigned to investigating the theoretical aspects of beam dynamics in accelerators was set up on the occasion of the renovation of Saturne,then continued to exist and to make essential contributions to accelerator development for a period of more than twenty years.

In view of its numerous achievements, one can say that this group has always fulfilled its task with success. Its capacity of mastering the essential features of a project was known outside LNS and was thus exploited very often through technical collaborations for the design of many facilities. The best way of summarizing its activity in a few words is simply to recall that altogether it has published about 300 reports in the form of internal notes,contributions to international conferences and workshops or articles in scientific journals and that it has participated in studies concerning most of the major projects constructed or contemplated in France during these last twenty years: Saturne 2, MIMAS, Super-Aco, ESRF, SISSI, SPIRAL, ELFE, SOLEIL.

In connection with these projects,analytical methods and numerical tools were developed for modelling and simulating phenomena occuring in circular accelerators (nonlinear resonances, coherent instabilities,synchrotron radiation,depolarizing resonances). As a result of these works,computer codes were elaborated .They are now known and used in numerous laboratories in the world.

At this point, it has to be stessed that this considerable amount of work was of course and at first the result of the efforts of the very talented accelerator physicists who composed that group but also resulted directly from the atmosphere, habits and working conditions which were created year after year and favoured a strong interaction between individuals together with a great complementarity of their competences.

In the short time at my disposal, it is difficult to review all the activities of this accelerator physics group from 1974 until today. I shall thus exercise drastic choices and concentrate on few issues which were of importance for the accelerators studied at LNS.

I cannot close this introduction without saying a few words on the role of Henri Bruck in the promotion of accelerator physics in France and particularly at Saclay.He was one of these pioneers who designed and constructed the first high energy synchrotron in France and who established the bases of accelerator theory at a time when numerous new concepts were arising. An essential element of his activity was certainly the education of many students. Some of them occupied later important positions of responsability. The knowledge accumulated by Henri Bruck and his collaborators formed the bases that made it possible the construction of Saturne 2 and therefore the creation of LNS.

2 Design studies of Saturne 2

The optimization of the basic parameters of a circular accelerator which has to fulfil a series of specifications is always a difficult task because the importance of most phenomena which can affect the beam performances is directly determined by the ring optics.

This is evidently the reason why there were numerous and active discussions about the magnetic structure of a new machine when it was decided to renovate the old weak focusing synchrotron Saturne. Initially, another weak focusing machine capable of producing 10^{12} protons per cycle at 2 GeV was proposed but it was rapidly abandoned when H. Bruck and his collaborators showed that a machine based on the strong focusing principle would allow the energy to be increased to 3 GeV and the intensity to 2.5 10^{12} protons per cycle for about the same total cost. [1]

Nevertheless, novel preoccupations arose and had to be taken into account in the design of the future machine for obtaining the foreseen performances. Some examples of the theoretical studies directed towards this end are outlined hereafter.

2.1 Design of a transitionless lattice for Saturne 2.

In strong focusing proton synchrotrons, there is an energy at which the derivative of the revolution frequency with respect to particle momentum becomes zero and changes sign. This energy is called the transition energy. In the neighbourhood of this energy the longitudinal beam dynamics is profoundly disturbed: the sychrotron frequency goes to zero and radiofrequency manipulations are required to maintain phase stability. In first alternating-gradient synchrotrons, transition was a great source of anxiety but solutions were found to cross it without serious difficulties. Later, it became one of the most important restrictions for high intensity beams because several intensity dependent phenomena occur during the crossing and lead to longitudinal emittance blow-up and beam losses.

To avoid these unpleasant problems in the case of Saturne 2, methods to raise the transition energy above 3 GeV were investigated.

Two very simple remarks helped to find rapidly an elegant solution:

-the horizontal tune ν_x is approximatively equal to the Lorentz factor γ at transition in regular lattices. Therefore, it had to be of the order of 4.

-four straight sections were needed to accomodate injection magnets, R.F cavities and 2 extraction systems. Thus, a missing-magnet structure with a four-fold symmetry had to be considered.

A machine satisfying these criteria was designed. While detailed optimization of its lattice parameters were in progress, it appeared that when ν_x was set above 3.7, the momentum compaction became negative and therefore the transition energy an imaginary number. [2]

One of the first, if not the first synchrotron having an imaginary transition energy was born. In recent years, with the advent of high intensity machines, it became attractive to design transitionless lattices but around the mid-seventies, Saturne 2 was probably considered a curiosity.

2.2 Crossing of spin resonances

The production of polarized proton beams was one of the requests formulated in the specifications of Saturne 2. Therefore, the effects of depolarizing resonances had to be investigated.In circular accelerators, these resonances occur when the energy dependent spin tune $\nu=\gamma G$ (G being the gyromagnetic anomaly) is equal to one of the frequencies contained in the spectrum of the horizontal fields seen by particles.

These resonances cannot be avoided because γ increases continuously during the acceleration cycle. For instance, the so-called imperfection resonances are crossed whenever the total proton energy is a multiple of 523.35 MeV.

In 1975, acceleration in a synchrotron of polarized protons across numerous resonances with negligible loss of polarization had been already successfully demonstrated at the Argonne ZGS. Nevertheless, there were big differences between a weak focusing machine like the ZGS and a strong focusing synchrotron like Saturne 2. Resonance strengths in Saturne could be orders of magnitude larger than in the ZGS and an important correction system might be required. A detailed theoretical analysis of this problem was thus indispensable.

Analytic expressions suitable for numerical calculations were elaborated and used to estimate the stength of the main resonances. [3,4]

Results confirmed that natural spin flip could be expected after the crossing of every imperfection and intrinsic resonances because of the strong vertical focusing which had been adopted. If one excepts the case of the first imperfection resonance crossed at 108 MeV and thus strongly influenced by large synchrotron oscillations, the experimental results observed in 1981-82 were found totally consistent with these expectations.[5]

3 Other accelerator design studies

As said in the introduction, LNS has been involved in the design or the upgrading of many new national and international projects. It is evidently not possible to detail here all the theoretical work which has been done on these occasions. I shall thus only review rapidly the main contributions and mention the relevant results which were obtained.

3.1 Super-Aco

Super-Aco is the first third-generation light source built in the world. In this type of machine,the very strong focusing required to produce a small equilibrium emittance causes a large variation of tunes with the relative momentum deviation of particles. This effect, called chromaticity, has to be corrected by means of sextupoles inserted in the accelerator's lattice in order to avoid too large tune spreads and transverse beam instabilities. Unfortunately, these sextupoles excite numerous nonlinear resonances which limit drastically the so-called dynamic aperture that is the boundary of stable oscillations in the transverse plane. Since large dynamic aperture is needed for the injection process and for long beam lifetime, sophisticated sextupole corrector schemes have to be found and carefully optimized.

At the time when the construction of Super-Aco was in progress, difficulties were encountered for the correction of its chromaticity because most existing simulation codes were not suited to calculate the natural chromaticity of machines having a small bending radius. In addition, there was no established strategy or analytical method providing criteria or guidelines for optimizing the sextupole arrangement and maximizing the dynamic aperture.

To overcome these problems, the help of LNS was requested. Its contribution can be summarized as follows:[6]

-an analytic method based on aberration theory was developed to calculate the exact first order chromaticity of Super-Aco,

-the position of the working point in the tune diagram and the sextupole corrector scheme were optimized in order to minimize the influence of the most dangerous resonances and to achieve a large dynamic aperture.

Results obtained during the commissioning in 1987 were in good agreement with the predictions and thus confirmed the validity of the analytical models which had been developed.

3.2 European Synchrotron Radiation Facility (ESRF)

ESRF, the 6 GeV low emittance storage ring optimized to produce high brilliance X-rays has successfully achieved and even surpassed its design performances in the course of commissioning in 1992. Nevertheless, as soon as 1993, a series of accelerator studies were planned in order to evaluate the limits of the facility in terms of brilliance and stability.

As a result of numerous improvements that have not to be listed here, a dramatic gain of 100 in brilliance was obtained in less than two years. Through a collaboration with ESRF, LNS contributed to this great success in proposing a new storage ring optics [7] which provided a natural emittance almost 2 times smaller than the design emittance of 7.nm together with a large dynamic aperture.

3.3 ELFE, an Electron Laboratory For Europe

From 1990 until 1996, LNS has been heavily involved in the design and feasibility studies of the different machine concepts which have been contemplated for ELFE, a 15 to 30 GeV high luminosity continuous electron beam accelerator for Nuclear Physics. Actually, first studies of a high duty cycle electron accelerator started in France as soon as 1987 and already at that time LNS contributed to the elaboration of two projects based on the use of a superconducting linac.[8,9]

At the beginning of 1990, the idea of a european machine with an energy above 10 GeV was evoked by a Committee of the French Academy of Science. An european research program was thus organized with the objective of investigating the fundamental beam dynamics issues and the technical difficulties that could influence the choices to be made for designing such an accelerator. In the course of these studies, accelerator physicists at LNS concentrated on the effect of synchrotron radiation on beam emittances in the transport system of a recirculating linac. It resulted a better understanding of the role played by the optics in the beam performances degradation.

In 1992, after NuPECC recommandations and from precise Physics requirements, an european working group composed of representatives from France, Germany, Italy and The Netherlands was set up in order to develop a technical proposal. Three different types of machine were examined, but after careful studies concerning their feasibility, the scheme initially envisaged at LNS, a three pass recirculating superconducting linac, was considered as the most adequate solution and selected by a committee of european experts.[10]

Two years later, a completely different approach arose from DESY in Germany and triggered a renewed activity on the ELFE project. The idea was to integrate ELFE into the TESLA facility under study at DESY by using the existing HERA storage ring as a pulse stretcher fed by a part of the TESLA linac. As a consequence, the total cost of ELFE would be greatly reduced.

The technical feasibility of ELFE at DESY was investigated by a group composed of accelerator physicists from Bonn, DESY, Frascati, Grenoble, NIKHEF and LNS. LNS participated in the following studies:

-modification of the HERA lattice for the needs of slow extraction,

-analysis of possible extraction methods, simulation of the extraction process and definition of the extraction channel.

Overall results obtained by the machine group showed that HERA could be used in association with TESLA to produce a beam with properties required for the ELFE experimental program.[11]

3.4 SOLEIL

SOLEIL is a very low emittance, medium energy synchrotron light source intented to provide high brilliance radiation in the 5 eV to 100 KeV energy range from a large number of insertions and bending magnets. The project which is presently under definition is the result of a long evolution.

Very first studies started towards the end of 1991 with, in mind, the objective of building a source to replace DCI and Super-Aco while retaining the potentialities of these two machines for users. As a consequence, the choice was made to design a positron storage ring with a moderate natural emittance of 35.nm at 2.15 GeV in order to ensure good Touschek lifetime for operation with intense bunch currents. A complete proposal was developed by LURE in collaboration with LNS and was presented in January 1994.

In the following years, the project evolved substantially towards much more ambitious performances in terms of brilliance, beam lifetime and stability. Extensive studies were undertaken to optimize a lattice aiming at exceptional objectives:

-very low emittance of about 2.nm at 2.15 GeV,

-very large dynamic aperture and momentum acceptance to maximize beam lifetime at high brilliance,

-very long straight sections in view to allow the installation of new types of insertion.

As a result of detailed theoretical studies, a four period storage ring that offered all the possibilities mentioned above was designed and presented at the EPAC 96.[12]

This accomplishment was the result of a wonderful collective effort of accelerator physicists from LURE and LNS.

4 Accelerator studies of general interest and development of computer codes

Besides specific works devoted to the design or the upgrading of a particular machine, the accelerator physics group at LNS undertook many theoretical studies of general interest. In parallel, sustained efforts were continuously directed towards the development of general-purpose computer codes for the analysis of particle beam dynamics in accelerators and transport systems. A few important examples of these works are listed below.

4.1 Theoretical aspects of coherent instabilities

An intense beam travelling in an accelerator induces electromagnetic fields which are determined by the boundary conditions imposed by the beam surroundings (vacuum chamber, cavities,...). These fields act back on the beam and may affect its stability or its properties. These phenomena are called collective effects because they are caused by the collective action of all the particles in the beam.

First studies of these collective instabilities were started at LNS towards 1977 in order to prepare the commissioning of Saturne and to define a stategy to control eventual problems.[13] In the following years, more general aspects were investigated in the spirit of answering some questions that were not completely understood at that time. These works found an intense echo in the accelerator physics community and became rapidly one of the most important references in this field.[14,15]

4.2 Spin dynamics

Another important topics extensively studied at LNS was spin motion in synchrotrons. Early investigations were made to understand the depolarizing mechanisms and to find correction schemes for maintaining the polarization of proton beams in Saturne 2 all along the acceleration cycle. Using a classical description of spin motion in magnetic fields, analytic expressions were derived and used to evaluate the strength of depolarizing resonances.

Later, a formalism based on the use of spinor algebra was analyzed and applied to the study of more general questions like the effect of synchrotron frequency sidebands or the snake resonances in large storage rings.[16]

4.3 Nonlinear beam dynamics

Nonlinear perturbations of particle motion have always been a great source of anxiety for accelerator physicists since the invention of strong focusing. In the case of Saturne 2, the great preoccupation was the tolerances to be respected for the magnets in order to minimize the strength of nonlinear resonances.[17]

For synchrotron light sources like Super-Aco, ESRF and SOLEIL, the dominant problem was the optimization of the dynamic aperture in the presence of the very strong sextupoles employed to correct chromaticity. Perturbation theory in the Hamiltonian formalism was applied to the analysis of motion when tunes are not close to one particular resonance. Analytic expressions were derived and used to define a strategy allowing the enlargement of dynamic aperture of storage rings.[18]

4.4 Development of computer codes

Development of computer codes has always represented an important part of the activities of the accelerator theory group at LNS. Three multi-purpose programs devoted to the design of accelerators and the study of beam dynamics are the result of a continuous effort of several years. They are now known in many laboratories in the world and have been used for the design of many types of accelerators.

-BETA [19], is a tool for the study of synchrotrons and transport lines. It handles large and small machines for electrons or hadrons and performs many tasks: linear lattice parameters calculation and matching, particle tracking, calculation of chromatic effects, study of dynamic aperture and nonlinear resonances, simulation of field errors and misalignments.[20]

-ZGOUBI[21], is specially adapted to the definition and adjustment of beam lines and large-acceptance spectrometers. It allows the study of complex systems containing several types of optical elements. Compared to other existing codes, it presents the following peculiarities: a numerical method based on Taylor series allowing a fast and very accurate integration of the equations of motion in magnetic or electric fields, the possibility of using a mesh , which allows raytracing from measured or simulated field maps, particle and spin tracking in circular machines.

-DYNAC [22], is a multiparticle simulation code that has been elaborated in collaboration with other institutions to simulate beam dynamics in linear accelerators. It allows the computation of particle motion not only in single accelerating gaps but also in long and complex cavities. The method used to calculate space charge effects is still under development. It should remove the cylindrical symmetry imposed in some other existing codes.

5 Conclusion

Accelerator theory groups exist only at a few big laboratories. LNS was an exception. Despite its relatively limited size,LNS has maintained such a group during twenty years. That is how the knowledge inherited from Henri Bruck and his collaborators has been preserved and enriched. That is one of the reasons why LNS could play an important role in accelerator development on the national and international level.

References

1. Groupe de travail pour la Physique Nucléaire avec des particules lourdes d'énergie intermédiaire, *Projet R2 à focalisation forte* (Juin 1973).
2. H. Bruck, J.L. Laclare, G. Leleux, G. Neyret, A. Tkatchenko, DSS-GERS 74-75/TP.02 (1974).
3. E. Grorud, J.L. Laclare, G. Leleux, GOC-GERMA 75-48/TP.28 (1975).
4. E. Grorud, J.L. Laclare, G. Leleux, in *Proceedings of the 1979 Particle Accelerator Conference* (1979).
5. T. Aniel,J.L. Laclare, G. Leleux, A. Nakach, A. Ropert, *Journal de Physique, Colloque C2, Tome 46* (1985).
6. D. Poirier, Thèse Université de Paris Sud (1984).
7. L. Farvacque, J.L. Laclare, P. Nghiem, J. Payet, A. Ropert, H. Tanaka,A. Tkatchenko, in *Proceedings of EPAC 94* (1994).
8. M. Gouttefangeas, F. Netter, Rapport DPh.N/Saclay 2443 (1987).
9. J.P. Didelez, M. Gouttefangeas,J.M. Loiseaux, M. Promé, in *Projet d'un accélérateur supraconducteur de 4.GeV* (1990).
10. J. Arvieux, E. De Sanctis in *Conference Proceedings of the Italian Physical Society Vol. 44* (1993).
11. R. Brinkmann, P. Bruinsma, J.M. De Conto, J. Faure, B. Frois, M. Gentner, D. Husmann, R. Kose, J. Maidment, P. Nghiem, J. Payet, F. Tazzioli, A. Tkatchenko, Y.Y. WU, *Nuclear Physics A622* (1997).
12. M.P. Level, P. Brunelle, M. Corlier, G. Flynn, C. Herbeaux, A. Nadji, P. Peaupardin, M. Sommer, J. Neel, P. Nghiem, J. Payet, J.P. Penicaud, A. Tkatchenko, M. Tkatchenko, in *Proceedings of EPAC 96* (1996).
13. J.L. Laclare, LNS 003, (1977).
14. J.L. Laclare, in *Proceedings of the 11th Int. Conf. on High Energy Accelerator* (1980).
15. T. Aniel, J.P. Garnier, J.L. Laclare, LNS 098 (1985).
16. P. Nghiem, A. Tkatchenko, *Nucl. Instr. and Methods A335* (1993).
17. J.L. Laclare, G. Leleux, A. Tkatchenko, DSS-GERS-74-91/TP-06 (1974).
18. P. Audy, G. Leleux, A. Tkatchenko, LNS/89-14 (1989).
19. L. Farvacque, J.L. Laclare, A. Ropert, ESRF-COMP-87.01 (1987).
20. J. Payet LNS/GT/93-06 (1993).
21. F. Méot, S. Valéro, LNS/GT/93-13 (1993).
22. P. Lapostolle, E. Tanke, S. Valéro, *Particle Accelerator, Vol. 44* (1994).

BEAM INSTRUMENTATION AT L.N.S.

Patrick AUSSET
Institut de Physique Nucléaire, Université Paris-Sud, 91406 Orsay Cedex

The L.N.S. facility provides a large variety of ions (protons and heavy ions up to Kr30+), energy range (up to 3 GeV for protons), as well as large intensity range (108 to 1012 elementary charges/bunch). The measurement of the beam properties and the diagnosis of its behavior are the key for a good running of the accelerators and improvement of performances. A comprehensive set of instruments is therefore required for monitoring the accelerators and the transfer lines. Most of instruments based on interceptive or non- interceptive beam monitors, realized for measuring beam current, position and size are described.

1 INTRODUCTION

It is evident that instruments which detect, locate and quantify the beam of accelerated particles play a vital role in the running of any accelerator. These instruments can be classed together as beam monitors and are essential constituents of an accelerator. They are our organ of sense that let us perceive the properties of the beam and its behavior in the machine.

At L.N.S., the continuous development of versatile measurement techniques became more and more important in the last years. The guiding principles to bring the beam instrumentation system to a high level of efficiency were :

- Minimization of the disturbance to the beam during the measurement process.
- Remotely controllable movable sensors which position are known by computers.
- Failures of mechanism detectable by computer.
- Beam parameters measurements are converted to electrical signals for computer processing and cover a wide range of intensity, energy and kind of particle.

The main beam parameters which have to be acquired for daily operation can be reduced to :

- Beam intensity and shape of the bunch : Distribution of the accelerated particles over the longitudinal coordinates.

- Beam position in the vacuum pipe.

- Beam profiles : Distribution of the accelerated particles over the transverse coordinates.

These beam parameters can be measured by very different methods. The beam instrumentation was designed not only from the viewpoint of the producer of beam diagnostics equipment, but also from the consumer's one of the control room staff. That means that some basic instrumentation is able to work 24 hours a day and 7 days a week when the facility operated.Beam instrumentation was used for :

- Checking and recording the accelerator parameters.

- Optimization of the coupling from one machine to the next.

- Verification of the quality of the beam when delivering to the experimental areas.

Moreover beam instrumentation helped to a better understanding of the running of the machine, and to improve her performances. Therefore, other measurements were carried : transverse emittance, bunch stability, Q value.

2 DESCRIPTION OF OPERATION - BEAM CHARACTERISTICS

2.1 Sources and preinjectors

Three specialized sources of ions can produce three classes of particles for the various physics requirements at L.N.S. :
AMALTHEE is a duoplasmatron source designed to produce light ions at high intensity : protons, deuterons, He 3+, He4+. The intensity reaches 100mA, the pulse duration can be tuned from 0.1ms to 1.5 ms and the repetition rate is 1Hz. Energy is 750 keV.
HYPERION produces polarized protons and deuterons since 1981 with a polarization better than 90%. The intensity is 0.7mA. The spill is 1.5ms long , the repetition rate reaches 2Hz. Energy is 400 keV.
DIONE is an E.B.I.S. source installed in 1987 on a 25 kV platform in order to inject heavy ions into a R.F.Q. Depending on the ion specie, the intensity can reach several tens of μA.

2.2 *Injector*

The injection in the synchrotron MIMAS is a multiturn injection process : The particles are slowly decelerated by a betatron core while the magnetic field is kept constant. The resulting coasting beam, stored on the mean radius of the machine, is then adiabatically captured and accelerated from 187 keV to 47 MeV (protons) by large frequency swing R.F. cavity (0.15 MHz to 2.7 MHz). Finally, fast dynamic beam transfer (300 ns) occurs at flat top of the magnetic field (0.9T) between MIMAS and SATURNE.

2.3 *Accelerator*

The beam is once again kicked and accelerated up to 3GeV (proton) in the synchrotron SATURNE II. Thanks to a betatron acceleration performed by a magnetic flux variation in a core named gephyrotron , the beam is pushed into resonance. In this way the extracted particle energy is maintained constant during the extraction process without modulation of the intensity at accelerating R.F. voltage. Then the particles are extracted during 500ms into the beam lines of the experimental area.

3 POSITION BEAM MEASUREMENT.

Beam Position Monitors (B.P.M.) with accompanying electronics is one of the most useful instrument found in an accelerator because B.P.M. signals carry information not only on the beam position in the vacuum chamber, but also about the beam intensity and any accelerator parameters which are position dependent.

3.1 *MIMAS ring*

Sixteen electrostatic pick-up electrodes, one for each quadrupole, measure the transverse position of the beam. Electrodes for measurement in the horizontal and vertical plane are combined in the same unit. A diagonally cut cross section produces a linear response to a displaced beam. The sensitivity to beam current of this monitor depends on the electrode capacity to ground and is controlled with mechanical tolerances of the assembly. They are all made of stainless steel and their capacitance is 115 pF. For ultra-high vacuum requirements, the electrodes can be baked at 300° C. Their associated amplifiers cover a 10^4 dynamic range from several 10^7 charges/s to several 10^{11} charges/s. The precision of the position measurement reaches 1mm.

3.2 SATURNE ring

Twenty four metallized alumina electrodes (12 horizontal and 12 vertical) have been developed for position measurement. The associated electronics circuits allow observation of high intensity beam (10^{10} particles /s to several 10^{12} particles /s). A set of carefully shielded electrodes has been installed in a straight section for the position measurement of low intensity beam down to 10^8 particles/s.

Either B.P.M. set of MIMAS or SATURNE is able to perform a Q value measurement : A short kick of few mm is given to the beam once per cycle and the lowest frequency of the spectrum of the induced signal on a P.U. is measured.

High frequency analog signals delivered by the amplifiers associated with the B.P.M. are transmitted to the main control room for analog processing (peak detection) . The electronics devices for the B.P.M. of the 2 rings are computer controlled by an auxiliary CAMAC crate controller which performs the gain switching, the position calculation and the closed orbit display. A P.C. gives additional information such as betatron amplitude, in order to adjust the position and the angle of the beam injected in SATURNE II.

3.3 Transfer line

Eight B.P.M. electrostatic P.U. have been placed in the beam transfer line between MIMAS and SATURNE to adjust the beam position in the vacuum pipe. They are of the same type of the MIMAS ones and their sensibility is the same . An integrator replaces the peak detector of the previous electronics system. At last, the control of the position can be performed down to 10^8 particles /s.

In the low energy beam transfer line from the 3 ion sources to MIMAS, and for the 11 beam lines of the experimental area , the position of the beam is controlled by beam profilers.

4 INTENSITY MEASUREMENT.

Perhaps the most basic measurement on a beam is that of its intensity and its bunch shape monitoring. A relative or absolute measurement of the intensity and a monitoring of the spill of the beam is performed on each source, transfer line or accelerator.

4.1 Source and low energy beam transfer line.

A commercial beam current transformer monitors the pulse and measure the intensity produced by HYPERION and DIONE with a 1 μA resolution. A home made beam transformer placed in the vacuum pipe monitors the number of charges produced by DIONE with a residual noise corresponding to about 10^8 charges/ bunch.

Monitoring of the beam current in the low energy beam transfer line is achieved by 4 Faraday cup associated with a current to voltage converter. The signals are transmitted to the main control room and displayed on oscilloscope. Moreover, after digitalization, a numerical integration give the number of charges in the range of 10^7 charges/s to several 10^{11} charges/s. To suppress secondary ndary emission , Faraday cup are polarized to 150 V. Obviously they are totally destructive of the beam.

4.2 MIMAS ring, high energy transfer line , SATURNE ring.

For the commissioning in 1987, a commercial beam current transformer (sensitivity : 100mA /V) monitored the intensity of the beam accelerated in MIMAS . In connection with the heavy ion program, a new sensitive beam current transformer, the core of which is made of high permeability amorphous material, has been developed . Specific problem arose because of the large diameter (290 mm), the baking of the vacuum pipe (300° C), and the low level of the intensity of the heavy ion beams. Finally injection of Kr 30+ was observed (600 nA).

A commercial beam current transformer monitors the intensity of the beam in the high energy beam transfer line to SATURNE and another in the SATURNE ring.

4.3 Absolute measurement of the extracted beam of SATURNE.

A Secondary Emission Chamber monitors the spill of SATURNE II in the two extraction sections. Beams from 10^8 to 10^{12} can be ejected during a typical 500ms duration spill. Absolute measurement of the ejected charges leads to detect very low currents ($< 10^{-7}$ A). Unfortunately the S.E.C. whose efficiency depends on the nature of the extracted particle and on their energy cannot provide the number of charges.

These beam characteristics have needed to build a high sensitivity beam current transformer (noise : down to 80. 10^{-9} ; bandwidth : 1Hz ; 3KHz) whose function is to calibrate a S.E.C. during a special fast extraction process (3ms). Therefore, the ejected beam current increases up to 10^{-6} A and the

S.E.C. can be calibrated down to 5. 10^9 charges/spill with an accuracy less than 5% . When coming back to the standard slow extraction process, the number of charges is provided only by the S.E.C.

In addition, a servosystem has been added to produce a spill of intensity as constant as possible. The signal of the S.E.C. is compared to a reference and the resulting signal modulates the input command of the power supply of the Gephyrotron.

4.4 Longitudinal measurement - Observation of bunches

In SATURNE II, a wide-band R.F. electrostatic P.U. designed for intensity measurement, is assigned to the bunches observation and to the adjustment of the phase of the accelerating frequency. She works in the high intensity range (10^{10} to 10^{12} charges /s). The associated amplifiers allow a 125 MHz bandwidth. For lower intensities, a special carefully shielded (1 m long) allows observation of beam down to 10^8 charges /s.

In MIMAS a special set of 3 electrostatic P.U. is also specially assigned to the observation of longitudinal distribution of bunches.

5 TRANSVERSE BEAM PROFILE MEASUREMENT.

B.P.M. and beam profilers with accompanying electronics are an active field of development since no one is really satisfied with what he has. The most important development direction was not with the instruments themselves but with their integration into the machine operation. As the sophistication of the machine increases, it becomes essential that measurements are not only displayed on a screen for the operators but also fed back to the control of the machine and into changes in the magnet current.

5.1 Scintillator

The first sheer diagnostic to measure the transverse dimension of the beam was the scintillator screen placed in the path of the beam. In spite of the extraordinary development of signal digitization, processing and displaying, the screen remains as the most persuasive beam size or profile detector. The success of this kind of detector is universal.

At L.N.S. these screens consisted of thin (2mm) plate of Cr doped Al2O3 directly available from industry. The precision can reach 1mm down to few 10^9 protons $/s.cm^2$ at GeV energies. Troubles arise from electronic T.V. cameras which were not enough radiation resistant (except the old ones equipped with vacuum valves). At each point where the position of the beam is crucial, is placed a scintillator screen. At last, about ten scintillators have been working on the beam transfer lines.

5.2 Beam profiler

They are the universal detector to determine the distribution of the intensity of the beam in the transverse plane. They allow the position and focusing adjustments in the beam transfer lines and the emittance measurements. At L.N.S. many kinds of sensor, specially fitted to the local beam characteristics, are used. The most representative of them can be listed as follow :

- Sources and low energy line 30 m (long) :

  In these sections, where a baking of the vacuum chamber is not needed, the profilers are plane of 32 gold-plated tungsten wires ($\oslash = 20\mu$m) on a teflon-glass frame; The usable surface is about 70mm.70mm and they can support a maximum temperature of 80° C.

  When a baking is needed the profilers are aluminium plates, 1 mm thick upon which 32 or 48 strips of gold are printed by a silk-screen process. The spacing of the strips ranges from 250 μm to 2.5 mm and the thickness of the layer of gold is 20 μm. They are destructive of the beam and are moved by pneumatic devices which are under manual or computer control. Double detector have also been developed : they provide simultaneously the horizontal and the vertical profile with two cross layers, each of 32 gold wires electrically isolated .

- Experimental area beam line :

  the eleven beam transfer lines of the experimental area are equipped with about 60 beam profilers. These ones were developed from traditional multiwire ion chamber. Each monitor consists mainly of a harp of 48 golden plate tungsten wires ($\oslash = 20\mu$m). The spacing of the wires is either 1 mm or 2 mm depending on the total width of the beam to measure . A compressed air actuated high vacuum feed through moves in and out the harps.

  The chamber is air filled if the intensity of the beam is less than 10^8 charges /s and filled with a mixture of argon and carbon gas for intensity ranging from several 10^4 charges/s to 10^8 charges/s. The charges resulting from the ionization process are converted to a voltage and represents the beam profile. A program named ENVEL performs :

- optics calculations,

- acquisitions and control on beam line elements of various types,
- combines these two skills to provide the physicists with procedures for :
  - initially setting their beam line for correct focusing,
  - checking vacuum valves and obstacles state,
  - automatically tuning the beam line dipoles.

5.3 Overview of the control system :

The synchrotrons and the beam transfer lines are operated by a fully distributed control system based upon a network of identical microcomputers integrated in a serial branch CAMAC interface system. Each subset of the machine can be accessed either from touch panel operator consoles or locally. Each computer is equipped with an autonomous peripheral environment which allows an automatic start.

The network has a star topology. Each branch of the star is a serial CAMAC branch. At the center of the star, a communication system named SAMU provides transmission and CAMAC handling. It enables to drive up to 256 CAMAC crates organized in 16 branches of 16 crates. The SAMU system is a eight MOTOROLA 68000 processors system built in a VME crate : eight MOTOROLA 68000 processors.

The operating system used in the SAMU and the crate controller is a fast multi-tasking rela time system.

This versatile control system has been operational since the very start of MIMAS and gives since then full satisfaction to its users.

6 CONCLUSION

At the end of 1997 the L.N.S. had two groups of accelerators : the 20 years old synchrotron SATURNE and the 10 years old booster storage ring MIMAS. The beam monitors and control system have got considerable improvement and modification and they do not keep the original style in order to converge on a homogeneous beam diagnostic system.

Acknowledgments

This instrumentation has been the work of a large number of people of the L.N.S. staff and also of students who have contributed over this 20 years to this task. It is very difficult to pay fair tribute to each of them. We want to thank them all.

References

1. L. DEGUEURCE, P. AUSSET: Beam diagnostics for heavy and polarized ions in the range of 10 keV /amu to 1 GeV /amu, proceedings of the 1988 E.P.A.C. Rome.
2. L. DEGUEURCE: Beam diagnostics for the EBIS ion source at SATURNE, Letter to the editor, N.I.M. A260 (1987) p538-542.
3. P. AUSSET, J.M. LAGNIEL, P. LEBOULICAUT, P. MATTEI, D. URIOT: Low intensity charges measurements of the 1 GeV/amu SATURNE II extracted beam, proceedings of the 1992 E.P.A.C. Londres.
4. P. KNAUS, S ROSSI, L. BADANO, J. BOSSER, P. AUSSET, P.A. CHAMOUARD: A betatron core for optimized slow extraction in a proton/ion medical accelerator, proceedings of the 1998 E.P.A.C. Stockholm.
5. L. DEGUEURCE, A. NAKACH, J. SOLE: The SATURNE beam measurement system for orbit corrections and high and low beam acceleration, 11th International Conference on High Energy Accelerators (1980) GENEVE.

MIMAS

J.-L. LACLARE
Groupe de projet SOLEIL, Délég. Rég. Ile de France du CNRS, 91198 Gif sur Yvette, FRANCE

1 Status of SATURNE in the early 80's

To introduce the subject of the presentation, it is interesting to recall the status of SATURNE at the time when we started discussing the concept of MIMAS.

1.1 Polarised Particles

At that time, SATURNE had just started to add polarised protons and deuterons to its series of projectiles. The polarisation of the beams was excellent in the full range of energy : (up to 3 GeV for protons or 1.15 GeV/n for deuterons) and this despite the dramatic number of depolarising resonances to be crossed.

For Nuclear Physics at intermediate energy, it was an exceptional window at energies below and above the meson factories. This window was largely exploited at SATURNE. About 40 % of the experiments were performed with polarised beams.

For Accelerator Physics, given the strong focusing of SATURNE lattice that enhances by extremely large factors the strength and width of resonances, it was a "première" which success was confirming the possibility to accelerate polarised particles in high energy machines. Why not much higher than SATURNE ?

As a matter of fact, the AGS in Brookhaven had started a programme to accelerate polarised protons up to 15 GeV at the beginning of 1984. A similar project was being developed at KEK (Japan). At CERN, the pressing need for polarised beams to be accelerated into the PSB-PS-SPS complex was being expressed.

At SATURNE, the unique weak point of the facility was the too low intensity of the extracted beams. In principle, it could not exceed the 10^9 particles per pulse, although the intensity at the source was to a certain extent remarkable between 100 and 200 μA.

If this performance could be considered adequate for a certain number of experiments, it was considered nevertheless as the major obstacle in view of the full development of Nuclear Physics with polarised beams at SATURNE.

1.2 Heavy Ions

As far as heavy ions are concerned, CRYEBIS and its platform were to be installed to allow for the acceleration in SATURNE of the first heavy ion beams between 100 et 1150 MeV/amu. Up to that time, the BEVALAC in Berkeley and the SYNCHRO-PHASOTRON in Dubna were sharing the exclusivity of that type of beams. The GSI in Darmstadt was nearly to join them.

With relativistic heavy ions well above GANIL energies, a totally new generation of experiments was made possible. Nevertheless, the performances were certainly to remain relatively modest. As a matter of fact, as it was already the case for polarised particles, for heavy ions as well, with the 20 MeV Drift Tube proton Linac between the source and SATURNE, the facility was intrinsically limited.

On the one hand, it was impossible to accelerate masse over 40 since the linac was opaque for low charge to mass ratios. On the other hand, for light ions, the intensity could only be very moderate (around a few 10^8 charges extracted per second with CRYEBIS 82). This corresponded to a total charge in SATURNE slightly above the detection threshold and merely adequate to control the beam all along the acceleration cycle. We were confronted with a certain contradiction.

The pending question was how to better serve our users in the future ?

2 The Concept of MIMAS

It must be reminded that SATURNE had been originally designed to accelerate intense beams of protons at the space charge limit. If in accordance with predictions, the request for heavy ion beams would get stronger and reach about 40 would be operated 80 % of the time with weak beams (40 % polarised particles 40 % heavy ions).

As a consequence, for the two most requested types of beams, intensity would have been a limitation blocking certain experiments. With a view to better serve the most demanding users and prepare the next generation of experiments, we opted for an increase in beam intensity. The concept of a new injector was born and we proposed at once :

- first to augment significantly the intensity of polarised particles and light heavy ion beams.

- then to have access to the heaviest heavy ions.

The solution to this double ambition consisted in the construction of a new injector specifically adapted to the operation of SATURNE with specialised sources producing weak beams. In the new scheme with MIMAS for polarised particles and heavy ions, the linac which had been constructed at the time when the elementary Particle Physics was dominant at SATURNE would be used only for the preacceleration of light particles from intense sources : the exact reason for which it had been built.

MIMAS was a circular machine of the synchrotron type naturally better matched than the Drift Tube Linac with the synchrotron SATURNE for :

- the accumulation of very long pulses of polarised particles from HYPERION,
- the stacking of several pulses of heavy ions extracted from CRYEBIS.

With MIMAS, we finally would again fill SATURNE with high intensity beams.

3 A few Milestones

The study started in 1979. Within a couple of years, it had been sufficiently detailed to allow for an immediate start of construction as soon as the project would be accepted. Le principle of MIMAS construction was agreed by the end of 1982. At the beginning of 1983, MIMAS was scheduled and financially supported by IN2P3 and CEA. The shares were 1/3 IN2P3, 2/3 CEA. The construction was scheduled to last for 3 and a half year and the connection to SATURNE was therefore expected in the course of 1987.

4 MIMAS operation principle

MIMAS is a small synchrotron with a circumference of 37 m exactly 1/3 of SATURNE. Upstream, it was connected with the 3 preinjectors :

- HYPERION for 375 keV polarised protons or 187.5 keV per nucleon polarised deuterons,
- The preaccelerator composed with CRYEBIS and the RFQ for 187.5 keV/amu ions with q/A's ranging from 0.25 to 0.5,
- AMALTHEE for standard 750 keV protons and 375 keV per nucleon deuterons.

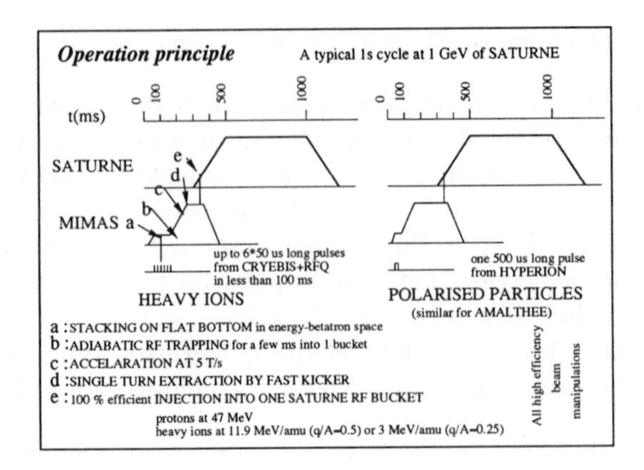

Figure 1: The operation principle.

The operation principle was simple. To illustrate the way MIMAS was functioning, we take the example of a 1second cycle of SATURNE at 1 GeV in the Fig 1. The slow extraction flat top could be 500 ms long. In the interval between two injections into SATURNE, MIMAS is operated as a low energy (a few hundreds of keV/n) accumulator on a flat bottom of the magnetic field.

During the first phase, one injects

- either a long pulse of polarised particles from HYPERION of about 400 to 600 μs (similar for AMALTHEE),
- or several short pulses of heavy ions of about 50 μs at the confinement rate of the source ; that is to say about 10 ms. In this way, one can stack up to about 6 pulses in less than 80 ms.

This accumulation provides already a significant gain. In a second phase, in comparison with the linac injector, we cumulate a series of high efficiency beam manipulations. After injection, the continuous beam is adiabatically captured to form a single bunch by applying a sawtooth law that lasts for some 1 ms to the RF voltage. The efficiency for trapping particles is excellent and nearly 100 %. The third phase corresponds to acceleration with a 5 Tesla/s raise of the magnetic field slightly superior to that of SATURNE.

Finally, for MIMAS comes the flat top at the energy for transfer and the fast extraction in synchronism with the RF of SATURNE. After it has been kicked out by a the fast switching extraction magnet, the MIMAS beam is guided through the transfer line that connects the two synchrotrons to the interior of straight section N° 3. It is then kicked into SATURNE by a second fast magnet to fill one third of the circumference. Particles are then, on the wing, accelerated to their final energy and extracted in the direction of the experiments.

5 Presentation of MIMAS

5.1 The lattice

The chosen magnetic structure was extremely simple. It combined the three basic magnet components on a single girder :

- a quadrupole focusing the horizontal plane,
- a dipole to bend the trajectories in the mid plane,
- a quadrupole focusing the vertical plane.

To constitute the circumference 8 such modules were evenly distributed along a circle with a 5.85 m radius.

5.2 The chamber aperture

With a view to stack side by side in the horizontal plane the many turns of injected beam, and then trap them with the RF, the useful physical aperture for the beam had to be made very large ±13 cm horizontally and ±8 cm vertically, although we had chosen a strong focusing. Such dimensions were definitely well above standard.

5.3 The Magnets

This chamber aperture had to be accommodated inside the magnets poles which had the main characteristics to be short and bulky. The fringing field of the magnets was a major concern since its contribution to the field integral was far from being negligible.

The dipoles of rectangular shape had complicated extremities to minimise eddy currents. We had to face all types of unforeseen events during construction. Among these, a truck loaded with one of the 8 dipoles fell in a ravine and the manufacturer had to restart the construction line to produce one extra unit.

The quadrupoles were as long as wide and asked for specific analytical and experimental developments for the search for 3D pole shapes satisfying the high required quality with regards to the gradient integral. The campaign of end pole shimming was extremely successful.

5.4 *The Radio Frequency system*

The required voltage of 4 kV could be achieved with a pair of cavities. The RF had to match the revolution time of the very slow heavy ions with Z/A =.25 at injection and the already rather fast 47 MeV protons at extraction.

Accordingly the RF swing was larger than 15 between 0.16 and 2.48 MHz. The cavities were very heavily loaded with ferrites which polarisation was varied along the acceleration cycle to obtain the change in resonant frequency and remain in synchronism with particles. This unprecedented swing performance constituted a record and, since then, the know how of SATURNE in this respect was exported to many places.

5.5 *The kicker magnets*

Initially, SATURNE laboratory had no experience in design and construction of ultra fast magnets. At that time CERN had developed a solid expertise and they kindly accepted to form our team. Two kicker magnets were necessary to the transfer of the beam between the two machines. MIMAS beam was occupying about 600 ns (that is to say most part of the circumference) and after acceleration, little space (less than 200 ns) was left unoccupied between the tail and the head of the single bunch for the firing of the kicker.

On the other side, SATURNE kicker had more than 2/3 of the circumference $\sim$ 950 ns to decrease its field. The kicker project was very successful.

5.6 *Vacuum aspects*

With a view to accelerate the heaviest incompletely stripped heavy ions, the ring had to be operated under ultra high vacuum conditions in the few 10^{-11} torr. This was superbly achieved by generalising the thermal preparation at $900°C$ of all the parts constituting the vacuum chamber walls. Special pieces (kickers ferrites for instance) under vacuum had specific treatment. In addition a system for a $300°C$ in-situ bake out was installed. In the Fig 2, the result of the injection of 10 fully stripped N7+ pulses from the source repeating at 10 Hz is shown for different residual gas pressures. For 10^{-10} torr, the maximum stored current corresponds to 3 effective pulses. From that point, a new injected pulse just compensates for the losses between two successive injections. This example shows that an excellent vacuum was a prerequisite to envisage the effective stacking of several heavy ions pulses.

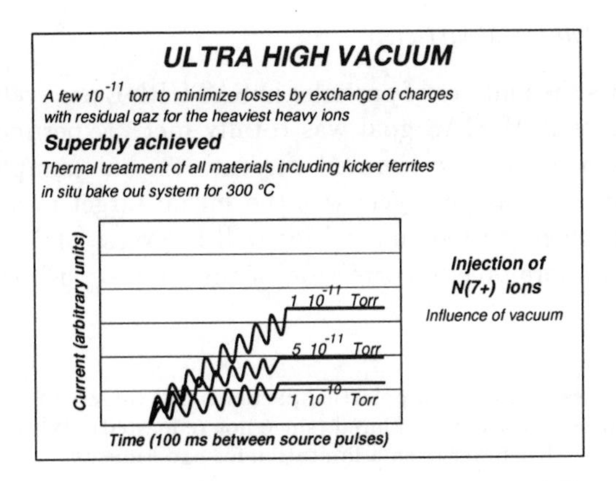

Figure 2: Vacuum effect.

5.7 *Other pieces of equipment*

On MIMAS, many pieces of equipment were specific. Among others let us mention :

- A betatron (an original idea from Pr H Brück) was used to stack in momentum the injected beam. It consisted in a tore of laminated steel excited by a winding playing the role of the primary circuit of a transformer. The beam path was the secondary circuit. The corresponding decelerating voltage was created by the flux variation excited by a change of current in the primary circuit. The performances of the power supply were quite difficult to meet.

- The ring was equipped with all sorts of correctors made with printed circuit, a speciality formerly developed on SATURNE.

- Good pulsed power supplies with an excellent tracking between quadrupole and dipole currents were necessary.

- Diagnostics were also of a special design due to the large chamber size and also the wide range of currents to be accelerated (several orders of magnitudes between the heaviest ions and the protons).

6 Commissioning time and achieved performances

It took a few months to test the new injector and to successfully connect it to SATURNE. The achieved performances are summarised in the next two tables.

6.1 *Light and Polarised particles*

The table 1 shows the number of particles in SATURNE per cycle. A-posteriori, it can be stated that MIMAS goal was totally met. Experiments have never been limited by a too low accelerated current. As an example, the achieved $8.\ 10^{11}$ figure for protons per cycle was the initial target performance. However, if it had been requested, intensities well in excess of $1.\ 10^{12}$ could have been served by a more precise correction of the ring to push the space charge limit.

Table 1: Performances for light and polarised particles. All targeted figures were achieved. Maximum possibilities not always explored (since not requested). With MIMAS, intensity has never been a limitation for experiments.

Performances for light and polarised particles.						
	A	Q	Q/A	Particles/cycle	Min Energy (MeV)	Max Energy (MeV)
Light particles						
protons	1	1	1	$8.\ 10^{11}$	100	2950
deuterons	2	1	0.5	$5.\ 10^{11}$	26	1145
Helium 3	3	2	0.66	$2.\ 10^{11}$	46	1730
Helium 4	4	2	0.5	$2.\ 10^{11}$	26	1145
Polarised Particles						
protons	1	1	1	$2.\ 10^{11}$	100	2950
deuterons	2	1	0.5	$3.\ 10^{11}$	26	1145
Lithium 6	6	3	0.6	$1.\ 10^{10}$	26	1145

MIMAS also proved to be an excellent injector for polarised particles with unprecedented figures for currents and polarisation. All predictions were met.

6.2 *Heavy Ions*

The table 2 summarises the performances for heavy ions. Achieved intensities were as expected. As it had already been the case for polarised particles, it can be stated that intensity of heavy ions has never been a limitation.

However, we were wrong when we had anticipated a demand around 40%. Heavy ions were requested for only 10% of the beam time.

6.3 *Beam availability*

In comparison with the previous Drift Tube Linac injector, MIMAS proved to be more stable and more robust. This was highly visible right from the start of MIMAS since beam availability (percentage of achieved divided by scheduled beam time) increased immediately from 81% to 92% as shown in the Fig 3.

Table 2: Performances for Heavy Ions. All targeted figures were achieved. Requests for H.I. were $\sim$ 10% less than anticipated 30-40%.

Performances for heavy ions.						
	A	Q	Q/A	Particles/cycle	Min Energy (MeV/amu)	Max Energy (MeV/amu)
Heavy Ions						
Carbon	12	6	0.5	$1.\ 10^9$	26	1145
Nitrogen	14	7	0.5	$1.\ 10^9$	26	1145
Oxygen	16	8	0.5	$1.\ 10^9$	26	1145
Neon	20	10	0.5	$1.\ 10^9$	26	1145
Argon	40	16	0.4	$1.\ 10^8$	17.5	825
Krypton	84	30	0.36	$2.\ 10^6$	13	690
Xenon	129	33	0.26	$3.\ 10^6$	7	400
Gold	197	50	0.26	$1.\ 10^6$	7	400

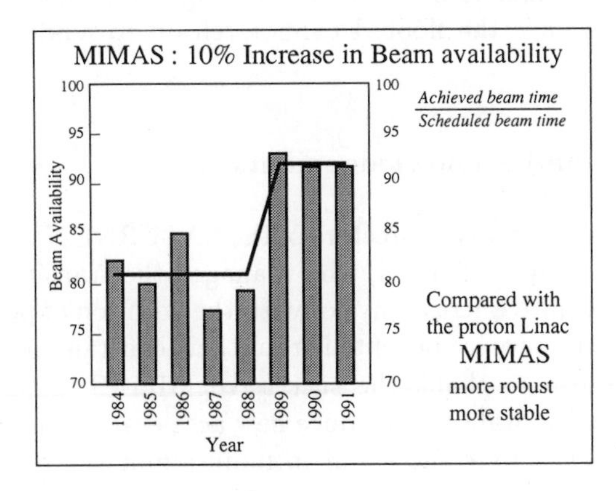

Figure 3: Beam availability.

7 An unexplained phenomenon : Spontaneous deceleration of the beam at injection

In general, MIMAS accelerator Physics was extremely well understood. The unexpectedly "too good" injection efficiency was the exception. Due to other priorities and a limited number of machine studies, it will remain a puzzle forever. Normally, to get radially stacked, the beam from the preinjector had to be decelerated by means of the betatron. In this way, the excess of horizontal betatron amplitude (which would normally limit the lifetime of the particles to the few turns it takes to return to its initial conditions in the injection septum) is compensated for by a drift of the chromatic closed orbit. In practice, it was repeatedly observed that (whatever the current, the source and the type of particles) injection was possible without deceleration.

8 An Homage to the pioneers

Starting in the 60's a tradition of excellence in Accelerator Physics had been established at Laboratoire National SATURNE by Professor Henry Bruck. He wrote a famous blue book which was considered as the bible for several generations. He organised courses in this discipline for the French "3eme Cycle" through which were educated a large number of Doctors of my generation. To a large extent, he can be considered as the father of the accelerator field in France and at the origin of most of the existing competences at both CNRS and CEA. Since I have the floor, I cannot refrain to render homage to this great pioneer.

9 Conclusions and acknowledgements

I want to thank the successive Directors of the SATURNE National Laboratory and in particular Jacques Thirion, who managed the transition from Particle Physics to Intermediate Energy Physics with the addition of polarised particles and heavy ions. The project benefited from their constant support.

I would like also to thank the successive MIMAS managers and group leaders who did not hesitate to introduce new accelerator concepts and develop new techniques. They were successful in maintaining the motivation of staff, although every time a new accelerator was to be constructed, SATURNE had to pay a heavy tribute in personnel (GANIL, ESRF, etc....).

Above all, I want to thank the laboratory staff who demonstrated that they could develop of a big project like MIMAS while continuing in parallel to operate the facility : a remarkable performance.

This presentation benefited from the many status reports made on the occasion of the "Journées d'Etude SATURNE":

JL Laclare: Le nouvel injecteur MIMAS, JES 3 Fontevraud, 25-29 avril 1983, p 433

E Grorud, L Farvacque: Equipements et techniques à mettre en oeuvre pour MIMAS, JES 3 Fontevraud 25-29 avril 1983, p 703

PA Chamouard et A Tkatchenko: MIMAS, SATURNE et leurs equipements, JES 4 La Londe les Maures, 10-14 Nov. 1986, p 283

PA Chamouard: Bilan et perspectives un an après le démarrage de MIMAS, JES 5 Piriac, 16-20 Mai 1989, p 288

PA Chamouard: Evolution des accélérateurs du LNS depuis 1989, JES 6 Mont Ste Odile, 18-22 Mai 1992, p 165

PA Chamouard: SATURNE 2: Résultats et bilans 8 ans après la mise en service de MIMAS, JES 7 Ramatuelle, 29 Janv.-2 Fev. 1996, p 196

The heavy ion source DIONE

Jean FAURE
CEA/DSM/UGP SATURNE 91 191 Gif sur Yvette CEDEX FRANCE

1 INTRODUCTION

The operation of DIONE, a source of highly ionized heavy ions, has been described in detail previously [1-5]. In brief, an external electron gun emits a stream of electrons which are focused and conducted through a series of cryogenically cooled drift tubes by the strong magnetic field (B = 6 T) of a superconducting solenoid. A schematic of the source is shown in fig. 1. For normal operation, three different sets of axial potentials are used, as shown in fig. 2. First a set of potentials is applied so that ions produced in an external ion source are allowed to enter the central trapping region. Then the injected ions are contained in the trap for a time long enough to reach high charge states by multiple electron collisions. Finally, the ions produced are expelled from the source.

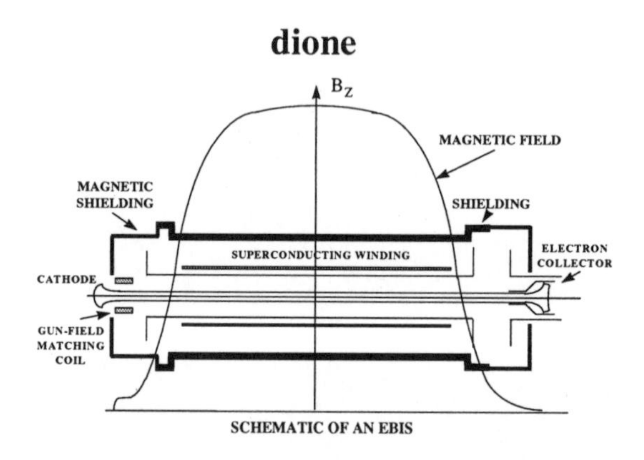

Figure 1: Schematic of an EBIS source.

External ion source has been developed to make DIONE operation independent of the particular atoms used, as well as to facilitate the changeover from one material to another.

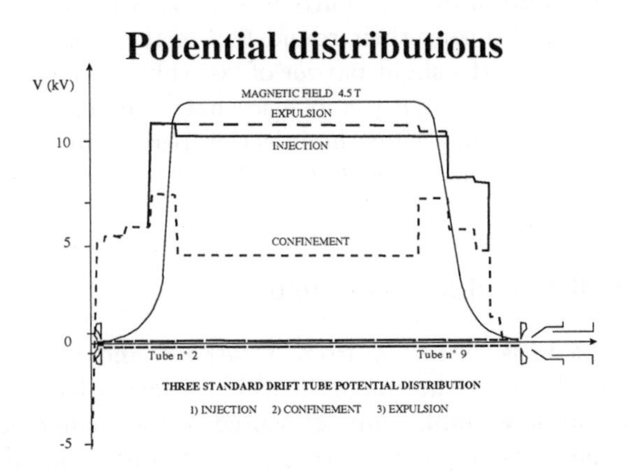

Figure 2: The three standard drift tube potential distributions.

2 External primary source and drift tubes potential distributions

A standard duoplasmatron has been used as the external ion source until 1993. Then, a hollow cathode ion source has been built to inject gaseous and metallic elements [7].

The hallow cathode ion source is a self-sustained glow discharge in a buffer gas between an anode and a cathode. The cathode is a cylindrical cavity to increase the gas ionization in the negative glow area of the discharge. Sputtering phenomenon by ions striking the cathode produces material sublimation. Therefore, this source is adapted for gaseous or metallic ions, as well.

At the exit of the external source, the ions are selected by the means of a Wien filter. As can be seen, a series of electrostatic lenses, steering plates, and a cylindrical deflector is used to bring the ions into DIONE. Some of these elements, in particular the steering plates near the electron collector, and the deflector, must be pulsed to allow for different voltages during the injection of ions as opposed to during the extraction of ions from DIONE.

The potentials of the drift tubes during ion injection from the external source are shown in fig. 2. These values were chosen so that the low charge state ions produced by the external source would slow down as much as possible once they enter DIONE. Therefore, the potential value for injection corresponds to the external source potential. This will increase the round trip time for ions injected in the source and therefore increase the quantity of ions in the source for a given injected current. The ions will also have a chance of being further ionized during their round trip in the source and therefore will be unable to overcome the slight barrier of last tube. The potential of the tube near the electron gun is slightly above so that to maintain the injected ions in the confinement region. After the injection period, the potentials revert to the standard containment values and then the expulsion potentials, as shown in fig. 2.

3 Brillouin flow and ions confinement

The electron beam is extracted from a cathode and accelerated at 10 kV. During this acceleration the magnetic field variation has to be adjusted so that the Brillouin flow equilibrium is insured in the confinement region. This configuration provides the best electron beam compression and density.

The density is a few 1000 A/cm. The ions are contained in the space charge potential trap of the electrons.

An excellent efficiency of injection into DIONE has been measured [8]. But, during the ionization process the ions are heated. The electron-ions elastic collision between the electron beam and the working ions is involved. So that the ejected beam intensity decreases rapidly when the confinement time is large for highly charged ion like Kr^{30+}, for instance.

It has been shown that a reliable method to prevent the ions heating is to provide an effective ion cooling by continuous He-gas injection. Nitrogen-gas should be better but the helium temperature pumping system made this process very unstable by successive adsorption and desorption on the drift tubes [6].

For this reason a warm EBIS project started up, but due to the leak of heavy demand from users, is has been abandoned. Nevertheless a prototype has been built to prove the feasibility of a 10^{-11} torr vacuum system without cryogenic pumping system [9].

4 DIONE performances

The extracted heavy ions pulse length is as short as possible (less than 50 μs) to be injected in MIMAS and SATURNE. Depending upon the ions species and the MIMAS vacuum 1 to 8 DIONE ejected pulses could be injected.

The next table gives the obtained performances.

HEAVY IONS PRODUCTION PERFORMANCES OF DIONE

Ion species	Source exit (number of nuclei)	Number of injected pulses in MIMAS	Number of nuclei accelerated in Saturne
Li^{3+} polarized	6μA ; 25ms	8	1 10^9
$^{14}N^{7+}$	10^9	1	7 10^8
C^{6+}	10^9	4	10^9
Ne^{10+}	10^8	4	2 10^8
Ar^{16+} Ar^{17+} Ar^{18+}	10^8 3.5 10^7 2 10^6	3	1.2 10^8
$^{84}Kr^{30+}$ $^{84}Kr^{26+}$	8 10^6 2 10^6	1	2 10^6 8 10^6
$^{56}Fe^{20+}$	4.2 10^6	1	3 10^6
$^{129}Xe^{33+}$	2.9 10^6 (MIMAS entrance)	1	2 10^6 (MIMAS only)
$^{197}Au^{50+}$	3 10^6	1	10^6

The beam emittance measured at the DIONE exit is 3.5 10^{-7} m.rad norm.

1. J. Faure, B. Feinberg, A. Courtois and R. Gobin, *Nucl. Instrum. Methods* A **219**, 243 (1987)
2. J. Faure, B. Feinberg, A. Courtois and R. Gobin, *Nucl. Instrum. Methods* A **219**, 449 (1984)
3. B. Gastineau, J. Faure and A. Courtois, *Nucl. Instrum. Methods* B **9**, 538 (1985)
4. L. Degueurce, *Nucl. Instrum. Methods* A **260**, 538 (1987)
5. A. Courtois, J. Faure, R. Gobin, P.A. Leroy, B. Visentin and P. Zupranski, *Nucl. Instrum. Methods* A **311**, 10 (1992)
6. B. Visentin, J. Faure, R. Gobin, P.A. Leroy, *Nucl. Instrum. Methods* A **313**, 23 (1992)
7. R.S.I. 65 (4), APRIL 1994 A hollow cathode ion source as an EBIS injector for metallic elements
8. B. Visentin, P. Van Duppen, P.A. Leroy, F. Harrault and the ISOLDE collaboration, *Nucl. Instrum. Methods* B **101**, 275 (1995)
9. VACUUM 46/ N8-10/ 77-780 (1995) Ultra high vacuum technology applied to the design of warm EBIS

PHYSICS WITH HEAVY IONS: CENTRAL COLLISIONS AND MULTIFRAGMENTATION

J. GOSSET
DAPNIA/SPhN, CE de Saclay, F-91191 Gif-sur-Yvette, France

The temperature-density phase diagram of nuclear matter has been explored at Saturne-2 in two regions. At energies per nucleon of a few hundred MeV, central collisions between heavy nuclei bring some heat and compression to the nuclear matter formed with nucleons from the interacting region. They have been studied with 4π detectors, mainly Diogène but also the Miniball/Miniwall array. Evidences for collective flow perpendicular to the reaction plane (squeeze-out) and for a correlation between nuclear flow and pion emission have been reported for the first time from Diogène results. In the same range of energy per nucleon, light nuclei (^{3}He and ^{4}He) impinging on heavy targets bring to the target nucleus a large thermal excitation without too much compression or angular momentum. Under these conditions nuclei preferentially emit several intermediate-mass fragments, a phenomenon called multifragmentation that has been studied with more and more sophisticated means, including the 4π detector ISiS.

Introduction

At the beginning of the seventies the properties of dense and hot nuclear matter became an important issue, both in fundamental nuclear physics to explore the temperature-density phase diagram of nuclear matter far away from the point corresponding to normal nuclei (normal nuclear density ρ_0 and zero temperature), and in astrophysics to understand the core of neutron stars and the explosion of type II supernovæ. From the experimental point of view, central collisions between heavy nuclei at an energy per nucleon of about 1 GeV were expected to compress the interacting region between the projectile and the target nuclei up to a few times ρ_0 and to bring it to temperatures as high as 50-100 MeV. From the theoretical point of view exotic phenomena were expected like the production of shock waves, pion condensates or density isomers. First experiments were performed in Berkeley and Dubna. Inclusive experiments were soon realized not to be sufficient to sort out the events according to the impact parameter of the collisions, and to disentangle the interesting compression effects from more trivial ones like the geometry. Experiments evolved towards more and more exclusive ones. In the transformation from Saturne-1 into Saturne-2, the possibility to accelerate heavy ions up to a few hundreds of MeV per nucleon was foreseen at the very beginning. Right after the first Saturne-2 workshop in 1977, the Diogène collaboration was formed, with physicists from Saclay, Clermont-Ferrand and Strasbourg, and the aim to build an electronic 4π detector devoted to exclusive measurements of central collisions between heavy nuclei at energies per nucleon of a few hundred MeV. The results obtained with the Diogène detector in the eighties will be reviewed in the second section under the label "central collisions", together with results obtained more recently with another 4π detector, the Miniball/Miniwall array.

At a smaller energy per nucleon, nuclei are not expected to overlap completely. Nucleus-nucleus collisions lead to the formation of a nuclear system with high excitation energy and many experiments have been devoted all around the world to infer the properties of this excited nuclear system from its decay characteristics. In such studies it is important to disentangle the relevant parameters. With symmetric collisions between heavy nuclei, a rather high angular momentum is also brought to the excited nuclear system. With light nuclei (^{3}He and ^{4}He) impinging on heavy targets at energies per nucleon of a few hundred MeV, it is possible to get a high excitation energy without too much angular momentum or compression. This is a rather clean situation to explore the temperature-density phase diagram of nuclear matter at densities close to or below ρ_0 and temperatures smaller than 10 MeV. At low excitation energy, the system evaporates a few nucleons and there remains one heavy residue or two fission fragments. At higher excitation energy, but not high enough for the system to be vaporized into its constituent nucleons, it preferentially emits several intermediate-mass fragments, a phenomenon called multifragmentation. The results obtained from such studies at Saturne-2, with more and more exclusive detection systems, will be reviewed in the first section under the label "multifragmentation".

Both experimental programs are related together by the ambitious aim to derive the equation of state of nuclear matter. This is a very difficult task for several reasons. Most theoretical predictions are made within theories that deal with infinite nuclear matter, whereas all experiments are using finite systems only. In order to be able to extract an equation of state one would prefer that the system be in some kinetic equilibrium for at least some time during the whole path it follows in the density-temperature plane as the collision proceeds. In the final state which is observed, one has to look for robust variables that reflect precisely enough the state reached by the system at the highest density. In the Saturne-2 energy range one basic model to which data are usually compared consists in intranuclear cascade (INC) simulations, eventually followed by some evaporation code that takes into account the excitation energy brought into the system by the fast cascading process. These INC calculations are supposed to take into account off-equilibrium effects, but they do not include any equation of state explicitly. In more sophisticated models, like BUU (Boltzmann-Uehling-Uhlenbeck) or QMD (Quantum Molecular Dynamics), one tries to take into account an equation of state in the transport equations through a density dependent mean field with more or less compressibility. This mean field can be momentum dependent too. In-medium nucleon-nucleon cross sections can also be used instead of the values in free space.

The present review is an experimental one and tries to cover extensively all results obtained about the two subjects at Saturne-2. The general physics context of all these works can be followed with more details in the proceedings of the successive international conferences on nucleus-nucleus collisions. The last ones took place in Gatlinburg [1] in 1997 and Taormina [2] in 1994.

1 Multifragmentation

First experiments [3,4] by a group from Saclay dealt with the measurements of angular correlations between fission fragments resulting from the bombardment of the fissile ^{232}Th nucleus with light projectiles (p, d, α) at energies between 70 and 1000 MeV. The fission fragments are detected with position sensitive ionization chambers. The distributions of linear momentum imparted to the target are deduced from the angular correlations. From these data and data from other experiments at lower energies one can see a clear evolution of the reaction mechanism as a function of incident energy. At low energy, almost full momentum is transferred. From 10 to 70 MeV per nucleon, the momentum transfer is incomplete but proportional to the projectile mass. At higher energies, the momentum transfer saturates, a phenomenon which is tentatively explained in terms of an optimum excitation energy of the fissioning nucleus. This excitation energy would be too high in central collisions and the fission mainly occurs from peripheral collisions, in agreement with results from INC calculations.

Next experiments, [5,6] by the same group from Saclay and physicists from Darmstadt, focused on the bombardment of lighter targets (Au, Ho, Ag) with α particles. By selecting targets lighter than Th, the yield from peripheral collisions is reduced by the increase in the fission barrier, thus allowing events with the highest linear momentum transfer and excitation energy to be favoured. With the new setup fragments are measured in a very broad mass range. Emphasis is placed on determining the properties and the production mechanism of the intermediate-mass fragments (IMFs). From fragment yields and fragment-fragment angular correlations, the competition between fission and multifragmentation is examined. Even though the linear momentum transfer saturates at high energy, the IMF emission continues to increase, and the fission to drop, till 800 MeV per nucleon. At this energy, the IMF contribution is

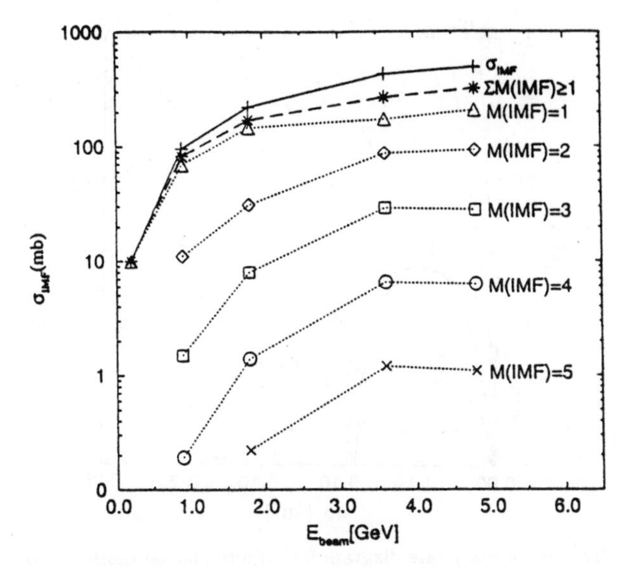

Figure 1: Cross sections for IMF multiplicity M as a function of bombarding energy for the ^{3}He+Ag system (figure taken from ref. 14).

about the same for the three targets. The relative velocities between IMFs exceed strongly the values of binary fission. Monte Carlo calculations show that these relative velocities exclude a sequential emission from the recoil nucleus and support a simultaneous breakup emission.

In the last experiments a beam of ^{3}He nuclei was used to bombard Ag and Au targets at energies between 480 and 4800 MeV. The experimental setup used by the same group from Saclay and physicists from Indiana allowed for more and more exclusive measurements of multifragmentation. An array of 36 particle-identification telescopes covering 8% of total solid angle was first used.[7–10] The most complete results[11–15] were obtained with the Indiana Silicon Sphere 4π detector array (ISiS)[16] consisting of 162 gas ionization/0.5 mm silicon/28 mm CsI telescopes covering 74% of 4π. Both light charged particles (LCP), i.e. H and He isotopes, and IMFs ($3 \leq Z \leq 20$) are fully Z identified with low thresholds (energy per nucleon as low as 0.8 MeV). The maximum telescope stopping power is E/A = 96 MeV. Complete particle identification (both Z and A) is provided for energetic H, He, Li and Be isotopes with energies per nucleon larger than 8 MeV.

The results of the experiments may be summarized as follows. Maximum

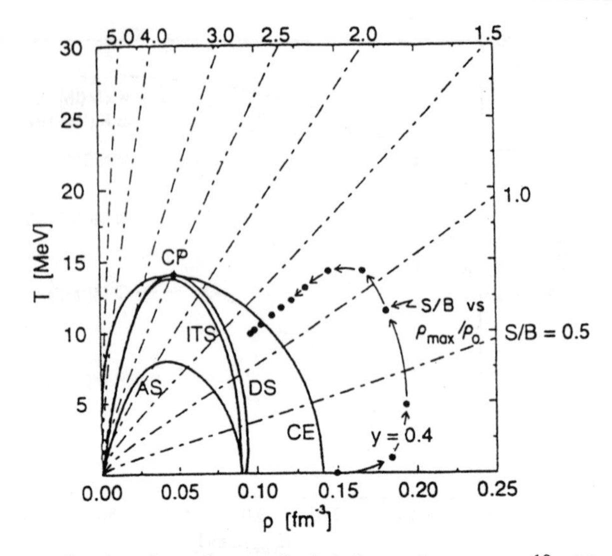

Figure 2: Temperature-density phase diagram for infinite nuclear matter [19] with $y = Z/A = 0.4$; entropy per baryon is indicated by dashed lines. Areas are defined as CE for liquid-gas coexistence, DS, ITS, and AS for diffusive, isothermal, and adiabatic spinodal instabilities. CP is the critical point. Reaction trajectory for the 4.8 GeV ^{3}He+Ag system [17] is plotted in average entropy (S/A) versus maximum density (ρ_{max}/ρ_0) coordinates in 4 fm/c steps. The trajectory is for impact parameter $b = 1.8$ fm and is based upon BUU calculation. [18] The apparent increase in density at early times reflects the localized effect of the projectile as it passes through the Ag nucleus. The whole figure is taken from ref. 14.

deposition energies of about 950 MeV and 1600 MeV are inferred from the data for Ag and Au nuclei respectively, [13] with a saturation for the Ag target at the highest bombarding energies, visible for example in the excitation functions for IMF events (Fig. 1). The results demonstrate the importance of accounting for fast cascade processes in defining the excitation energy of the targetlike residue. Correlations between various observables and the average IMF multiplicity indicate that the total thermal energy and the total observed charge provide useful gauges of the excitation energy of the fragmenting system. Comparison of the experimental distributions with INC predictions shows qualitative agreement. Rapidity, moving source, and sphericity-coplanarity analyses are consistent with near-simultaneous emission of IMFs from a source in approximate kinetic equilibrium. [14] For the most dissipative collisions, the spectral peaks are broadened and shifted to very low energies, indicative of emission from an extended nuclear system with density of about $\rho_0/3$. Predictions of an INC expanding, emitting source model compare well with experimental multiplicity distributions and the evolution of fragment spectral shapes. The time dependence of multifragmentation has been derived from the analysis of large-angle correlations between IMFs emitted in ^{3}He + Au collisions at 4800 MeV. [15] The results suggest that IMF emission occurs from an evolving system in which lighter fragments are emitted preferentially from a hotter, more dense source prior to breakup of a dilute residue that also produces heavier fragments. Finally, as an illustration of the possible paths followed in the (density-temperature) plane, the central collision trajectory for the 4.8 GeV ^{3}He+Ag reaction, [17] based on BUU calculations for finite nuclei, [18] is plotted on Fig. 2 in the phase diagram for infinite nuclear matter with $Z/A = 0.4$. [19]

2 Central collisions

Most of this section is devoted to the results obtained with the Diogène 4π detector. The experiments performed with the Miniball/Miniwall array are simply mentioned in the subsection devoted to triple differential distributions and flow analyses.

2.1 The Diogène 4π detector and its experimental program

Electronic detectors were preferred to visual ones like streamer chambers already in use in Berkeley and Dubna. The emphasis was put on good measurements of charged pions in addition to H and He nuclei, which was not possible with a detector like the Plastic Ball [20] that was being developed for experiments at the Bevalac in Berkeley. Our choice [21] was finally a combination between a pictorial drift chamber for the central detector at large angles and a scintillator hodoscope at forward angles.

The pictorial drift chamber covers the angular range from 20° to 132°. It is contained in a cylindrical barrel (hence the name Diogène) located inside a solenoid with a magnetic field of 1 T parallel to the beam direction. It samples three-dimensional points and energy loss along particle trajectories. From the magnetic rigidity and the energy loss, light charged particles (π^+, π^-, p, d, t, ^{3}He and ^{4}He) are identified. The energy threshold is 15 MeV for pions and 30 MeV for protons. The momentum resolution $\sigma(P)/P$ varies between 10 and 15%. The angular resolution is about 2° for both the polar and the azimutal angles. This central detector is able to handle particle multiplicities as high as about 30. The scintillator hodoscope (plastic wall) covers the angular range from 0° to 6°. Only the charge of the light fragments is resolved through energy loss and time of flight measurements.

Experiments started with ^{4}He projectiles, from 1982 to 1984, on targets of

C, Cu and Pb at energies per nucleon of 200, 400, 600 and 800 MeV. Heavier ions were used as soon as they became available. Between 1984 and 1986, targets of C, NaF, Nb and Pb were bombarded with ^{20}Ne projectiles at energies per nucleon of 200, 400, 600, 800 and 1000 MeV. In 1987 and 1988, targets of Ca, Nb and Pb were bombarded with ^{40}Ar projectiles at energies per nucleon of 200, 400 and 600 MeV. Finally, proton beams were also used to bombard targets of C, Nb and Pb at 800 and 1600 MeV. About 150 000 events were recorded for each projectile/target/energy combination.

At the end of the experimental program, two parallel plate avalanche chambers [22] were installed in the beam line upstream from the Diogène target and used with the argon beam. Events corresponding to interactions of the beam halo could be more efficiently rejected. The knowledge of vertex position could help to improve the performance of the track reconstruction program.

Because of its large solid angle coverage, the Diogène pictorial drift chamber was also used to study quite different subjects, like the inelastic proton-proton scattering with a polarized proton beam [23] and the decay and absorption of the Δ resonance in nuclei with (^{3}He, t) reactions. [24,25]

2.2 Overview of results and analyses

As already mentioned, it is essential to sort out the collisions according to their impact parameter, in order to disentangle the interesting compression effects from more trivial ones like the geometry. All Diogène results have been analyzed as a function of the multiplicity measured in the central detector, assumed to be monotonously correlated with the impact parameter. [26] Successive slices of about equal magnitude in the inclusive multiplicity distribution are made, that correspond to successive slices in the estimated impact parameter.

The Diogène detector does not have a full angular coverage, the angular and momentum resolutions are not negligible, and the track reconstruction program does not have 100% efficiency. In order to make comparisons between Diogène results and results from model calculations, it is important that events produced by any model calculation go through an experimental filter that takes into account all these imperfections. Such an experimental filter has been applied to INC simulations, [27,28] to which most of Diogène results have been compared.

For the impact parameter selection, the quality factors of various global variables had been compared within INC calculations, [29] and no global variable seemed more appropriate than the multiplicity. The reliability of the method mentioned above for estimating the impact parameter has been checked within the framework of INC calculations. [26] Even though the multiplicity distributions do not agree perfectly between model and experiment, the same procedure can be applied to both distributions for estimating the impact parameter. Experimental results can then be compared consistently to model predictions as a function of this estimated impact parameter.

In the following subsections, results concerning multiplicities and double differential distributions will be presented first. The production of composite particles might have something to do with the entropy of the system, the pion multiplicity with the compression energy, and the slopes of the energy spectra with the temperature of the system. Then two-particle correlations will be reviewed. At small relative momentum between identical particles, their interpretation through interferometry leads to some information about the size of the emitting source. With (π^+, p) correlations one can look for Δ-resonance production. Finally, triple differential distributions will be summarized. The reaction plane is determined event by event and collective flow effects, related to the pressure build-up, are discussed both for baryons and pions.

2.3 Multiplicities and double differential distributions

The proton multiplicities, or more precisely the pseudoproton multiplicities or the multiplicities of protonlike particles, i.e. of free protons plus protons that are bound in detected composite fragments (d, t, ^{3}He and ^{4}He), have been used for all systems to select impact parameter. This is interesting for easier comparison to results from INC simulations that do not take into account the formation of composite particles.

π^+ and π^- multiplicities have been measured in proton-nucleus, [30] α-nucleus [31,32] and Ne- and Ar-nucleus [33] collisions as a function of the impact parameter, projectile energy and target mass. Multiplicities predicted by INC calculations are generally smaller than the experimental ones. Such a difference had been already observed on results obtained with the streamer chamber in Berkeley and used to extract an equation of state of the nuclear matter. [34] The fact that some difference already exists with projectiles as light as ^{4}He, and even with protons on heavy targets, where no compression is expected to occur, is rather an indication that the INC misses some pion absorption.

In order to summarize the double differential distributions for the emission of any ejectile as a function of the centrality of the collision, it is better to use variables that have a simple behaviour under Lorentz transformation than to use energy or momentum and angle that would be natural for measurements from particle telescopes or spectrometers. For the longitudinal dimension the rapidity y is such a variable. For the transverse dimension, instead of the transverse momentum p_T, one can also use a dimensionless quantity γ_T equal to the ratio between the transverse mass $m_T = \sqrt{(m^2 + p_T^2)}$ and the rest mass m. Using the invariant double differential distribution $\mathrm{d}^2N/(\gamma_T^2\mathrm{d}\gamma_T\mathrm{d}y)$ presents the advantage that, for isotropic thermal emission from a single source with Boltzmann spectrum, it is an exponential function $N_e(y)\exp(-\gamma_T/\gamma_e(y))$ for any fixed value of the rapidity, with a simple dependence of the coefficients $N_e(y)$ and $\gamma_e(y)$ upon the temperature T_0 of the source, and also upon the number of emitted particles for the coefficient $N_e(y)$. In particular, $\gamma_e(y)$ is close to T_0/m. Moreover the product $N_e(y)\gamma_e(y)$ is a constant.

Very often, the invariant double differential distributions $\mathrm{d}^2N/(\gamma_T^2\mathrm{d}\gamma_T\mathrm{d}y)$ measured with Diogène could be quite well fitted with such an exponential function for each rapidity bin. Even if measurements are performed within some acceptance limits, the previous fit allows for a simple extrapolation to the whole acceptance in transverse momentum for a given value of the rapidity in order to get a rapidity distribution $(\mathrm{d}N/\mathrm{d}y)_e$, which can be used together with $\gamma_e(y)$ to summarize the double differential distributions. Such a procedure has been extensively applied to our results for π^+ and π^- distributions [33] and also to proton and deuteron distributions in some cases. [35] For both pions and protons, the product $N_e(y)\gamma_e(y)$ is far from being constant, and the departure from being constant is stronger for protons than for pions.

The proton spectra measured in α-nucleus and Ne-nucleus collisions, with or without selection of most central collisions, have been compared with the predictions of a thermodynamical model using collective velocity distributions combined with statistical thermodynamics in local rest frames. [36] The spectra measured in Ne-NaF and Ne-Pb collisions at E/A = 400 and 800 MeV for charge 1 and 2 fragments detected at forward angles in the plastic wall show evidence for the existence of two emitting sources, [37] one from the projectile spectators and the other one from the participants. The contribution from both sources has been examined as a function of the centrality of the collision.

Deuteron emission has been studied for α-nucleus collisions. [38] The coalescence relation between protons and deuterons is examined for the inclusive part of the spectra. The size of the interacting region is evaluated from the coalescence coefficients. The rms radius is typically 4-5 fm, depending on the target mass. The proton and deuteron spectra corresponding to central collisions are fitted assuming emission from a single source moving with intermediate velocity between that of the projectile and that of the target. The extracted "temperatures" are independent of the emitted particle, indicating that the fragments have a common source.

For Ne + Nb collisions at energies per nucleon of 400 and 800 MeV, the experimental results for the emission of protons and light nuclear fragments (d, t, ^{3}He and ^{4}He) have been compared with INC calculations complemented by a simple percolation procedure.[35] The model reproduces quite well global experimental observables like nuclear fragment multiplicity distributions, and nuclear fragment to proton ratios. For rapidity distributions, the best agreement occurs for peripheral reactions.

2.4 Correlations between two particles

Two-proton correlation functions at small relative momentum have been systematically studied for Ne-nucleus and Ar-nucleus collisions.[39,40] Information on the space-time structure of the emitting source is derived as a function of the centrality of the collision. The Koonin's formula[41] is used, and special attention is devoted to take into account all experimental biases in order to get the distorted theoretical correlation curves before comparison to the experimental data. Interesting conclusions are also obtained when comparing the extracted source radii to the dimensions of the overlapping volume between the projectile and the target in a cylindrical clean cut model.

The resolution on the fragment momentum is not accurate enough that, from (p, α) and (d, α) correlation functions,[40] one can extract quantitative values like source sizes, or temperatures from the relative production yield of two excited states of a given species.

Invariant mass distributions have been examined for (π^+, π^-) pairs[42] from p-nucleus collisions measured with Diogène, and from p-nucleus and ^{12}C-nucleus collisions[43,44] measured with the Dubna propane bubble chamber. An enhancement in the low invariant-mass region is observed only for heavy targets. A shadowing effect due to pion absorption inside the nuclear medium is considered as a possible source of this pattern. Azimuthal (π^+, π^-) correlations and pion yield dependence on the impact parameter, as well as transport calculations, seem to confirm this assumption.

Attempts to extract Δ production from invariant mass distributions of (p, π^+) pairs have been made for p-nucleus[45,46] and Ne-nucleus[33] collisions, using the event mixing method. Clear signals around the Δ mass are visible in the difference between invariant mass spectra for pairs from the same event and for pairs where the proton and the π^+ are taken from different events, but in the same slice of impact parameter. For p-nucleus collisions, the signal is observed at a mass smaller than the free Δ mass by 10 to 60 MeV, and this mass difference increases with the centrality of the collision. For Ne-nucleus collisions, the signal is centered at about 1200 MeV. It decreases with the centrality of the collision. It is always smaller than the signal predicted by INC calculations. In the event mixing method the difference is not so easy to interpret because spurious terms can distort its shape and magnitude as compared to the true signal.[47] Anyhow, Diogène results about Δ production are to be considered as a pioneering work, since two recently published letters are still dealing with this subject. The first one[48] concerns results obtained with the EOS time projection chamber[49] in Berkeley. The event-mixing method is used to extract Δ production from (p, π^+) and (p, π^-) pairs in ^{58}Ni + Cu collisions at an energy per nucleon of 1.97 GeV. In the second one,[50] charged pion spectra measured in ^{58}Ni-^{58}Ni collisions at energies per nucleon of 1.06, 1.45 and 1.93 GeV with the FOPI detector[51] in Darmstadt are interpreted in terms of a thermal model including the decay of Δ resonances. Thermal pions are dominant at high transverse momentum, whereas pions originating from Δ resonances are mostly visible at low transverse momentum. About 10 and 18% of the nucleons are excited to Δ states at freeze-out for beam energies per nucleon of 1 and 2 GeV, respectively.

2.5 Triple differential distributions and collective flow

In the field of relativistic nucleus-nucleus collisions, a lot of work has been devoted to the extraction of collective flow that would result from pressure build-up effects. This flow may occur in the reaction plane (side-splash and bounce-off) or perpendicularly to it (squeeze-out). It has been thoroughly investigated in the framework of INC calculations by D. L'Hôte, from the Diogène collaboration, together with J. Cugnon.[52,28]

In the Diogène flow analyses, the azimuth of the reaction plane is determined event by event from the transverse momentum method,[53] as the azimuthal angle for which there is a maximum correlation between the rapidity y and the transverse momentum $\vec{p}_T$ of all detected baryons. More specifically it is given, for each particle (index ν) to be considered in the flow analysis, by the direction of the vector

$$\vec{Q}_\nu = \sum_{\mu \neq \nu} (y_\mu - < y >) Z_\mu \vec{p}_{T\mu} / m_\mu .$$

The sum runs over all baryons reconstructed in the event (index μ), with mass m_μ and charge Z_μ, except eventually the particle ν itself, in order to remove autocorrelations. $< y >= \sum_\mu Z_\mu y_\mu / m_\mu$ is the average rapidity, calculated for each event over all charges carried by the baryons. It has to be evaluated for each event because we are not dealing only with symmetric systems. Such an analysis leads to triple differential distributions of the emitted particles, the three variables being the rapidity y for the longitudinal component and the two components $p_{x'}$ and $p_{y'}$ of the transverse momentum $\vec{p}_T$, in the estimated reaction plane and perpendicular to it, respectively. In all flow analyses it is very important to consider and take into account the distorsions of the extracted quantities due to the uncertainty in the determination of the azimuth of the reaction plane. It has been done carefully for the various kinds of flow analysis applied to Diogène measurements. These distorsions will be simply called $\Delta\phi$ distorsions in the following.

The dependence of $p_{x'}/m$ upon rapidity has been analyzed as a function of the centrality of the collision for pseudoprotons as well as for composite particles (d, t, ^{3}He, ^{4}He).[54,35] Even for asymmetric systems, it is always linear, at least in the rapidity range where the contribution of spectators from the target or the projectile is negligible. Flow effects in the reaction plane can be summarized by the slope of this linear dependence, corrected for $\Delta\phi$ distorsions. This flow parameter increases with the mass of the considered baryon. It does not depend very much on the centrality of the collision. It is always larger than the value predicted by INC simulations. Finally, it is clear that this flow parameter is affected by the detector acceptance and does not reflect the true flow of nuclear matter. The mass dependence of this flow parameter has also been measured at Saturne-2 with another 4π detector, the Miniball/Miniwall array, in ^{84}Kr+^{197}Au collisions at an energy per nucleon of 200 MeV,[55] using techniques that are free of reaction plane dispersion.[56] The fragment flow per nucleon increases with mass, following a thermal or coalescencelike behaviour. Comparisons to BUU calculations demonstrate a definite preference for a momentum dependent mean field and offer support for a 20% density dependent reduction in the nucleon-nucleon cross section from the value in free space, with very little sensitivity to the compressibility parameter.

For each value of the rapidity, the azimuthal distribution of the emitted particles with respect to the reaction plane can be fitted by a Fourier series up to second order. The first order coefficient is related to the usual flow in the reaction plane. It gives rise to peaks at 0° and 180° in the azimuthal distribution of particles emitted at positive, respectively negative, values of the rapidity with respect to the center-of-mass rapidity. The second order coefficient, which gives rise to peaks at ±90° with respect to the reaction plane, most visible in the rapidity range where the usual flow is zero, is related to a new kind of flow, or squeeze-out, perpendicular to the reaction plane. This squeeze-out was observed for the first time in Diogène results (Fig. 3a) concerning pseudoprotons emitted in Ne-nucleus collisions at an energy per nucleon of 800 MeV.[57] The dependence upon transverse momentum, rapidity and impact parameter, of the second-order coefficient corrected for $\Delta\phi$ distorsions, has been systematically studied.[54]

In order to try and get rid of the influence of the detector acceptance, to summarize the triple differential distributions more completely than with the

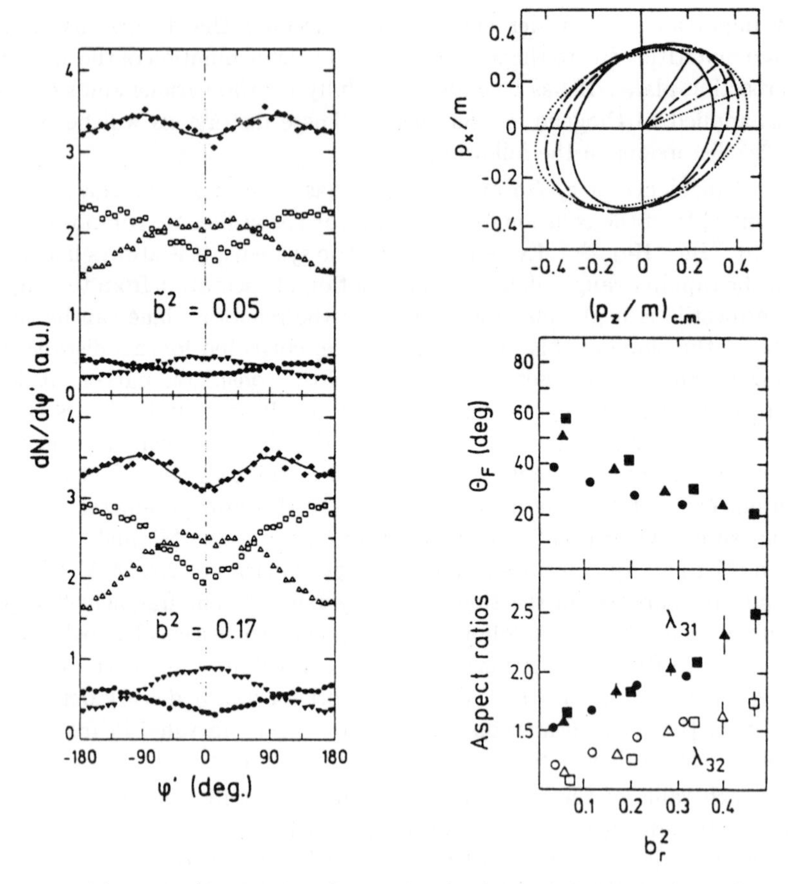

Figure 3: (**a**) LEFT (figure taken from ref. 57): Azimuthal distributions of proton-like particles emitted in different rapidity bins for Ne+Pb collisions at 800 MeV per nucleon. The upper and lower parts correspond to the two multiplicity cuts (18, ∞) and (13, 17), and the corresponding estimates for the squared reduced impact parameter $(b/b_{max})^2$ are indicated. The lines correspond to the Fourier series fits in the intermediate rapidity bin. The symbols correspond to the following five rapidity bins (y_{CM} is the estimated center-of-mass rapidity): $y < y_{CM} - 0.3$ (full circles); $y_{CM} - 0.3 < y < y_{CM} - 0.1$ (open squares); $y_{CM} - 0.1 < y < y_{CM} + 0.1$ (full lozenges); $y_{CM} + 0.1 < y < y_{CM} + 0.4$ (open triangles); $y_{CM} + 0.4 < y$ (full triangles). (**b**) RIGHT (figures taken from ref. 58): TOP: Ellipses representing the two-dimensional Gaussian fits (contours at $1/e$ of the maximum) to the in-plane distributions of proton-like particles, for Ar+Pb collisions at 400 MeV per nucleon and for increasing values of the squared reduced impact parameter $(b/b_{max})^2$=0.06 (full line), 0.20 (dashed), 0.33 (dot-dashed), 0.46 (dotted); BOTTOM: Impact parameter dependence of the flow angle θ_F, and of the aspect ratios λ_{31} (full symbols) and λ_{32} (open symbols), for Ar+Ca (circles), Ar+Nb (triangles) and Ar+Pb (squares) at 400 MeV per nucleon.

flow parameter and the squeeze-out coefficient, and to enrich our information on the flow from participant nucleons only, fits to a simple anisotropic Gaussian distribution were successfully tried within an acceptance which removes most of the spectators contribution. Instead of applying a three-dimensional Gaussian fit on the triple differential distribution - which is numerically difficult to handle with only about 20 000 events per multiplicity bin - two-dimensional distributions $(p_z/m, p_{x'}/m)$ and $(p_z/m, p_{y'}/m)$ distributions are fitted with two-dimensional anisotropic Gaussian distributions that represent the in-plane and out-of-plane distributions obtained by integrating over the third dimension. The p_z component is calculated in the center-of-mass of the emitting source. In any given class of multiplicity, the rapidity of the source is estimated as the rapidity at which the average in-plane momentum is zero. The adjusted parameters - one angle θ and three variances, σ_3^2 and σ_1^2 in the reaction plane ($\sigma_3 > \sigma_1$), and σ_2^2 perpendicularly to the reaction plane - are corrected for $\Delta\phi$ distorsions, which leads to the true flow angle θ_F and variances f_3^2, f_1^2 and f_2^2, closely related to the flow angle and the eigenvalues obtained from a sphericity analysis. First results (Fig. 3b) were obtained for Ar-nucleus collisions at an energy per nucleon of 400 MeV.[58] Complete results are available in a thesis.[54] The flow angle is increasing with decreasing impact parameter and is larger for heavier targets. The out-of-plane aspect ratio $\lambda_{32} = (f_3/f_2)^2$ is always smaller than the in-plane aspect ratio $\lambda_{31} = (f_3/f_1)^2$, in agreement with the squeeze-out effect. Both aspect ratios decrease when going to more and more central collisions. The large values of the flow angle (20°-40°) obtained for the symmetric Ar+Ca system at an energy per nucleon of 400 MeV could appear as contradicting the Plastic Ball results,[59] where no peak at finite angle in the $dN/d(\cos\theta_F)$ distributions was observed. However, their event-by-event sphericity analysis is subject to finite number distorsions[60] which explain the disappearance of the peak under certain conditions.[28] Moreover, it is not able to resolve the contributions from participants and spectators, as stressed by the authors themselves.[59] This clearly demonstrates the potentiality of the Gaussian fitting method. Predictions from INC simulations differ drastically from experimental values. For Ar-nucleus collisions at energies per nucleon of 400 and 600 MeV, the flow angle θ_F is 2-3 times larger in the experiment than in the INC.

Flow analyses may also be performed for produced particles. The correlation between nuclear flow and pion emission was observed for the first time in Diogène results (Fig. 4) concerning Ne-nucleus collisions at an energy per nucleon of 800 MeV.[61] Complete results have also been published.[62] The pions exhibit a non-zero flow. For asymmetric systems, the average transverse momentum projected in the reaction plane is always positive. In other words

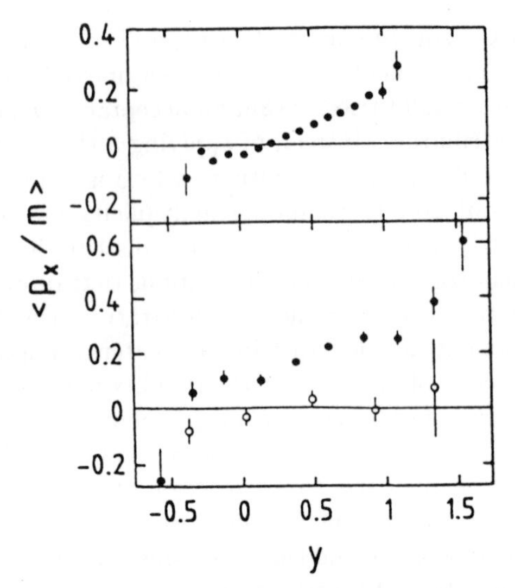

Figure 4: Average in-plane transverse momentum versus rapidity for protons (top) and π^+ (bottom) from central Ne+Pb collisions at E/A=800 MeV (figure taken from ref. 62). The open circles (bottom figure) are INC results.

there is a preferential pion emission in the direction of the lighter (projectile) nucleus. Such a behaviour contradicts the predictions of INC calculations and is underestimated by QMD calculations which include mean field effects, whereas a full BUU calculation [63] was able to reproduce the data. It can be interpreted within a simple geometrical model in terms of absorption by the spectator nuclear matter. The target shadowing explanation has a remarkable consequence on the sign of the baryon flow, which cannot be determined from the transverse momentum method. From the pion absorption hypothesis and the observed correlation between pion and baryon flows, it follows that the baryon flow is directed towards the light projectile at forward rapidities, which is in agreement with the sign predicted by the different models.

A review about collective flow in heavy-ion collisions has been published recently [64] and may help the reader to get more perspective on that subject.

2.6 Summary and outlook

Specific Diogène contributions to the field of relativistic nucleus-nucleus collisions may be summarized as follows. The study of asymmetric systems leads to paths in the temperature-density phase diagram that are different from the path followed by symmetric systems, which were mostly studied with the Plastic Ball. Diogène measurements are not based on energy measurements through total absorption like in the Plastic Ball. This led to good results on π^+ and π^- production. The Δ resonance production could be seen from (π^+, p) invariant-mass distributions. About collective flow, evidence for out-of-plane effects (squeeze-out) was first reported from our results. A new method, based on fits to anisotropic Gaussian distributions, has been applied for the first time, which allows to see more clearly the flow of participants by eliminating most of the spectators contribution. Correction for distorsions had to be developed for all different flow analyses. A correlation between nuclear flow and pion emission was also observed for the first time.

The Diogène detector was not built to handle high multiplicities that occur for symmetric collisions between heavy nuclei at energies per nucleon around 1 GeV. Detectors similar to Diogène, but better for that respect, had to be built to exploit fully the collisions between heavy nuclei that were accelerated with the Bevalac in Berkeley and with SIS in Darmstadt: the EOS time projection chamber [49] and the FOPI 4π detector, [51] respectively. Results obtained with these two detectors continue to appear in the literature.

Finally I would like to thank - this is surely the last opportunity - all people that worked for the success of the Diogène experimental program. That took more than ten years and involved many people, from the design of the detector to the last publications. All these people showed a lot of passion and enthousiasm, and were eager to transmit them to the eighteen students that got a PhD with Diogène results. After the completion of the experimental program, Diogène collaborators went to work on many different subjects, including solid-state physics and climatology. A few of them are now involved in what seems the most natural continuation of their work, the study of nucleus-nucleus collisions at much higher energies in a quest for the quark-gluon plasma, a new state of matter predicted by quantum chromodynamics, the fundamental theory of strong interactions.

Acknowledgements

I am grateful to Marie-Claude Lemaire for carefully reading the manuscript.

References

1. Proceedings of the Sixth International Conference on Nucleus-Nucleus Collisions, June 2-6, 1997, Gatlinburg, Tennessee, USA, *Nucl. Phys.* A **630** (1998).

2. Proceedings of the Fifth International Conference on Nucleus-Nucleus Collisions, May 30-June 4, 1994, Taormina, Italy, *Nucl. Phys.* A **583** (1995).
3. F. Saint-Laurent et al., *Phys. Lett.* B **110**, 372 (1982).
4. F. Saint-Laurent et al., *Nucl. Phys.* A **422**, 307 (1984).
5. G. Klotz-Engmann et al., *Phys. Lett.* B **187**, 245 (1987).
6. G. Klotz-Engmann et al., *Nucl. Phys.* A **499**, 392 (1989).
7. S.J. Yennello et al., *Phys. Lett.* B **246**, 26 (1990).
8. S.J. Yennello et al., *Phys. Rev. Lett.* **67**, 671 (1991).
9. S.J. Yennello et al., *Phys. Rev.* C **48**, 1092 (1993).
10. K. Kwiatkowski et al., *Phys. Rev.* C **49**, 1516 (1994).
11. K. Kwiatkowski et al., *Phys. Rev. Lett.* **74**, 3756 (1995).
12. K.B. Morley et al., *Phys. Lett.* B **355**, 52 (1995).
13. K.B. Morley et al., *Phys. Rev.* C **54**, 737 (1996).
14. E. Renshaw-Foxford et al., *Phys. Rev.* C **54**, 749 (1996).
15. G. Wang et al., *Phys. Lett.* B **393**, 290 (1997).
16. K. Kwiatkowski et al., *Nucl. Instrum. Methods* A **360**, 571 (1995).
17. G. Wang et al., *Phys. Rev.* C **53**, 1811 (1996).
18. P. Danielewicz, *Phys. Rev.* C **51**, 716 (1995).
19. H. Müller and B.D. Serot, *Phys. Rev.* C **52**, 2072 (1995).
20. A. Baden et al., *Nucl. Instrum. Methods* A **203**, 189 (1982).
21. J.-P. Alard et al., *Nucl. Instrum. Methods* A **261**, 379 (1987).
22. R. Burgei et al., *Nucl. Instrum. Methods* A **287**, 481 (1990).
23. C. Comptour et al., *Nucl. Phys.* A **579**, 369 (1994).
24. T. Hennino et al., *Phys. Lett.* B **283**, 42 (1992).
25. M. Roy-Stephan, invited talk at this meeting.
26. C. Cavata et al., *Phys. Rev.* C **42**, 1760 (1990).
27. J. Cugnon, D. Kinet and J. Vandermeulen, *Nucl. Phys.* A **379**, 553 (1982).
28. J. Cugnon and D. L'Hôte, *Nucl. Phys.* A **452**, 738 (1986).
29. J. Cugnon and D. L'Hôte, *Nucl. Phys.* A **397**, 519 (1983).
30. M.-C. Lemaire et al., *Phys. Rev.* C **43**, 2711 (1991); double-differential distributions of charged pions are tabulated in M.-C. Lemaire et al., Report CEA-N-2670 (1991), unpublished.
31. D. L'Hôte et al., *Phys. Lett.* B **198**, 139 (1987).
32. D. L'Hôte, Thèse d'Etat, Université de Paris-Sud, 1987.
33. C. Cavata, Thèse, Université de Paris-Sud, 1989, Report CEA-N-2629 (1990), unpublished.
34. R. Stock et al., *Phys. Rev. Lett.* **49**, 1236 (1982).
35. G. Montarou et al., *Phys. Rev.* C **47**, 2764 (1993).

36. M.-J. Parizet et al., *Int. J. Modern Phys.* **4**, 3689 (1989).
37. N. Bastid et al., *Nucl. Phys.* A **506**, 637 (1990).
38. G. Montarou et al., *Phys. Rev.* C **44**, 365 (1991).
39. P. Dupieux et al., *Phys. Lett.* B **200**, 17 (1988).
40. P. Dupieux et al., *Z. Phys.* A **340**, 165 (1991).
41. S. Koonin, *Phys. Lett.* B **70**, 43 (1977).
42. J. Pluta et al., *Nucl. Phys.* A **562**, 365 (1993).
43. I. Angelov et al., *Sov. J. Nucl. Phys.* **30**, 824 (1979).
44. G.N. Agakishiev et al., *Sov. J. Nucl. Phys.* **37**, 938 (1983).
45. M. Trzaska et al., *Z. Phys.* A **340**, 325 (1991).
46. M. Trzaska, GSI-91-14.
47. D. L'Hôte, *Nucl. Instrum. Methods* A **337**, 544 (1994).
48. E.L. Hjort et al., *Phys. Rev. Lett.* **79**, 4345 (1997).
49. G. Rai et al., *IEEE Trans. Nucl. Sci.* **37**, 56 (1990).
50. B. Hong et al., *Phys. Lett.* B **407**, 115 (1997).
51. A. Gobbi et al., *Nucl. Instrum. Methods* **324**, 156 (1993).
52. J. Cugnon and D. L'Hôte, *Phys. Lett.* B **149**, 35 (1984).
53. P. Danielewicz and G. Odyniec, *Phys. Lett.* B **157**, 146 (1985).
54. M. Demoulins, Thèse, Université de Paris-Sud, 1989, Report CEA-N-2628 (1990), unpublished.
55. M.J. Huang et al., *Phys. Rev. Lett.* **77**, 3739 (1996).
56. P. Danielewicz et al., *Phys. Rev.* C **38**, 120 (1988).
57. M. Demoulins et al., *Phys. Lett.* B **241**, 476 (1990).
58. J. Gosset et al., *Phys. Lett.* B **247**, 233 (1990).
59. H.Å. Gustafsson et al., *Phys. Rev. Lett.* **52**, 1590 (1984).
60. P. Danielewicz and M. Gyulassy, *Phys. Lett.* B **129**, 283 (1983).
61. J. Gosset et al., *Phys. Rev. Lett.* **62**, 1251 (1989).
62. J. Poitou et al., *Nucl. Phys.* A **536**, 767 (1992).
63. B.A. Li, W. Bauer, and G.F. Bertsch, *Phys. Rev.* C **44**, 2095 (1991).
64. W. Reisdorf and H.G. Ritter, *Annu. Rev. Nucl. Part. Sci.* **47**, 663 (1997).

SATURNE - AN INTENSE SOURCE OF COSMIC RADIATION

R. MICHEL

Center for Radiation Protection and Radioecology, University Hannover
Am Kleinen Felde 30, D-30167 Hannover, Germany

This work gives an introduction to cosmic ray particles and to their interactions with matter as well as to some of their widespread applications in astrophysics, cosmic ray physics, cosmophysics, planetology, astronomy and space technology. It shall serve as a source of references of experiments performed at SATURNE II dealing with the various aspects of cosmic ray interactions with matter. These experiments covered the determination of cross sections for the production of residual nuclides by medium-energy particles, the simulation of the cosmic ray exposure of meteoroids and of planetary surfaces and the investigation of radiation hardness of materials, radiation detectors and electronic devices for space, aviation and accelerator technology. Major scientific achievements have been and still will be obtained from this work at SATURNE II which was excellently suited to simulate cosmic radiation in the laboratory.

1 Cosmic Ray (CR) Particles and CR Interactions with Matter

In the solar system, one observes two natural types of medium- and high-energy corpuscular radiation, the galactic cosmic radiation (GCR) and the solar cosmic radiation (SCR). SCR particles are emitted during solar flares from the sun and consist on the average of 98 % protons and 2 % α-particles[1]. Their spectral distributions and intensities vary from flare to flare, typical energies going up to a few hundred MeV/A. The SCR is a special case of the general phenomenon of stellar cosmic radiation which is to be expected in the proximity of each star and which can be much more intense and probably also of higher energies in certain active stars such as WR and T-Tauri stars.[2]

GCR particles come from outside the solar system. They are injected into the interstellar medium by supernova explosions and are accelerated stochastically by complicated processes[3] and partially show extreme energies up to 10^{21} eV.[4] GCR consists out of 87 % protons, 12 % α-particles and 1 % heavier ions which show similar energy spectra if looked at as function of energy per nucleon.[5] Mean GCR energies are a few GeV/A. GCR spectra are modulated by interaction of GCR particles with the solar magnetic field and thus depend on the solar activity. SCR and GCR proton spectra at 1 A.U. are shown in Figure 1.

CR particles undergo a wide variety of interactions. Heavy GCR particles are fragmented by collisions with interplanetary hydrogen. Stellar CR particles interact with matter which is injected back from very active stars into interstellar matter.[6] These processes occurring outside the solar system are important nucleosynthetic processes which, on the one hand, strongly influence the abundance patterns of heavy GCR nuclides[7] and, on the other, are increasingly discussed as a source of nucleosynthesis during galactic evolution[8] and as a possible source of short-lived radioactivity in the early solar system giving rise to isotopic anomalies in exotic phases of pristine solar system matter.[9]

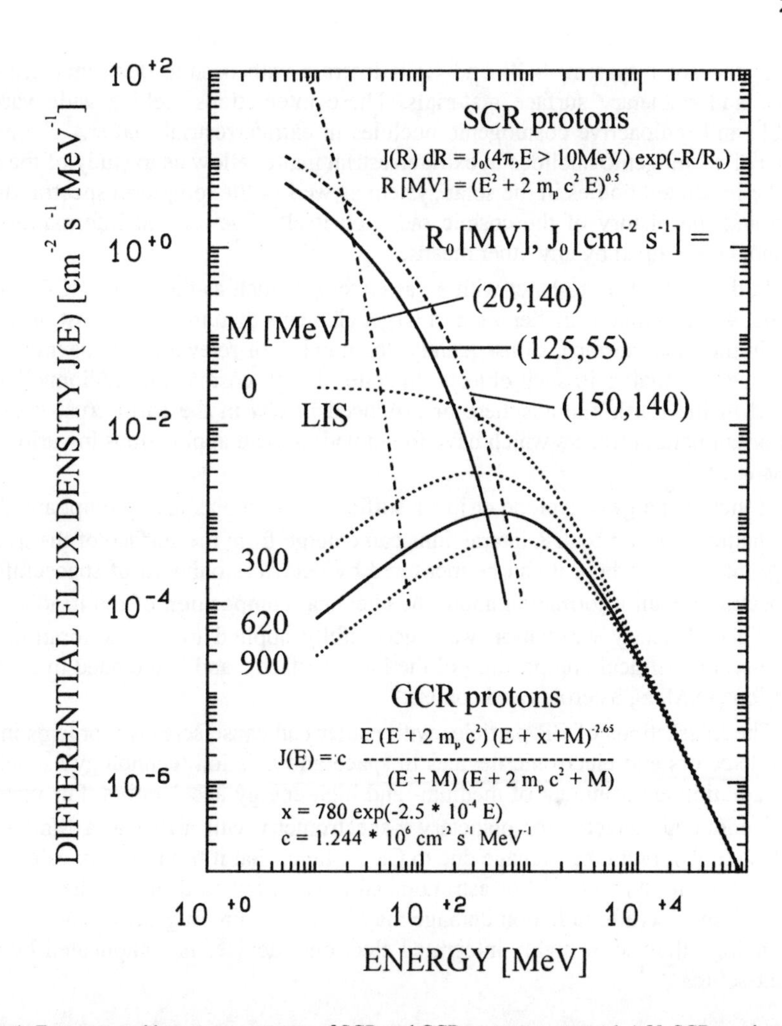

Figure 1: Extremes and long-term averages of SCR and GCR proton spectra at 1 A.U. SCR can be described as exponentially falling rigidity spectra with characteristic rigidities R_0 and 4π integral flux densities J_0 of protons with energies above 10 MeV.[10] Here, SCR spectra are plotted covering the observed range of R_0-values, 20 - 150 MV.[1] GCR spectra can be parameterized using a modulation parameter M.[11] GCR spectra are shown for times of an active sun (1969, M = 900 MeV) and of a quiet one (1965, M = 300 MeV). In addition, the local interstellar GCR spectrum (LIS)[11] and mean SCR and GCR proton spectra during the last 10 Ma are shown. The long-term spectra were derived from an analysis of cosmogenic nuclides in lunar rocks and meteorites for which simulation experiments performed at SATURNE II provided the basis.[12]

In the solar system, SCR and GCR interact with cosmic dust, meteoroids and lunar and planetary surface materials. These interactions yield a wide variety of stable and radioactive cosmogenic nuclides in extraterrestrial matter.[13-17] Investigations of cosmogenic nuclides in extraterrestrial matter allow us to study of the history of the irradiated bodies in the solar system as well as the long-term spectral distribution and the history of the cosmic radiation itself. They reveal information which cannot be obtained by any other means.

In the surface of a planet with a gas envelope, such as the Earth, SCR and GCR particles find a much smaller suite of target elements than in cosmic dust, meteoroids and in the lunar surface. Consequently, the number of relevant cosmogenic nuclides produced is smaller in such objects. In spite of that, cosmogenic radionuclides produced in the Earth's atmosphere or produced *in situ* in the earth crust they act as important natural tracers which have found wide spread applications in various fields of science.[18]

If there is no gas envelope or just a sufficiently thin one, also prompt and delayed γ-radiation induced by CR interactions can emerge from the surface of the irradiated body into space where it can be measured by detectors onboard of spacecrafts. The γ-spectra contain information about the chemical composition of the irradiated matter. This planetary γ-radiation was successfully applied to the exploration of the large-scale chemical composition of the lunar surface[19] and is intended to be used in missions to Mars, asteroids and comets.[20]

The interactions of CR particles with matter can cause serious problems in radiation detectors and electronic devices in space and aviation technology as well as in the radiation environment of medium- and high-energy accelerators. For instance, a semiconductor detector for planetary γ-spectrometry will arrive at a Martian orbit with so high a radiation damage due to CR particles that it has to be annealed in orbit before starting its work.[21] For astronomical devices onboard of satellites, space stations or space crafts radiation damage may become a limiting factor and in aviation technology the use of highly integrated electronic devices is complicated by single-event-setups.[22,23]

2 Cosmic Ray Studies at SATURNE II

Though accelerators are usually installed to solve problems of fundamental nuclear and particle physics, they play an important role for a wide variety of scientific and technical applications. For an advanced understanding of CR interactions with matter accelerator experiments are fundamental. Three groups of tasks can be defined which only can be solved using accelerators. They can be summarized under the headings *cross sections measurements*, *simulation experiments* and *material testing*. The goals of accelerator experiments for cosmic ray studies are to provide the nuclear reaction data needed, to validate models of CR interactions on the basis of simulation experiments under completely controlled conditions and to test the feasibility of technical solutions for space missions, aviation and accelerator technologies in a laboratory on Earth.

SATURNE II was outstandingly well suited to study CR interactions with matter. With its capability to accelerate protons, ^{4}He-particles and heavy ions it allowed to study the transition from preequilibrium reactions to spallation and fragmentation for all relevant CR particle types. The particle energies were variable enough to cover energies where many relevant cross sections depend strongly on energy. The beam intensities were high enough to simulate millions of years CR exposure within reasonable experiment times. During two decades, 13 out of 305 approved experiments at SATURNE II, were dedicated to the various aspects of CR interactions with matter, most of them being performed in recent years when medium-energy nuclear data for applications were increasingly recognized as important work items.[24,25]

2.1 Thin-Target Cross Sections of CR Interactions

Cross sections are the basic nuclear quantities on which interpretations or model calculations of GCR interactions have to be based. They have to be known from threshold up to about 10 GeV if CR interactions are considered. In contrast to basic nuclear science, where exclusive data provide the wealth of information, for applications inclusive data are needed.[26] Integral cross sections for the production of residual nuclides as function of particle energy, dominate the modeling of CR interactions in the interstellar medium as well as in solar system matter. Presently a successful and reliable modeling of cosmogenic nuclide production is only possible by combining experimental thin-target cross sections of the underlying nuclear reactions with calculated spectra of primary and secondary particles inside the irradiated matter.[12,26,27]

H. Sauvageon et al.[28-30] were the first to use SATURNE II to measure production cross sections. LNS 35 was part of a series of experiments by the Bordeaux group[28-39] aimed on the basic understanding of spallation reactions[28-32] and on the measurement of nuclear reaction data for models of CR interactions.[33-39] The CR related experiments were interrupted by the early death of S. Regnier. Later on, however, B. Lavielle and E. Gilabert from Bordeaux measured in collaboration with the author the production cross sections of krypton and xenon isotopes from all cosmochemically relevant target elements in the course of experiments LNS 167, LNS 172, LNS 232 and LNS 275[40-42] (see below).

A systematic and comprehensive approach to measure all relevant cross sections for the production of cosmogenic nuclides in extraterrestrial matter was made by the author and various collaborators (experiments LNS 167, LNS 172[43], LNS 232[44,45] and LNS 275[46-48]). After an early phase of initial experiments at KFA/Jülich, UCL/Louvain La Neuve and IPN/Orsay,[49] and references therein, at CERN/Geneve[50] and at LANL/Los Alamos[51], a general experimental routine evolved to determine complete excitation functions from thresholds up to 2.6 GeV by combining experiments at LNS/Saclay, TSL/ Uppsala and PSI/Villigen covering the energy ranges from 2.6 GeV to 200 MeV, from 180 MeV to 70 MeV and below 70 MeV, respectively, [52] and references therein.

The target elements C, N, O, Mg, Al, Si, Ca, Ti, Mn, Fe, Co, Ni, Cu, Nb, Sr, Y, Zr, Nb, Ba and Au were investigated. Residual radionuclides and stable rare gas isotopes were measured after irradiation by AMS or X- and γ-spectrometry and by conventional mass spectrometry, respectively, covering all relevant cross sections for the production of cosmogenic nuclides. A collaboration of up to 13 institutes at Ahmedabad, Bordeaux, Köln, Hannover, Jülich, Mainz, New Brunswick, Philadelphia, Uppsala and Zürich contributed to these investigations.[40-42,49-70] In 1997, the database covered nearly 550 target/product combinations with about 15,000 cross sections.[52,70] It is available in EXFOR format[71] (EXFOR number *O0276*) at the Nuclear Data Bank of NEA/OECD/ Paris. Large numbers of targets have been irradiated and stored for the measurement of long-lived and stable nuclides in the future. Recent analyses covered the production of long-lived radionuclides such as ^{14}C[72], ^{36}Cl[73], ^{41}Ca and ^{129}I[74,75] and of stable rare gas isotopes.[76]

By this collaboration, the production of cosmogenic nuclides by α-particles was also investigated for α-energies up to 180 MeV (KFA/Jülich and PSI/Villigen) and at 2.4 GeV (LNS/Saclay) resulting in a data base for 485 reactions with 6368 individual cross sections[77,78] thus providing a first basis for an adequate consideration of α-particles in model calculations of cosmogenic nuclide production rates.

The cross sections measured to satisfy the data needs of cosmic ray interactions with matter likewise are needed feasibility studies of accelerator-driven energy systems for the transmutation of nuclear waste[79] and energy amplification[80] which have caused a huge demand of medium-energy cross sections[81] (see also the paper by S. Leray in this proceedings). The CR related data provide a basis to estimate nuclide production in accelerator components, shielding materials and ambient air.[82] However, more data and a larger target element coverage is needed. In a collaboration of institutes from Bordeaux, Brueyere le Chatel, Köln, Hannover, Studsvik, Uppsala and Zürich such experiments were performed (LNS 275) investigating the target elements Mo, Rh, Ag, In, Te, La, Ta, W, Re, Ir, Pb, Bi, Th and U. These experiments are not yet finally evaluated. Only some preliminary data were reported so far.[83-89]

The final thin-target activation experiment to be mentioned is LNS 170. Sümmerer and coworkers measured the target fragmentation of Au and Th by 2.6 GeV protons and derived a semi-empirical formula for the estimation of production cross sections .[90] Their results were used for lay-out and design of the Projectile-Fragment Separator, FRS, at GSI/Darmstadt.[91]

All these cross section measurements also had consequences for the theoretical interpretation of nuclear reactions at medium energies. Many individual tests of various nuclear reaction models and codes were performed, *e.g.* refs.[49-52,58], and the cross sections were also used to develop systematics of medium-energy reactions described by semi-empirical formulas.[32,90] But, only recently a comprehensive test of a large variety of models and codes was performed on request of the Nuclear Science Committee of NEA/OECD. The comprehensive and consistent database of cross sections obtained in the experiments LNS 167, LNS 172 and LNS 232 provided a corner stone of experimental evidence in this International Model and Code Intercomparison of Medium Energy Activation Yields[92] in which calculated results from 29 contributions by 18 participants or participating collaborations were compared with nearly 6000 experimental cross sections for 202 target/product. 22 different models or codes were applied giving a representative survey on today's modeling capabilities. After this first general and systematic survey[92,93], now detailed analyses of individual models have to be performed[89,94] with the goals of defining the necessary ingredients of optimal reaction models and codes and of developing global best parameter sets for such models.

All the above experiments widely satisfied the data needs for model calculations of cosmogenic nuclide production with respect to charged-particle-induced reactions. However, the situation was much worse for neutron-induced reactions and the production by GCR particles of cosmogenic nuclides in thick-targets, e.g. meteoroids and planetary surfaces, is dominated by reactions of secondary neutrons.[95] Experimental cross sections for n-induced reactions above 14 MeV are generally sparse and for all relevant cosmogenic nuclides practically not existent.[96] Therefore, frequently the assumption of equal cross sections of p- and n-induced reactions is made in model calculations of production rates. However, this assumption is generally not valid and may lead to severely wrong production rates.[97]

The required cross sections of neutron-induced reactions can be derived from experimental thick-target production rates.[98] Provided the cross sections of the proton-induced production are well known, the neutron excitation functions can be extracted from the thick-target production rates by unfolding techniques starting from theoretical guess functions. From production rates measured in artificial stony and iron meteoroids irradiated isotropically with 600 MeV at the CERN SC[99,100,101] and with 1.6 GeV protons at SATURNE II (experiment LNS 172[102-105]; chapter 2.2) a consistent set of neutron excitation functions for more than 500 target/product combinations derived.[98,105] These neutron cross sections provided the basis for physical model calculations describing all aspects of cosmogenic nuclide production in extraterrestrial matter and for nuclide production at medium energies in general.[12,82]

All these experiments used activation techniques. However, the database thus obtained cannot satisfy the data needs of model calculations describing the interactions between the heavy GCR component and interstellar hydrogen. Such calculations shall model the nuclide abundance patterns of the heavy GCR component and shall extract the source composition of heavy CR particles, e.g. refs.[106-108]. For this purpose, systematic cross sections for a wide variety of stable nuclides are needed which cannot be obtained by activation experiments. Therefore, another approach with inverse kinematics was used in the experiment LNS 257[109,110] and fragmentation cross sections were determined from experiments using a 1.52 g/cm^2 thick liquid hydrogen target and on-line detection of product nuclides by a detector system (HISS) including a magnetic spectrometer.

This inverse kinematics was used earlier by Webber and coworkers who irradiated CH_2 and C targets and derived fragmentation cross sections for interactions with hydrogen by a CH_2-C subtraction technique.[111,112] The cross sections derived in this way were partially in contradiction to those measured by activation techniques[51] and the discrepancies have not yet been resolved.[113]

In a collaboration of A. Soutoul and coworkers from the Service de Astrophysique (SAp) at DAPNIA/CEA with W.R. Webber and his group this technique was improved by making use of SAp's hydrogen target and the above mentioned detector system. Starting in 1991, experiments were performed with beams ranging from neon to iron at the BEVALAC. The final analysis of these data is still underway. After the final shut-down of the BEVALAC, e.g. ref.[112], this collaboration continued their experiments at SATURNE II as experiment LNS 257 which took data in 1993 and 1994. The charge changing cross sections from 10 beams of isotopes of 8 different nuclei between Be and Ni which were accelerated to energies from 400 to 650 MeV/A were measured after interaction with a 1.52 g/cm^2 liquid hydrogen target. This is the first measurement of cross sections using a pure hydrogen target with a thickness approximating the amount of hydrogen traversed by cosmic rays in our galaxy. Beams of ^{9}Be, ^{11}B, ^{15}N and ^{22}Ne were investigated for the first time, others (^{12}C, ^{14}N, ^{16}O, ^{20}Ne, ^{56}Fe and ^{58}Ni) allow a comparison with earlier measurements in which the CH_2-C subtraction technique was applied. Isotopic cross sections of 130 secondary fragmentation isotopes were measured from the 10 different beams. Only preliminary results were published so far[114] and the final analysis is still underway.

2.2 Simulation of CR Interactions with Dense Matter

The exposure conditions in space are usually not well defined - often even unknown or subject to investigation. Therefore, the primary and secondary particle fields inside an irradiated object have to be derived by model calculations, e.g. ref.[95], describing particle production and transport from high and medium energies down to thermal ones. Simulation experiments are needed for the validation of such model calculations of the spectra of primary and secondary CR particles interactions in space. At an accelerator this can be done under completely controlled conditions leaving no room for invention of free parameters as consequence of the unknown irradiation conditions.

There were two large simulation experiments in which GCR interactions with extraterrestrial matter was mimicked in the laboratory in order to obtain „ground truth" for interpretative models which cannot be verified by information obtained from space missions or space-born objects. In two experiments at SATURNE II (LNS 172 and LNS 236) the bombardment by GCR protons of stony and iron meteoroids and the GCR-induced γ-ray production in planetary surfaces were simulated, respectively.

For the planetary γ-spectrometry, in LNS 236 a thick-target arrangement (1.5 m * 1.5 m * 1.8 m) was exposed to 1.5 GeV protons at intensities of 10^6 to 10^7 protons per spill.[115,116] Within an outer iron sleeve the thick-target arrangement had a surface target hit by the proton beam and an inner core. As surface target and inner core could be exchanged in order to get information on targets with different chemical compositions. γ-ray spectra were measured with a high purity germanium detector in the energy range from 0.1 MeV to 10 MeV. The experimental data will serve as tests of model calculations describing the γ-ray production by medium-energy proton bombardment.[117-120] Due to failure of the Mars Observer mission and due to preference to other space missions the evaluation of experiment LNS 236 is delayed and only preliminary data were published so far.[121-123] The final results, however, will have important impact on future space missions for which this type of γ-spectrometry is foreseen.

As has been explained in detail elsewhere,[99,100] such stationary thick targets cannot simulate realistically the exposure to galactic protons of meteoroids in space. As had been demonstrated in thick-target experiments at the CERN SC[99-101,124,125] the irradiation conditions of meteoroids in space can be simulated by isotropic irradiation of artificial meteoroids. However, theses experiments in which 3 artificial stony meteoroids made of diorite and gabbro were isotropically irradiated by 600 MeV[100] protons failed to reproduce the production rates observed in meteorites by a factor of two and more. This was attributed to 600 MeV being too low an energy to be representative for GCR protons.

The first realistic simulation of the GCR exposure of meteoroids was achieved in experiment LNS 172[43] using a proton energy of 1.6 GeV which is equivalent to the mean GCR proton energy during times of a quiet sun such as the Maunder minimum. In two separate irradiation experiment artificial stony and iron meteoroids with radii of 25 cm and 10 cm respectively were isotropically irradiated by 1.6 GeV protons. The isotropic irradiation was achieved by four simultaneous movements during irradiation. By two translational movements (up/down and left/right) the artificial meteoroids were exposed to a parallel proton beam covering their total cross sections. In this parallel proton beam the artificial meteoroids were rotated around two perpendicular axes. For a photograph of the artificial stony meteoroid in the irradiation machine see the cover of nouvelles des saturne.[43] During effective beam-on-target times of 282.1 h and 125.7 h, respectively, the artificial stony and iron meteoroid received proton fluences of $1.3 * 10^{14}$ cm^{-2} and $2.4 * 10^{14}$ cm^{-2} equivalent to 1.6 and 3.0 Ma exposure in space of real meteoroids. The observed production rates matched those observed in real meteorites within 10 % and justified the headline „*Les cosmos en accelere*" of *Le Monde* about SATURNE´s capabilities.[126]

Inside the artificial stony and iron meteoroids, several thousand of thin targets of all cosmochemically relevant target elements were irradiated and production rate depth profiles of all relevant cosmogenic nuclides (> 500 target/product combinations) were measured. A theoretical analysis of the experimental production rates was performed using depth- and size-dependent spectra of primary and secondary particles calculated by the HERMES code system[116] and experimental and theoretical thin-target cross sections of the underlying nuclear reactions. The results of all these thick-target experiments performed at LNS[68,102-105,128-133] and at CERN[99-101,124,125] were used to extract information about the excitation functions of the neutron-induced reactions by least-squares adjustment. Thereby, a consistent set of excitation functions of the underlying neutron-induced reactions was established.[98] With the new neutron cross sections it is possible to describe simultaneously all data from the simulation experiments with an accuracy better than 10 %.[12,82,105,133]

2.3 CR Particles and Material Science

Devices to be flown onboard of extremely costly space missions cannot be justified without prior testing in the laboratory. In order to investigate radiation damage to and radiation hardness of detectors and electronic devices material testing has to be performed at accelerators. A total of 6 experiments was dedicated to the problems of such material testing. These experiments cover the investigation of radiation damage to electronic devices and detectors. The borderline between CR investigations and terrestrial applications is not well defined. The terrestrial application include the radiation hardness of detectors at medium- and high-energy accelerators, electronic devices for aviation technology as well as devices for space missions or astronomical measuring systems to be used in earth orbits and space.

Experiment LNS 175[134,135] was the first experiment at LNS to study radiation damage in semiconductor detectors intended for use in space missions for remote sensing of large scale chemical composition by GCR-induced planetary γ-emission.[19] Such a detector onboard a space craft to mars will arrive damaged and have to be annealed in orbit prior to measurements. In LNS 175 systematically such effects were studied to develop the necessary techniques for space missions.[21,136-138]

In LNS 234[139], a detector for X-ray astronomy was tested by proton irradiations. Experiment LNS 261[140,141] simulated the GCR exposure of Si, BGO, CsI, CdTe and I_2Hg detectors.[142-146] In this experiment also cross sections for the residual nuclide production in CsI and CdTe were determined.[143,144] These experiments were dedicated to various space missions such as the proposed ESA mission INTEGRAL.[147] Also by LNS 294[148] detector materials for the ESA space mission INTEGRAL were tested. Finally, in LNS 263 silicon detector materials for use at the CERN LHC were investigated[149-152] and in experiment LNS 215 the dynamic effects of light ions on electronic components were studied.[153]

3 Achievements due to Saturne

There is a number of major achievements which have to be attributed to work performed at SATURNE II. Further achievements will be obtained in the future when the experiments done in the past will be finally evaluated and when they will have their impact on scientific missions to come. As a conclusion of this overview, highlights shall be emphasized where experiments at SATURNE II provided the basis for major scientific improvements.

Experiments at SATURNE II established a *culture of nuclear data for applications*. They provided the quantitative basis for the physical understanding of nuclear reactions at medium energies. For energies between 200 MeV and 2.6 GeV the worldwide largest database of cross sections for the production of residual nuclides was established which will be used for a long time in the various fields of application which range with respect to the experiments described here from cosmic ray physics, production of cosmogenic nuclides to various technical applications.

A *physical model of cosmogenic nuclide production* without free parameters was established on the basis of the thick-target experiments LNS 172. Combining calculated spectra of primary and secondary galactic particles in real meteoroids and planetary surfaces with experimental cross sections for p-induced reactions and the deduced cross sections for n-induced reactions, model calculations of cosmogenic nuclide production rates can be performed which explain all aspects of cosmogenic nuclide production in meteoroids and which all to calculate production rates with an accuracy of better than 10 % and production rate ratios better than 3 %.[12] More than 30 years after the formulation of the problem in 1961[154] experiments at SATURNE II provided the solution. The model calculations cover now the cosmogenic nuclides ^{10}Be, ^{14}C, ^{26}Al, ^{36}Cl, ^{41}Ca, ^{53}Mn, ^{60}Fe, ^{129}I as well as He, Ne, Ar, Kr, and Xe isotopes.[12,72,104,105,129,130]

From the analysis of the observed concentrations of cosmogenic nuclides in extraterrestrial matter, the model allows to determine long-term averaged spectra of SCR and GCR and to derive cosmic ray exposure ages, constraints on the preatmospheric size of the meteoroids and the shielding-depth of the samples in them and to distinguish multiple impacts and complex exposure histories, terrestrial ages are determined and the problem of pairing of different meteorites.

Finally, experiments at SATURNE II provided a *scientific and technical basis for future space missions.* If the detectors and electronic devices tested collect their data and give information originating from far beyond the orbit of Saturne it hopefully will be gratefully remembered by the scientific community and the public that SATURNE significantly contributed to these achievements.

Acknowledgments

The experiments LNS 167, LNS 172, LNS 232 and LNS 275 performed by the author and his collaborators were partially funded by the Deutsche Forschungsgemeinschaft, the Schweizer Nationalfond and by the EC in the framework of the Human Capital and Mobility Program. The generosity and assistance of the directorate of LNS and of the LNS staff is gratefully acknowledged. The author is grateful to the spokesmen of other experiments for providing helpful information.

References

1. J.N. Goswami *et al.*, *J. Geophys. Res.* A **93**, 7195 (1988).
2. M Arnould *et al.*, *Astron. Astrophys.* **321**, 452 (1997).
3. D.A. Bryant *et al.*, *Nature* **356**, 582 (1992).
4. K. Suga *et al.*, *Phys. Rev. Lett.* **27**, 1604 (1971).
5. R.G. Alsmiller *et al.*, ORNL-RSIC-35 (1972).

6. A.G.W. Cameron, in *Protostars and Planets III*, eds. E.H. Levy *et al.* (Univ. Arizona Press, Tucson, 1993), 47.
7. P.B. Price, *Space Sci. Rev.* **15**, 69 (1973)
8. D.D. Clayton, in *Meteorites and the solar system*, eds. J.F. Kerridge and M.S. Matthews, (University of Arizona Press, Tucson, 1988), 1021.
9. G.J. Wasserburg, in *Protostars and Planets II*, eds. D.C. Black and M.S. Matthews, (University Arizona Press, Tucson, 1983), 703.
10. R.E. McGuire and T.T. von Rosenvinge, *Adv. Space Res.* **4**, 117 (1984).
11. G.C. Castagnoli and D. Lal, *Radiocarbon* **22**, 133 (1980)
12. R. Michel *et al.*, *Nucl. Instr. Meth. Phys. Res.* B **113**, 434 (1996).
13. J. Geiss *et al.*, *Space Sci. Reviews* **1**, 197 (1962).
14. D. Lal, *Space Sci. Reviews* **14**, 3 (1972).
15. R.C. Reedy *et al.*, Ann. *Rev. Nucl. Part. Sci.* **33**, 505 (1983).
16. S. Vogt *et al.*, *Rev. Geophys.* **28**, 253 (1990).
17. R. Michel and B. Zanda, *Sixiémes Journées d'Études Saturne, 18 - 22 Mai 1992, Mont Sainte-Odile*, 337 (1992).
18. R.C. Finkel and M. Suter, Adv. *Anal. Chem.* **1**, 1 (1993).
19. W.V. Boynton *et al.*, *J. Geophys. Res.* **97**, 7681 (1992).
20. J.R. Arnold *et al.*, *AIP Conf. Proc.* **186**, 453 (AIP, New York, 1989).
21. J. Brückner *et al.*, *IEEE Transact. Nucl. Sci.* **38**, 209 (1991).
22. K. Johansson *et al.*, in *Conf. Proc. Vol. 59 „Nuclear Data for Science and Technology"*, eds. G. Reffo *et al.* (Società Italana di Fisica, Bologna, 1997), 1497.
23. H.H.K. Tang, in *Conf. Proc. Vol. 59 „Nuclear Data for Science and Technology"*, eds. G. Reffo *et al.* (Società Italana di Fisica, Bologna, 1997), 1492.
24. N.P. Kocherov, *Intermediate Energy Nuclear Data for Applications*, INDC(NDS)-245, IAEA, Wien (1991).
25. J.J. Schmidt, in *Nuclear Data Evaluation Methodolgy*, ed. Ch.L. Dunford (World Scientific, Singapore, 1993), 3.
26. R. Michel (1994) in *Proc. 8émes Journées Saturne, 5-6 May. 1994, Saclay, Accelerators Applied to the Nuclear Waste Problem,* eds. S. Leray and Y. Patin (LNS, Saclay, 1994, LNS/Ph/94-12), 37.
27. R. Michel, in *Proc. 7e Journées dÉtudes Saturne, Ramatuelle, 29.01. - 02.02.1996* (CEA/IN2P3, 1996), 432.
28. H. Sauvageon, *Phys. Rev.* C **24**, 2667 (1981).
29. H. Sauvageon *et al.*, *Phys.Rev.* C **25**, 466 (1982).
30. H. Sauvageon *et al.*, *Z. Phys.* **314**, 181 (1983).
31. M. Lagarde-Simonoff *et al.*, *J. Inorg. Nucl. Chem.* **37**, 627 (1974).
32. H. Sauvageon, *Z. Phys.* A **326**, 301 (1987).
33. F. Baros and S. Regnier, *J. Physique* **45**, 855 (1984).

34. M. Barbier and S. Regnier, *J. Inorg. Nucl. Chem.* **33**, 2720 (1971).
35. S. Regnier *et al.*, *Earth Planet. Sci. Lett.* **18**, 9 (1973).
36. S. Regnier *et al.*, *Compt. Rend. Acad. Sci. Paris (Serie B)* **280**, 513 (1975).
37. S. Regnier *et al.*, *Phys. Letters* **68B**, 202 (1977).
38. S. Regnier, *Phys. Rev.* C **20**, 1517 (1979).
39. S. Regnier *et al.*, *Phys. Rev.* C **26**, 931 (1982).
40. E. Gilabert *et al.*, *Meteoritics & Planet. Sci.* **30**, 510 (1995).
41. S. Neumann *et al.*, in *Conf. Proc. Vol. 59 „Nuclear Data for Science and Technology"*, eds. G. Reffo *et al.* (Società Italana di Fisica, Bologna, 1997), 1519.
42. E. Gilabert *et al.*, *Nucl. Instr. Meth. Phys. Res.* B, to be submitted (1998).
43. R. Michel, *nouvelles de saturne,* **15**, 74 (1991).
44. Anonymous, *nouvelles de saturne,* **15**, 104 (1991).
45. R. Michel, *nouvelles de saturne,* **17**, 62 (1993).
46. R. Michel, *nouvelles de saturne,* **18**, 118 (1994).
47. R. Michel, *nouvelles de saturne,* **19**, 73 (1995).
48. R. Michel: *nouvelles de saturne,* **20**, 65 (1996).
49. R. Michel *et al.*, *Nucl. Phys.* A **441,** 617 (1985)
50. R. Michel *et al.*, *Analyst* **114**, 287 (1989).
51. R. Michel *et al.*, *Nucl. Instr. Meth. Phys. Res.* B **103**, 183 (1995).
52. R. Michel *et al.*, *Nucl. Instr. Meth. Phys. Res.* B **129**, 153 (1997).
53. Th. Schiekel *et al.*, *Nucl. Instr. Meth. Phys. Res.* B **113**, 484 (1996).
54. S. Theis *et al.*, Lunar Planetary Science **17**, 887 (1986).
55. B. Dittrich *et al.*, *Nucl. Inst. Meth. Phys. Res.* B **52**, 588 (1990).
56. B. Dittrich *et al.*, *Radiochim. Acta* **50**, 11 (1990).
57. K. Prescher *et al.*, *Nucl. Inst. Meth. Phys. Res.* B **53**, 105 (1991).
58. R. Bodemann *et al.*, Nucl. *Instr. Meth. Phys. Res.* B **52,** 9 (1993).
59. R. Michel, in *Proc. Int. Conf. Nuclear Data for Science and Technology*, ed. J.K. Dickens (ANS, Inc., La Grange Park, Illinois, 1994), 337.
60. Th. Schiekel *et al.*, in *Proc. Int. Conf. Nuclear Data for Science and Technology*, ed. J.K. Dickens (ANS, Inc., La Grange Park, Illinois, 1994), 344.
61. Th. Schiekel *et al.*, *PSI Progress Report 1993, Annex IIIA*, 50 (1994).
62. K.J. Mathews *et al.*, *Nucl.Instr. Meth. Phys. Res.* B **94**, 449 (1994).
63. Th. Schiekel *et al.*, in *Progress Report on Nuclear Data Research in the Federal Republic of Germany for the Period April, 1st, 1994 to March, 31th, 1995*, NEA/NSC/DOC(95)10, INDC(Ger)-040, Jül-3086, 31 (1995).
64. Th. Schiekel *et al.*, *Nucl. Instr. Meth. Phys. Res.* **114**, 91 (1996).
65. B. Dittrich-Hannen *et al.*, *Nucl. Instr. Meth. Phys. Res.* B **113**, 453 (1996).
66. B. Dittrich, thesis, Universität zu Köln (1990).

67. Th. Schiekel *et al.*, *Nucl. Instr. Meth. Phys. Res.* B **113**, 484 (1996) and *Meteoritics & Planetary Science* **30**, 573 (1995).
68. M. Lüpke, thesis, Universität Hannover (1993).
69. Th. Schiekel, thesis, Universität zu Köln (1995).
70. R. Michel *et al.*, *Conf. Proc. Vol. 59 „Nuclear Data for Science and Technology"*, eds. G. Reffo *et al.* (Società Italana di Fisica, Bologna, 1997), 1458.
71. V. McLane, *EXFOR Sytems Manual: Nuclear Reaction Data Exchange Manual*, BNL-NCS-6330, July 1996.
72. U. Neupert, thesis, Universität Hannover (1996).
73. F. Sudbrock *et al.*, *Conf. Proc. Vol. 59 „Nuclear Data for Science and Technology"*, eds. G. Reffo *et al.* (Società Italana di Fisica, Bologna, 1997), 1534.
74. Ch. Schnabel *et al.*, *Radiocarbon* **38**, 107 (1996) and *PSI Annual Report 1996/Annex IIIA*, 40 (1997).
75. C. Schnabel *et al.*, *Conf. Proc. Vol. 59 „Nuclear Data for Science and Technology"*, eds. G. Reffo *et al.* (Società Italana di Fisica, Bologna, 1997), 1559.
76. H. Busemann *et al.*, *Meteoritics & Plan. Sci.* **30**, 494 (1996).
77. H.-J. Lange *et al.*, *Appl. Rad. Isotop.* **46**, 93 (1994).
78. H.-J. Lange, thesis, University Hannover (1994).
79. C.D. Bowman *et al.*, Nucl. Instr. Meth. Phys. Res. A **320**, 336 (1992).
80. F. Carminati *et al.*, CERN/AT/93-47(ET) (1993)
81. R. Michel (1994), *nouvelles de saturne*, **18**, 118 (1994).
82. R. Michel *et al.*, in *Proc. 2nd Int. Conf. on Accelerator-Driven Technologies and Applications,* ed. H. Condé (Uppsala University, Gotab, Stockholm, 1997), 448.
83. M. Gloris *et al.*, in *Progress Report on Nuclear Data Research in the Federal Republic of Germany for the Period April, 1st, 1995 to March, 31th, 1996*, NEA/NSC/DOC(96) 24, INDC(Ger)-042/LN, Jül-3246, 31 (1996).
84. M. Gloris *et al.*, *Nucl. Instr. Meth. Phys. Res.* B **113**, 429 (1996).
85. M. Gloris *et al.*, in *Progress Report on Nuclear Data Research in the Federal Republic of Germany for the Period April, 1st, 1996 to March, 31th, 1997*, NEA/NSC/DOC(97) 13, INDC(Ger)-043/LN, Jül-3410, 37 (1997).
86. M. Gloris *et al.*, in *Proc. 2nd Int. Conf. on Accelerator-Driven Transmutation Technologies and Applications,* ed. H. Condé (Uppsala University, Uppsala, 1997), 549.
87. M. Gloris *et al.*, *Conf. Proc. Vol. 59 „Nuclear Data for Science and Technology"*, eds. G. Reffo *et al.* (Società Italana di Fisica, Bologna, 1997), 1468.
88. X. Blanchard et al., in *Proc. 2nd Int. Conf. on Accelerator-Driven Transmutation Technologies and Applications,* ed. H. Condé (Uppsala University, Uppsala, 1997), 543.
89. M. Gloris, thesis, Universität Hannover (1998).
90. K. Sümmerer *et al.*, Phys. *Rev.* C **42**, 2546 (1990).

91. H. Geissel *et al.*, *GSI Scientific Report 1988*, GSI-89-1, 277 (1989).
92. R. Michel and P. Nagel, *International Codes and Model Intercomparison for Intermediate Energy Activation Yields*, NSC/DOC(97)-1 (NEA/OECD, Paris, 1997).
93. R. Michel and P. Nagel, *Conf. Proc. Vol. 59 „Nuclear Data for Science and Technology"*, eds. G. Reffo *et al.* (Società Italana di Fisica, Bologna, 1997), 879.
94. S.G. Mashnik *et al.*, LA-UR-97-2905 (1997)
95. R. Michel *et al.*, *Meteoritics* **26**, 221 (1991).
96. R. Michel, in *Intermediate Energy Nuclear Data for Applications*, ed. N.P. Kocherov, INDC(NDS)-245 (IAEA, Wien, 1991), 17.
97. R. Michel *et al.*, submitted to *Meteoritics & Planet. Sci.* (1998).
98. I. Leya and R. Michel, *Conf. Proc. Vol. 59 „Nuclear Data for Science and Technology"*, eds. G. Reffo *et al.* (Società Italana di Fisica, Bologna, 1997), 1463.
99. R. Michel *et al.*, *Nucl. Instr. Meth. Phys. Res.* B **16**, 61 (1986).
100. R. Michel *et al.*, *Nucl. Instr. Meth.* B **42**, 76 (1989).
101. D. Aylmer *et al.*, Jül-2130 (1987).
102. R. Michel *et al.*, *J. Radioanal. Nuclear Chem.* **169**, 13 (1993).
103. R.Michel *et al.*, Planet. Space Science **43**, 557 (1995).
104. I. Leya *et al.*, *Meteoritics & Plan. Sci.* **30**, 536 (1995).
105. I. Leya, thesis, Universität Hannover (1997).
106. W.R. Webber *et al.*, *Phys. Rev.* C **41**, 547 (1990).
107. W.R. Webber *et al.*, *Ap. J.* **476**, 766 (1997).
108. A. Lukasiak *et al.*, *Ap. J.* **488**, 454 (1997).
109. Anonymous, *nouvelles de saturne,* **17**, 117 (1993).
110. Y. Cassagnou *et al.*, *nouvelles de saturne,* **18**, 109 (1994).
111. W.R. Webber *et al.*, *Ap. J.* **348**, 611 (1990).
112. C.-X. Chen *et al.*, *Ap. J.* **479**, 504 (1997).
113. H. Vonach *et al.*, *Phys. Rev.* **C55**, 2458 (1997).
114. Y. Cassagnou *et al.*, *24th Int. Cosmic Ray Conf., Roma 1995*, **3**, 176. (1995).
115. Anonymous, *nouvelles de saturne,* **15**, 105 (1991).
116. J. Brückner, *nouvelles de saturne,* **18**, 67 (1994).
117. G. Dagge *et al.*, *Proc. Lunar Planet. Science*, **21**, 425 (1991).
118. J. Masarik and J. Brückner, *Proc. 25th Int. Cosmic Ray Conf., Durban,* **5**, 289 (1997).
119. J. Brückner and J. Masarik, *Planet. Space Sci.* **45**, 39 (1997).
120. J. Brückner *et al.*, *Lunar Planet. Sci.* **XXV** (LPI, Houston, 1994), 187.
121. J. Brückner *et al.*, *Lunar Planet. Science* **XXIII** (LPI, Houston, 1992), 169.

122. J. Brückner *et al.*, in *Proc. Int. Symp. on Capture Gamma-Ray Spectroscopy and Related Topics, Fribourg, Switzerland, Sept. 20-24* (World Scientific, London, 1993), 980.
123. U. Fabian *et al., Lunar Planet. Sci.* **XXVII** (LPI, Houston, 1996), 347.
124. P. Englert *et al.*, Nucl. Instr. Methods B **5**, 415 (1984).
125. B. Dittrich *et al.*, *Analyst* **114**, 295 (1989).
126. Le Monde, 19.6.1991.
127. P. Cloth *et al.*, Juel-2203 (1988).
128. R. Rösel, thesis, Universität zu Köln (1995).
129. I. Leya *et al.*, *Meteoritics & Planetary Science* **31**, A80 (1996).
130. I. Leya and R. Michel, *Meteoritics & Planetary Science* **32**, A78 (1997).
131. E. Gilabert *et al.*, *Meteoritics & Plan. Sci.* **31**, A50 (1996).
132. E. Gilabert *et al.*, *Meteoritics & Plan. Sci.* **32**, A47 (1997).
133. I. Leya *et al.*, to be submitted to *Meteoritics & Plan. Sci. (1998).*
134. J. Brückner, *nouvelles de saturne,* **18**, 60 (1994).
135. J. Brückner, *nouvelles de saturne,* **20**, 51 (1996).
136. M. Koenen *et al.*, *IEEE Transact. Nucl. Sci.* **42**, 653 (1995).
137. M. Koenen *et al.*, *IEEE Transact. Nucl. Sci.* **43**, 1570 (1996).
138. M. Koenen *et al.*, in *Conf. Record of the 1996 IEEC Nuclear Science Symposium, Anaheim, CA*, 882 (1996).
139. Anonymous, *nouvelles de saturne,* **15**, 104 (1991).
140. Anonymous, *nouvelles de saturne,* **17**, 111 (1993).
141. Anonymous, *nouvelles de saturne,* **18**, 111 (1994).
142. J.A. Ruiz *et al.*, *Ap. J. Suppl.* **92**, 683 (1994).
143. E. Porras *et al.*, *Nucl. Instr. Meth. Phys. Res.* B **95**, 344 (1995).
144. E. Porras *et al.*, Nucl. *Instr. Meth. Phys. Res.* B **111**, 315 (1996).
145. J.L. Ferrero *et al.*, *Astron. & Astrophys. Suppl. Ser.* **120**, 1 (1996).
146. J.L. Ferrero *et al.*, in *Proc. 4th Compton Symp.* eds. C.D. Dermer *et al.* (AIP, 1997), 1647.
147. A.J. Dean *et al.*, *INTEGRAL IMAGER EID PART B*, Proposal to ESA for the INTEGRAL M2 Mission, November 1994.
148. Anonymous, *nouvelles de saturne,* **18**, 124 (1994).
149. C. Arrighi *et al., nouvelles de saturne,* **17**, 113 (1993).
150. C. Arrighi *et al.*, *nouvelles de saturne,* **18**, 113 (1994).
151. Anonymous, *nouvelles de saturne,* **19**, 69 (1995).
152. Anonymous, *nouvelles de saturne,* **20**, 60 (1996).
153. J. Buisson, *Flux* **145**, 25 (1992).
154. J.R. Arnold *et al.*, *J. Geophys. Res.* **66**, 3519 (1961).

STUDY OF SPALLATION FOR TRANSMUTATION

S. LERAY
DSM/DAPNIA/SPhN,
CEA/Saclay,
F91191 Gif sur Yvette CEDEX, FRANCE

Abstract: SATURNE 2 was obviously the ideal tool to study the possibility of producing intense neutron flux in spallation sources. Although spallation reactions have been studied for many years, there has been a renewed interest in the last years because of applications of spallation sources to the transmutation of nuclear waste or energy production. New programs were devoted to the study of neutron and residual nuclides in spallation reactions.

1 Introduction

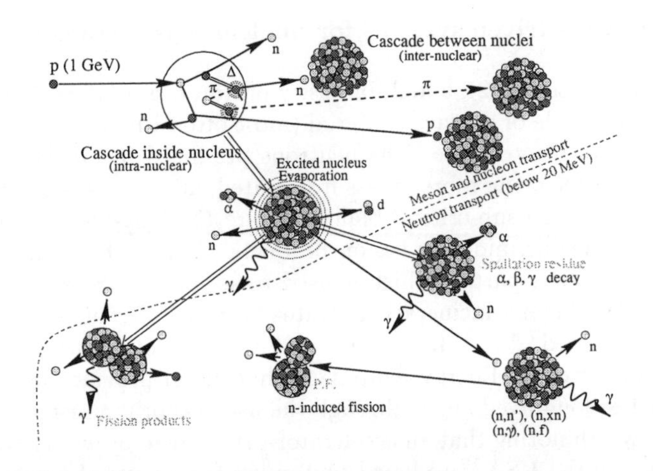

Figure 1: *Nuclear processes in a spallation target.*

Spallation is a nuclear reaction, involving a high energy - around one GeV - light particle and a nucleus, that leads to the emission of a large number of nucleons, mainly neutrons. This process has been known for many years [1] and studied from the beginning of SATURNE 2 mainly because this reaction mechanism is important to understand the cosmogenesis of nuclei in astrophysics (see the paper by R.Michel in this conference for studies at SATURNE in this field). Because some of the nucleons ejected during spallation reactions have sufficient energy to induce other spallation reactions with neighboring nuclei inside a thick target (see fig.1), it is possible to produce intense neutron flux with proton beams. With the progress achievable in high intensity accelerators (see the paper by J.M.Lagniel at this conference), it is now conceivable to design spallation sources that could compete with fission reactors in terms of neutron flux, while being more flexible. This has lead recently to a revival of interest for spallation physics.

Intense neutron flux are interesting for many applications such as solid state and material physics, for which several facilities are planned or under construction (SINQ at PSI [2], SNS [3] in the US, ESS project [4] in Europe ...). They could also be used to produce radioactive ion beams, isotopes for medicine or tritium for military purposes (APT project in the US [5] and TRISPAL [6] in France). However, the application that has actually been the dominant driving force for the recent revival of spallation physics is the transmutation of long lived nuclear waste in accelerator-driven systems [7,8,9,10,11].

2 Accelerator-driven systems for nuclear waste transmutation

In such systems (see fig.2), a high-intensity proton beam (10 to 250 mA) bombards a target made of a heavy material (most often Pb or Pb-Bi) surrounded by a sub-critical reactor core. The neutrons produced by spallation reactions in the target, after being more or less moderated, allow the maintenance of the chain reaction in the sub-critical blanket where the long-lived nuclear waste - plutonium, minor actinides or fission products - can be burned. The system generates electricity, part of which is used to supply the accelerator. Various schemes with different incineration strategies and different technical choices have been proposed [7,8,9,10]. In the proposal from Rubbia [11], accelerator-driven systems are viewed as Energy Amplifiers that could provide electricity while generating less long lived nuclei through the use of thorium instead of uranium cycle. It is worth noting that an accelerator-driven reactor was already studied around 1954 in the USA [12] to breed plutonium from natural uranium.

The main advantages of accelerator-driven systems are safety and versatility. The operation of the blanket in a sub-critical state is obviously a major safety advantage since it allows not to care about (small) reactivity insertions. Furthermore, operation well below criticallity would also permit the use of exotic fuels like large amount of Pu or minor actinides which is difficult in classical reactors because of the negative impact on reactivity parameters of these isotopes. Also, accelerator-driven systems are more flexible than reactors since the intensity of the accelerator can be adjusted to counteract the reactivity loss due to the growth of poisonous isotopes or to adapt to the elements to be transmuted. The more favourable neutron economy due to the extra

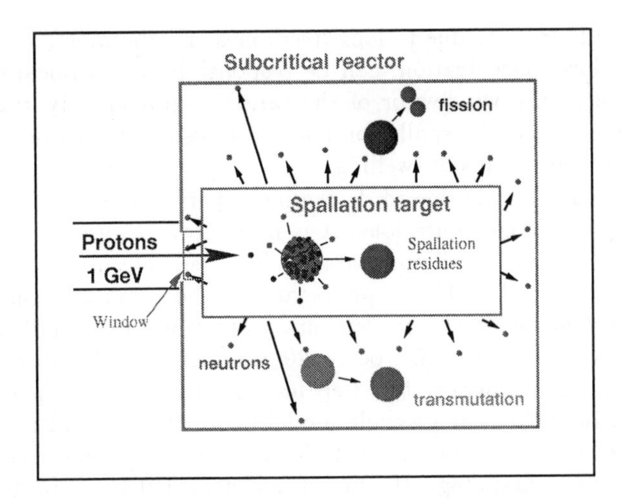

Figure 2: *Scheme of an accelerator driven system for nuclear waste transmutation.*

neutrons provided by the accelerator should allow the transmutation of significant amount of several long-lived isotopes, minor actinides or fission products. The main drawbacks of accelerator-driven systems are their complexity and the technological progress they imply concerning high intensity accelerators, the use of uncommon coolant as lead or molten-salt, the problems of radiation damage. Also the interface between the accelerator and the reactor which will be a thin window separating the accelerator vacuum and the target might be a source of concern for safety.

3 Spallation data needed for accelerator-driven systems

In the design of accelerator-driven systems, not only the spallation target has to be optimised in order to get the largest possible neutron flux with the desired spatial distribution, but it is also necessary to predict all the effects of the spallation reactions inside the target and in the surrounding media. For instance, it is important to know the energy spectrum and angular distribution of the high energy neutrons escaping from the target as they are essential for estimating radiation damage in structural materials and calculating required shieldings.

The isotopic distribution of the residual nuclides produced in the spallation target has also to be precisely predicted since some of the generated isotopes are radioactive and can induce problems of maintenance due to short lived nuclei or of radiotoxicity due to long lived ones. Furthermore, some elements, even in very low concentration, can be responsible for chemical corrosion or embrittlement of the window or of the target container. Hydrogen and helium, largely produced by spallation reactions, also matters since gas inside a material lead to problems of swelling.

Because it is not possible to determine experimentally all relevant quantities in all possible configurations, it is mandatory to use calculation codes to design and optimise spallation targets. The generally used codes are MonteCarlo transport codes able to propagate all the generated particles inside a complex system, in which the elementary reactions are described, above 20 MeV, by physics models, while, below 20 MeV, evaluated data files are read. Physics models are generally two-step models: the first step is described by the so-called Intra-Nuclear Cascade model in which the incident particle undergoes successive hard collisions with the individual nucleons inside the target nucleus. After this first stage, the target nucleus is left in an excited state and decays by evaporation or fission. Several codes containing different physics models are available. However, recent intercomparisons [13,14,15] of these codes have shown that many improvements are still needed and that there is a lack of experimental data to make a good validation, especially above 800 MeV. Namely, two types of measurement are necessary: the generation of fundamental nuclear elementary cross sections to improve basic nuclear models used in simulation codes, and thick target measurements to validate the transport part of the codes and determine the performance of various spallation targets.

SATURNE 2 was obviously the ideal place to study spallation physics: the available energy range was just suited, since in all the foreseen spallation sources the proton beam energy is ranging between 800 MeV and 1.6 GeV, and the beam intensity was large enough to perform double-differential cross-sections measurements or integral experiments. In the last years of SATURNE operation several programmes were devoted to this problematics and regarded both neutron and residual nuclide production studies.

4 Neutron studies

4.1 Neutron multiplicity measurements

- Integral thick target measurements

The most important factor in the design of a spallation target is of course the average number of neutrons produced per incident proton. Following experiments that were limited to 800 MeV at LAMPF [16], a collaboration between Bruyères-le-Châtel and Los Alamos has conducted average neutron multiplicity measurements in various targets at SATURNE between 800 MeV and 2 GeV, using the manganese bath technique. The cylindrical targets (Pb, W or a series of separated W plates) were placed inside a lead blanket surrounded by a water moderator tank containing 1% $MnSO_4$. The neutrons produced in the target are thermalized in the water and captured by the manganese to form radioactive ^{56}Mn. The manganese sulfate solution is forced to circulate to ensure mixing and its activity is measured by a high resolution germanium detector.

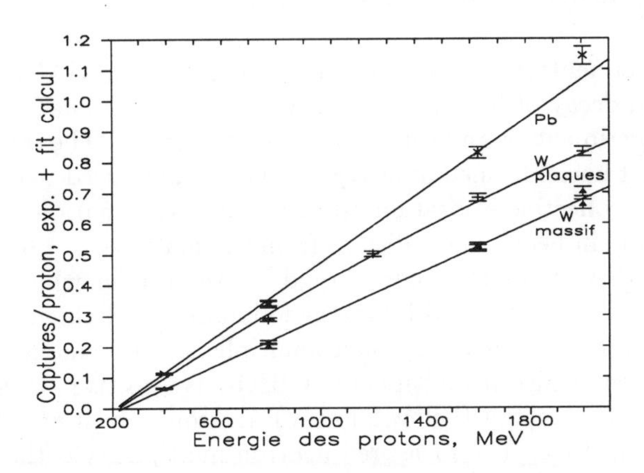

Figure 3: *Number of captures in the manganese per incident proton obtained by manganese bath technique (symbols) as a function of the beam energy for different targets compared with calculations performed with the code TIERCE from Bruyères-le-Châtel.*

Fig.3 from [17] shows the number of captures in the manganese per incident proton as a function of the beam energy for different targets ($\Phi = 25$ cm Pb, $\Phi = 15$ cm W and a series of W plates) compared with calculations performed with the high energy transport code TIERCE from Bruyères-le-Châtel [18]. The number of captures increases linearly with energy for the massive targets while a quadratic dependence better fits the W plates results. It can be noticed that in W the number of captures is much lower than in the Pb target although the numbers of neutrons produced in the tungstene and lead targets are of the same order of magnitude. This is due to the large capture cross-section of W for low energy neutrons. This can be solved by using separated plates as it can be seen in fig.3. It is also shown that the TIERCE calculations agree quite well with the experimental results. The maximum discrepancy is of the order of 8%. Similar results were obtained by Prael et al. [19] with the LAHET code system [20].

However, experiments without the lead blanket surrounding the targets were also performed. Then, the calculations overestimate the experimental results, especially at high energies and for smaller targets, by up to 24%. This could be ascribed to problems in the high energy transport code or to a bad prediction of the angular distribution of secondary particles produced in the intra-nuclear cascade model.

- **Multiplicity distributions on thin targets**

The ORION 4π neutron detector, which had been used in GANIL to study the formation and decay of hot nuclei in heavy ion collisions, was moved to SATURNE for experiments with proton and ^3He beams [21,22]. The original idea was to compare hot nuclei formed in heavy and light ion induced reactions in order to disentangle collective excitations such as spin, deformation or compression which arise only in heavy ion collisions from thermal effects. ORION is a $4\,m^3$ tank filled with gadolinium loaded liquid scintillator, in which neutrons are thermalized and then captured by the gadolinium. The light emitted from the radiative captures is detected by photomultipliers, thus enabling the measure of event-by-event neutron multiplicty. ORION is sensitive, with a large efficiency, only to low energy (below 20 meV) neutrons which mainly come from the evaporation stage. The measured neutron multiplicity is therefore related to the excitation energy of the hot nucleus after the intra-nuclear cascade.

The experiment was done using 475 MeV, 2 GeV protons and 2 GeV ^3He on Ag, Au, Bi, U targets. Results compared with an INC calculation from Cugnon [23] followed by GEMINI evaporation is shown in fig. 4. The agreement is rather good at 2 GeV but less at 475 MeV. Secondary reactions inside the tanks, evaluated to be of the order of 10 to 15 %, are not taken into account. During this experiment, also light charged particles and fission fragments were detected in coincidence with the neutron multiplicity. It was shown that the Z=2,3 particle spectra could not be explain by a pure evaporation model, suggesting that composite particle emission could arise from a pre-equilibrium mechanism. In the case of the uranium target, the fission probability was found to decrease at high neutron multiplicity i.e. large excitation energy, indicating the opening of other decay channels [22].

Following this experiment, the same collaboration GANIL-Berlin has been performing similar measurements on several facilities (CERN, COSY) to obtain a wide set of data on neutron multiplicity distributions both in thin and thick targets [24].

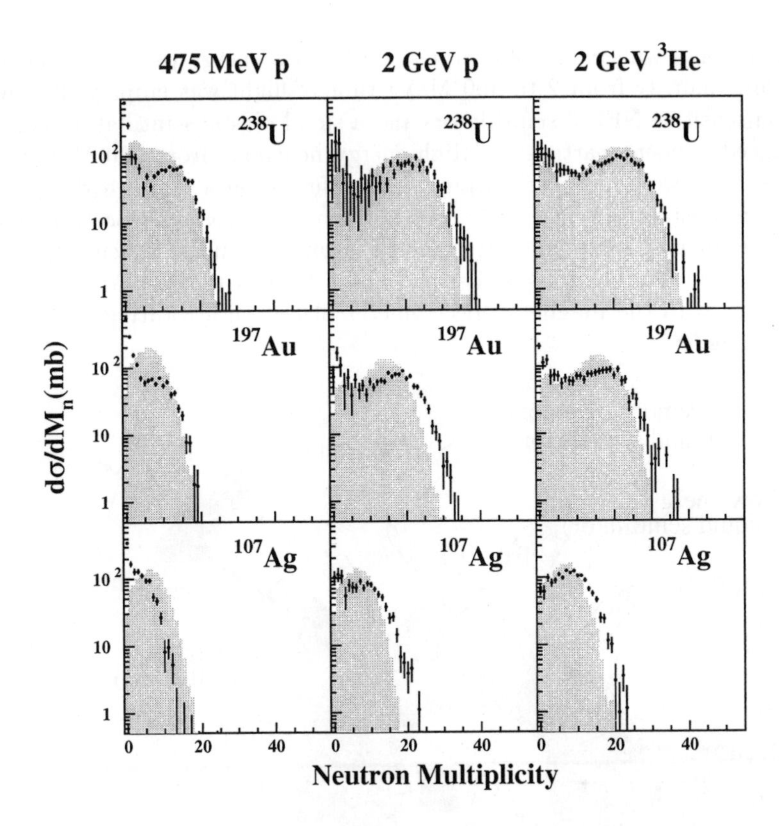

Figure 4: *Neutron multiplicity distributions obtained by the ORION collaboration. The shaded area represents the results of the calculation described in the text, folded with detection efficiency. From* [22]

4.2 Neutron energy and angular distribution measurements

The most extensive work that has been carried out at SATURNE on spallation was the so-called TRANSMUTATION programme (collaboration CEA-DSM, CEA-DAM, IN2P3). It aimed at measuring energy and angular distribution of spallation neutrons emitted in protons and deuterons induced reactions on various **thin and thick** targets. Double-differential cross-sections on thin targets are important to compare to and improve physics models while thick target measurements permits to test the transport part of the the calculation codes.

Two different experimental techniques were used in order to allow measurements on a wide range of energies: for the low energy part of the neutron spectrum (namely from 2 to 400 MeV) time-of-flight was employed between neutron-sensitive NE213-scintillators and a small plastic scintillator detecting the tagged incident particle [25]. High energy neutrons, from 200 MeV to the beam energy, were measured using (n, p) diffusion on a liquid hydrogen target and reconstruction of the proton trajectory in a magnetic spectrometer [26]. The experimental setup specially built to ensure the measurement of angular distributions from 0 to 160°, either by time-of-flight with several neutron scintillators or with the magnetic spectrometer placed on a rotating platform, is shown in fig.5 [27].

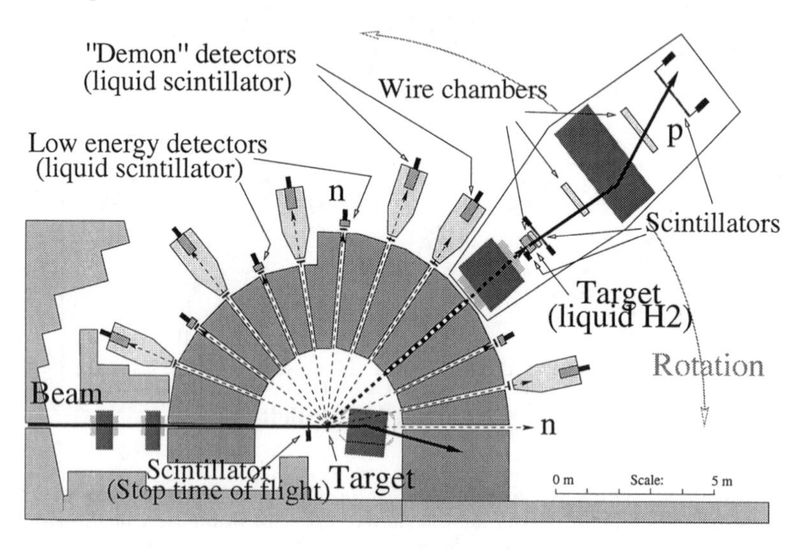

Figure 5: *Experimental setup for angular distribution measurements.*

A special attention was paid to the determination of detector efficiency which is a critical problem in neutron detection techniques. SATURNE was offering the possibility to accelerate deuterons and ^3He which could supply, by breakup reaction on a Be target, a quasi-monoenergetic neutron beam from 100 MeV to 1600 MeV. These neutron beams were used to directly determine the high energy detection system response function. This response function then served to unfold the measured proton energy spectra from the inelastic process contribution [26]. The neutron scintillator efficiency was obtained between 200 and 800 MeV from measurements with the quasi-monoenergetic neutron beam, relying on the $d + Be$ cross-section [25]. It has been completed at lower energies by measurements on the Tandem of Bruyères-le-Châtel (from 2 to 16 MeV) and in Uppsala by the associated particle method (from 22 to 120 MeV) [28]. The final efficiency curve of the DEMON detectors is shown on fig.6 and compared with calculations with O5S [30] and KSU [31] codes. The comparison between the spectrometer measurements (normalised through elastic $n-p$ cross-sections) and the Uppsala experiment allowed a new determination of the $d + Be$ cross-section [29] in the 30-120 MeV range.

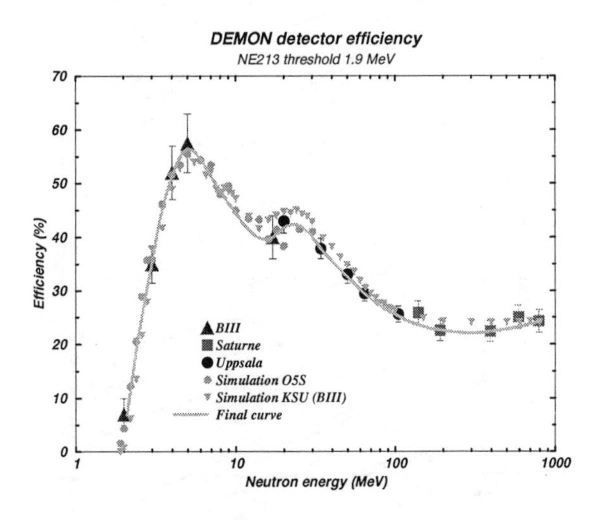

Figure 6: *Time-of-flight detector efficiency curve measured on Bruyères-le-Châtel Tandem from 2 to 17MeV, at Uppsala from 30 to 120MeV and at Saturne from 150 to 800MeV. The solid line shows the final efficiency curve used to obtain the results.*

- Thin targets

Measurements were performed with protons of 0.8, 1.2 and 1.6 GeV energy and deuterons of 1.2 et 1.6 GeV impinging on different targets. The neutrons were detected from 0° to 160° (see fig.5). The aim was to measure neutron spectra on nuclei representative of different parts of the periodic table of elements and corresponding to materials used in targets or in structures of accelerator driven systems: Al, Fe, Zr, W, Pb et Th. Deuteron, compared to proton measurements, are expected to give information on neutron induced reactions.

Fig.4.2 from X.Ledoux et al. [32] presents neutrons cross-section distributions in Pb(p,xn)X reactions at 1200 and 1600 MeV. The histograms are numerical simulations, taking into account the actual geometry and composition of the

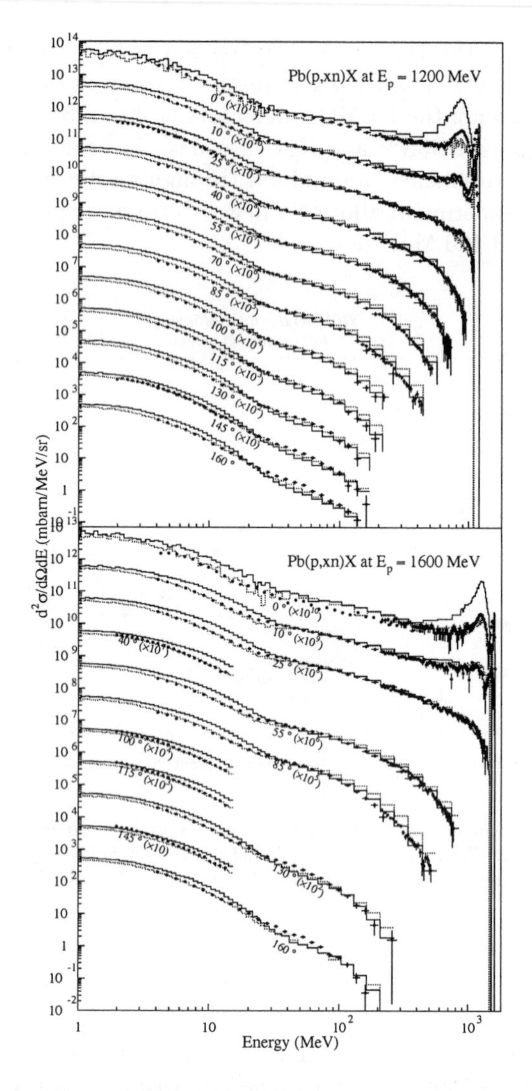

Figure 7: Production cross-sections measured on a 2 cm thick Pb target at 1200 and 1600 MeV. The histograms represent TIERCE simulations [18] using Bertini [33] (full line) or Cugnon [34] (dotted line) cascade model.

target, performed with the TIERCE [18] code system developped at Bruyères-le-Châtel within which two different intranuclear cascade codes, followed by the same evaporation model, have been used : Bertini cascade [33] (full line) and Cugnon code [23,34] (dotted line). At 0°, the Bertini cascade predicts a huge peak in the region corresponding to the Δ resonance excitation, which is not experimentally observed. This problem is actually known to be due to a too much forward peaked angular distribution of $NN \rightarrow N\Delta$ reaction in the Bertini INC code. A more realistic parametrisation of this angular distribution [35], obtained from the (n,p) measurements for determining the response function, was introduced in the code from Cugnon et al. [34] and now allows to get rid of this pathologic behaviour as it can be seen in the figure. Whatever the angle, calculations with the Bertini cascade overestimate the production cross-sections below 20 MeV. The Cugnon model generally leads to a much better agreement with the experimental results. This could be explained by the larger excitation energy left in the nucleus after the INC phase in the Bertini than in the Cugnon model, arising from a better treatment of the Pauli blocking by Cugnon.

- Thick targets

While double-differential cross-sections measurements on thin targets allow to test the validity of the physics models, thick target data provide information on the transport part of the codes.

The experimental setup, as described above, has also permitted the measurement of energy spectra of neutrons coming out from thick targets. In that case, because very few neutrons are expected to have energies higher than 400 MeV, only the time-of-flight detectors were used. A large programme of measurements with various lengths and diameters of cylindrical targets has been conducted with 0.8, 1.2 and 1.6 GeV protons and 0.8 and 1.6 GeV deuterons on lead, tungsten, iron and aluminium targets. The collimators and the possibility to longitudinally translate the target allowed to select neutrons coming from different emission zones, thus enabling to test the propagation of the cascade along the target. Because of the collimators, the target area seen, totally (full exposition zone) or partially (penumbra zone), by a detector depends on the angle of the collimator and on the target diameter and longitudinal position. This makes the conversion into neutron fluxes difficult. Therefore, data are presented as number of neutrons per incident proton, MeV and cm^2 of detector.

Fig.8 shows the results [36] obtained with a lead target of 20 cm diameter and 65 cm length at 800 MeV and 1200 MeV and 105 cm length at 1600 MeV for two different angles. It can be seen that, as expected, the number of outgoing neutrons increases with incident energy. The increase is proportionally larger for high energy back-scattered (angle 160) neutrons than for other neu-

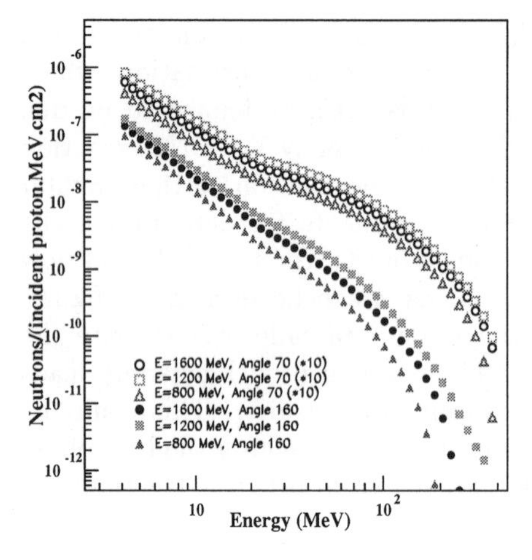

Figure 8: Results obtained on a lead target of 20 cm diameter, at longitudinal position 10 cm, at 800, 1200 and 1600 MeV for two different angles.

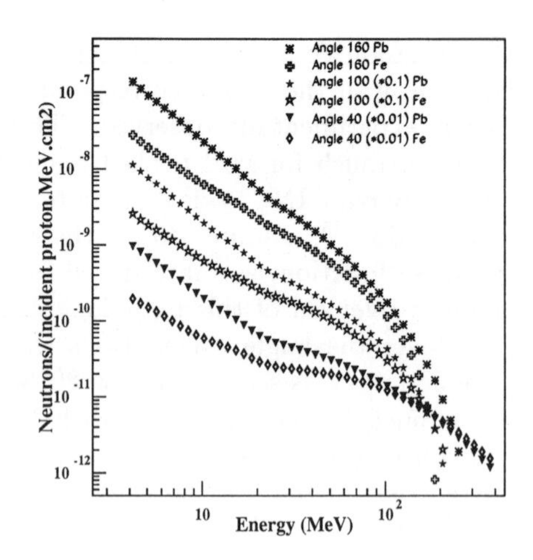

Figure 9: Comparison of results obtained with lead and iron targets of 65 cm length, 10 cm diameter, at longitudinal position 10 cm, at 1200 MeV.

trons. This indicates that, in the design of a spallation target, the choice of a high incident energy is not favorable in terms of back-scattered neutron fluxes towards the window separating the accelerator and the target. Fig.9 displays some of the comparison of results obtained with lead and iron cylinders of 20 cm diameter and 65 cm length, at longitudinal position 10 cm, for 1200 MeV and for different angles. While the number of low energy neutrons is much lower with iron than lead, it can be observed that the difference disappears at high energy. This is mainly due to the fact that the mean free path of high energy neutrons is the same in lead and iron. Comparisons with different transport codes are still under progress.

5 Residual nuclide studies

5.1 Thin target measurements

From the beginning of operation of SATURNE 2, a lot of measurements of residual nuclide production by radiochemical techniques have been performed. Most of this work, extensively described in the paper by R.Michel in this conference, was devoted to reactions relevant for nuclear cosmogenesis understanding and therefore limited to target nuclei lighter than gold. Data obtained on elements entering the composition of structure materials, like Fe or Ni, are of direct interest for spallation source design. Sümmerer et al.[37] studied the fragmentation of Au and Th targets at 2.6 GeV and, including analysis of previous data, has established an empirical formula, predicting the production cross-section of isotopes near to the target, still used today. The data collected for several years, at SATURNE and on other facilities by Michel et al.[38] has led to the determination of a large amount of excitation functions from a few tens of MeV to several GeV, which are important to test the energy dependence of isotope production in models. This data served as a basis of the international code intercomparison organised by the OECD/NEA[15].

More recently the same group in collaboration with Bruyères-le-Châtel[39] has extended their studies to heavier elements directly relevant for ADS as target or blanket materials. An example of the results obtained on a lead target at 211 Mev is shown on fig.10 with comparison with to different intranuclear cascade codes. As already pointed out in ref. [15], big discrepancies between model predictions and experimental data remains.

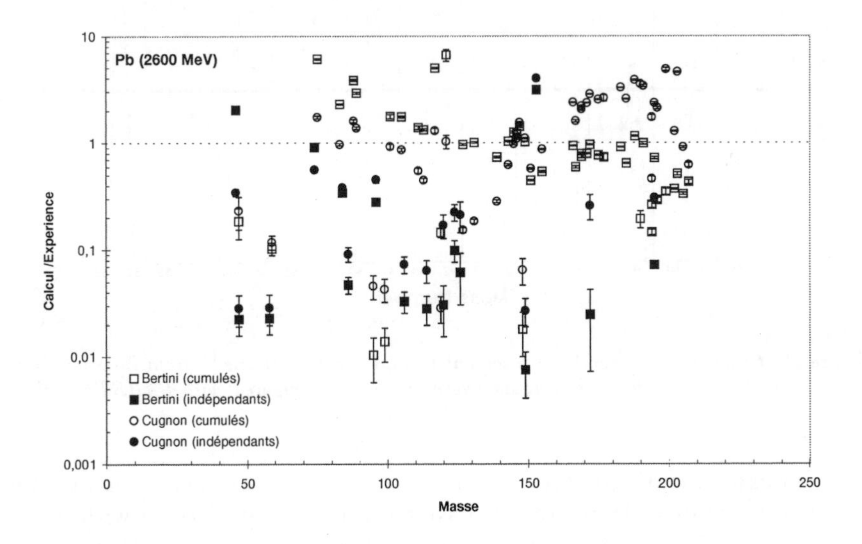

Figure 10: *Results obtained on a Pb target at 211 MeV. From*[40]

Experiments using radiochemical techniques suffers from the fact that the isotopes are identified after a radioactive decay chain, making the interpretation of the results rather difficult. A more direct measurement is possible using the heavy nucleus as a projectile on a hydrogen target in so-called reverse kinematic experiments in which the various produced isotopes are emitted in the forward direction with a velocity close to that of the beam and can thus be directly identified with detectors. This technique was used at SATURNE by Weber et al. [41] with light heavy ion beams. The isotopic cross-sections obtained using a ^{56}Fe beam have been compared in fig.11 with a calculation using the Bertini intra-nuclear cascade model which, contrary to what was above mentioned for Pb, gives a reasonable agreement with the data. These results are presently being used as input to time evolution code in order to evaluate embrittlement of an iron window in an ADS [42].

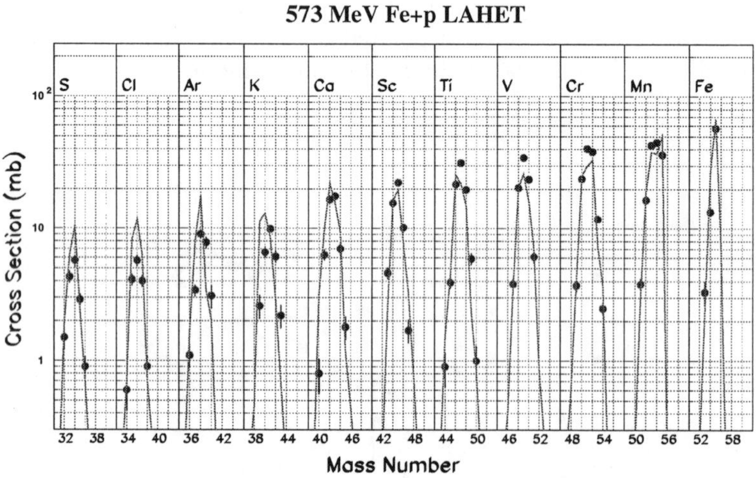

573 MeV Fe+p LAHET

Figure 11: *Experimental (symbols) fragment isotopic cross-sections* [41] *from interaction of a* ^{56}Fe *(573 MeV/A) beam with a liquid hydrogen target, compared with LAHET* [20] *calculations. From* [42]

New experiments using reverse kinematics and the fragment separator FRS at SIS accelerator in Darmstadt [43] have recently been performed with Au, Pb and U beams. They will allow the precise (N,Z) determination of all residues ranging from fission fragments to beam isotopes. The data measured at SATURNE will be necessary to get absolute values for the fission fragment isotopic production cross-sections because of the very low transmission of the spectrometer for these residues.

5.2 *Thick target measurements*

A group from Bruyères-le-Châtel [39] has also conducted at SATURNE thick target experiments, in which the production of the residual nuclides is studied by analyzing activation products by γ-spectrometry in thin foils placed at different longitudinal and radial positions inside a lead cylinder. These data compared to elementary cross-sections obtained on thin targets and to neutron production on thick targets will lead to further tests of the transport codes. The analysis is still under progress [40].

6 Conclusion

Although there has been a continous study of spallation reactions from the beginning of SATURNE 2, most of the work in this field was carried out during the last years of operation because of the new interest for applications of spallation to nuclear waste transmutation in accelerator driven systems. SATURNE was the one place where all the conditions were fulfilled to achieve important progress in the understanding of spallation mechanism and improve and validate calculations codes used in the design of spallation target: the accelerator energy range, the beam intensity high enough to perform double-differential or integral measurements in a reasonable time, the possibility to accelerate not only protons but also deuterons, to have access to neutron induced reactions, and light heavy ions, allowing reverse kinematics experiments, and finally the available experimental areas and efficient detection systems. The extensive program devoted to spallation neutron studies, covering multiplicity and energy and angular distribution measurements with both thin and thick targets, has led to a comprehensive set of data which extends the available information previously limited to 800 MeV, and, after the final completion of the analysis and comparison with models will undoubtedly lead to improved simulation codes for accelerator driven system design.

Acknowledgements

I am grateful to J.Fréhaut, X.Ledoux, R.Legrain, M.Ducauze and C.Volant for providing me with some of their results.

References

1. R.Serber, Phys. Rev. 72 (1947) 1114.
2. G.S.Bauer, IAEA, Vienna, TECDOC-836 (1995) p97.
3. Review of the Spallation Neutron Source (SNS), DOE/ER-0705 (June 1997).
4. G.S.Bauer, 2nd Int. Conf. on Accelerator Driven Transmutation Technologies, Kalmar, Sweden, June 3-7, (1996) 159.
5. J.C.Browne *et al.*, 2nd Int. Conf. on Accelerator Driven Transmutation Technologies, Kalmar, Sweden, June 3-7, (1996) 101.
6. J.L. Flament, 2nd Int. Workshop on Spallation Materials Technology, Acona, Italy, Sept. 19-22, 1997.
7. H.Takahashi and H.Rief, Proceedings of Specialists' Meeting on Accelerator-Based Transmutation, PSI Zurich, March 1992.
8. C. D. Bowman *et al.*, *Nucl. Instrum. Methods* A **320**, 336 (1992).
9. T.Takizuka, Proceedings of the Int. Conf. on Accelerator-Driven Transmutation Technologies and Applications, Las Vegas, NV, USA, August 1994, p64.
10. M. Salvatores et al., GLOBAL'93, Int. Conf. on Future Nuclear Systems, Seattle, Washington, USA, September 93.
11. C. Rubbia *et al.*, preprint CERN/AT/95-44(ET) (1995).
12. Status of the MTA process, California Research and Development Company, report LRL-102 (1954); C.M.Van Atta, A brief history of the MTA project, report UCRL-79151 (1977).
13. M. Blann *et al.*, *International Code Comparison for Intermediate Energy Nuclear Data*, OECD/NEA (1994).
14. D. Filges *et al.*, *OECD thick target benchmark*, Report NSC/DOC(95)2 (1995).
15. R. Michel and P. Nagel, *International Codes and Model Intercomparison for Intermediate Energy Activation Yields*, OECD/NEA (1997).
16. G. Morgan et al., Int. Conf. on Accelerator Driven Transmutation Technologies, Las Vegas NV, USA, July 1994, p682.
17. J. Fréhaut *et al.*, Rapport CEA-R-58/09.
18. O. Bersillon, *2nd Int. Conf. on Accelerator Driven Transmutation Technologies*, Kalmar, Suède, 3-7 Juin (1996).
19. R.E. Prael and M. Chadwick, Conf. Proceedings Vol. 59, Nuclear Data for Science and Technology, G.Reffo, A.Ventura and C.Grandi (Eds) SIF, Bologna (1997) 1449.
20. R.E. Prael and H. Liechtenstein, $LAHET^{TM}$ Code System, Report LA-UR-89-3014 Los Alamos National Laboratory (1989).

21. L. Pienkowski et al., *Phys. Lett.* B **336**, 147 (1994).
22. X. Ledoux et al., *Phys. Rev.* C **57**, 2375 (1998).
23. J.Cugnon, *Nucl. Phys.* A **462**, 751 (1987).
24. L. Pienkowski et al., Phys. Rev. **C56** (1997) 1909; D.Hilscher et al., *Nucl. Instrum. Methods* A **414**, 100 (1998).
25. F. Borne *et al.*, *Nucl. Instrum. Methods* A **385**, 339 (1997).
26. E. Martinez *et al.*, *Nucl. Instrum. Methods* A **385**, 345 (1997).
27. S. Leray *et al.*, Conf. Proceedings Vol. 59, Nuclear Data for Science and Technology, G.Reffo, A.Ventura and C.Grandi (Eds) SIF, Bologna (1997) 1426.
28. J. Thun *et al.*, in preparation.
29. J.F. Lecolley *et al.*, to appear in Eur. Phys. Journ.
30. R.E.Textor and V.V.Verbinski, report ORNL-4160 (1968).
31. R. A. Cecil, B. D. Anderson and R. Madey, Nucl. Instr. and Meth. **161**, 439 (1979).
32. X.Ledoux *et al.*, submitted to Phys. Rev. Lett.
33. H. W. Bertini *et al.*, *Phys. Rev.* **131**, 1801 (1963).
34. J.Cugnon, C.Volant and S.Vuillier, *Nucl. Phys.* A **620**, 475 (1997).
35. J.Cugnon *et al.*, PRC **56**, 2431 (1997).
36. S.Ménard, PhD Thesis, Orsay (1998).
37. K.Sümmerer *et al.*, *Phys. Rev.* **C42**, 2546 (1990).
38. R. Michel *et al.*, *Nucl. Instr. and Meth.* **B113** (1996) 429; *Nucl. Instr. and Meth.* **B113** (1996) 439.
39. P.Morel *et al.*, 2nd *Int. Conf. on Accelerator Driven Transmutation Technologies*, Kalmar, Sweden, June 3-7, 1996.
40. M. Ducauze, PhD Thesis, in progress.
41. W. R. Weber *et al.*, Ap. J. 508 (1998) 940; Ap. J. 508 (1998) 949.
42. R. Legrain and C.Volant, private communication.
43. F. Farget *et al.*, *AIP Conf. Proc. 447, Nuclear Fission and Fission Product Spectroscopy:* 2nd *Int. Workshop*, Seyssins, France, 1998, p11.

SATURNALIA

COLIN WILKIN
University College London, Gower St., London WC1E 6BT, UK

A personal choice is made of the highlights of the physics programme carried out at the Saturne-2 facility over the last twenty years.

It is the end, the end of the meeting and the end of Saturne, and the kind organisers have asked me to say a few words about both events. We have been gathered here in Paris for the last two days to celebrate the work of a whole collection of people and apparatus grouped around an accelerator Saturne-2, which closed down in December. I will not give a political speech but there is an important point which has to be made. The NSF facility at Daresbury was shut in the 1980's, at the height of its productivity, by a committee of physicists who were faced with a budgetary crisis caused by the need to support particle physics at a time of a rising Swiss Franc. When nuclear physicists queried the action, the administration replied that it was a decision taken on physics grounds, which is not the same as a decision taken physicists under pressure! As a result of this deliberate confusion, which was fed to the media, not only had the nuclear physicists lost their machine, but they felt devalued and insulted into the bargain. Do not let anyone tell you that Saturne could not have continued because there were not interesting problems still to be studied or that Saturne was not competitive for their study — both would be false.

The flexibility of the accelerator, with its wide choice of particle types and energy, and ease of energy changes, combined with an unrivalled collection of spectrometers and polarimiters, made it a splendid tool for unravelling problems in diverse branches of physics. Nevertheless the facility was closed. Let us agree simply to say that the authorities had different priorities to us.

Each of the excellent speakers has had to make a selection from the wealth of material in the subject area of his talk and, in my half hour, I will try to pick out some of the personal highlights of the whole programme of the Laboratory and show how, in many cases, these have influenced developments at other facilities. Though Saturne may be dead, its genes live on!

Before starting this though, let me make one more political point. It is much easier to get finance for doing new experiments at other facilities, than completing the analysis of data taken at Saturne. For the physicists involved it is also more appealing to throw themselves wholeheartedly into the new proposals. However we all know of many experiments, where there are still data on tape representing interesting phenomena, but which risk being left just gathering dust in cupboards. I will not embarass people by giving examples but I appeal to physicists and their laboratory directors to spend some time and money to extract the maximum from the data taken at Saturne.

It is exactly 30 years since I gave my first talk at Saclay, and in May '68 people had other things on their minds than the application of the Glauber model to the scattering of 1 GeV protons from carbon and oxygen, which was the subject of the talk. Though the Saturne accelerator was completely rebuilt 20 years ago, there has been a continuity in its physics programme and in some of the equipment so that I will stray across the 1977 frontier. Not being a "machine-man", there is of course a danger that I would not stress sufficiently the role of equipment in my valediction. To avoid this danger, I commend to you the excellent "vulgarisation" of the work carried out at Saturne [1]. Copies of the abridged English translation were given to all participants, although they will be disappointed not to find a description of the bonnes et mauvaises années there!

One should also not forget the Proceedings of the Journées d'Études Saturne, edited over many years by Pierre Radvanyi and Mme. Bordry. These volumes are mines of useful information about Saturne, its equipment and its physics and this is a true reflection of the meetings themselves, which were very important in developing the unique culture of Saturne. As an example, my first ideas on the possibility of quasi-bound $\eta\,^3$He states came to me at the Mont Ste. Odile meeting, and the first calculations were carried out on the train back from Strasbourg to Paris. Other people might remember rather the banquets and the difference between Roscoff and Cavalaire!

Elastic scattering proton-nucleus scattering and the excitation of nuclear levels was the first place where SPESI made an impact on the international stage. As discussed here by Vorobyov, this programme was an extension of the earlier Cosmotron work of Palevsky and friends. Vorobyov's talk reminded me of the calculations of my youth, but I had truly forgotten that the experimental programme of the Gatchina-Saclay colloaboration was so vast. Nevertheless it was a programme of classical nuclear physics where the only degrees of freedom were those of the nucleon. Within the Glauber model, they obtained matter densities in many nuclei, and also transition densities, without inserting specific forms for the nuclear densities. Thus they were able to look at changes in density from one isobar to another.

Since Pierre Radvanyi and others have mentioned the influence of Palevsky on the development of Saturne-1, allow me to add a couple of remarks to the history since I worked with Harry in Brookhaven in 1966-67. Saturne-1 was a clone of the Cosmotron, and Harry never understood why the French insisted on taking the original plans for the machine rather than introducing the improvements suggested by the American accelerator theorists who designed the Cosmotron. Secondly, even though Palevsky is now recognised for his work on elastic scattering, his real passion was for the comparison with the inclusive spectra where he hoped to derive nucleon-nucleon correlations from sum rules.

The first really innovative material was an investigation of intermediate energy pick-up reactions, where SPESI could easily measure the angular distributions of $^{12}\mathrm{C}(p,d)\,^{11}\mathrm{C}^*$ to half a dozen excited states of $^{11}\mathrm{C}$ at 700 MeV [2]. It is fitting that Jacques Thirion presented this at the same conference as Hoistad showed data on the (p,π^+) reaction, since the physics is closely related. Unlike elastic proton scattering at small angles, these large momentum transfer reactions are intimately related to mesonic or Δ degrees of freedom in the nucleus. This importance of virtual mesons or nuclear isobars, even for reactions where no mesons exist in either the initial or final states, is something that we had to bear in mind in many later experiments at Saturne-2. As a simple rule of thumb, rare events in hadronic reactions tend not to be as rare as would be indicated on the basis of purely nucleonic degrees of freedom.

Personally my most exciting moment at LNS was the night that we measured η production in the $\vec{d}p \to {}^3\mathrm{He}\,\eta$ reaction near threshold at SPESIV [3], though it had been studied previously at higher energies [4]. This was also an important moment for the laboratory as well since in this, and subsequent experiments at SPESII [5]:

- It showed a surprisingly high cross section, which is probably due to a virtual pion beam being created on one of the nucleons in the deuteron, with the pion producing an η-meson on the second nucleon. Such ideas are now used in the interpretation of threshold meson production in ion-ion collisions.

- The variation of cross section near threshold implies the existence of a quasibound $\eta\,^3\mathrm{He}$ state, a new form of nuclear matter at high excitation energy, whose existence was clearly confirmed for the $\eta\,^4\mathrm{He}$ system [6,7].

- The latter provided a natural explanation for the charge symmetry violation signal in the $dd \to {}^4\mathrm{He}\,\pi^0$ signal reported by the unforgettable ER54 group just below the η threshold [8].

- It allowed the most precise measurement of the mass of the η meson [9].

- It opened the possibility of studying the rare decays of the η with a tagged beam of such mesons [10]. Important results were found on the decays $\eta \to \gamma\gamma$ and $\eta \to \mu^+\mu^-$ and it is a matter of profound regret that the SPESII programme of rare decay studies had to be cut short for practical reasons. What a catastrophy.

- It lead to the study of heavier continuum states in $pd \to {}^3\mathrm{He}\,X$ under threshold conditions [11], as well as the isolation of heavier mesons X such as the ω [12], η' and ϕ [13].
- It also drove people to investigate threshold production of η [14,15], ω [16], η' [17] and ϕ [18] in nucleon-nucleon scattering using a variety of experimental techniques.

Almost all of these themes, born out of one passionate night of endeavour, will be carried on at other experimental facilities, although the diehard might well say with some justification that we could have done it still better at Saturne! Thus there is a big rare decay programme at Uppsala and searches for quasibound η-nuclear states at Jülich (and during the meeting Paul Kienle explained to me that there were similar proposals at GSI). Both laboratories will carry out production studies of the $\eta\,{}^3\mathrm{He}$ system and search for charge symmetry violation in $dd \to {}^4\mathrm{He}\,\pi^0$. I have even heard it said that by 2003 François Plouin will have finished his analysis of the mass of the η meson, but I have no confirmation of this.

Saturne was internationally recognised for the intensity and quality of its beams of polarised protons and deuterons [1], which was due to the combination of the Hyperion polarised ion source, the mini-synchrotron Mimas serving as injector, and a meticulous study of depolarising resonances in Saturne itself by the "machinistes". I was intitially confused to find that the "Group Théorie" at Saturne was composed purely of people who studied how the machine worked, whereas there were almost no nuclear/particle theorists in the laboratory. This showed the priorities of the founders of the laboratory and all other machines with which I am familiar have far less backing in this area. Though the successes in Indiana have been outstanding, it has been found that polarised particles are difficult to accelerate at both Uppsala and Jülich. In his talk Lagniel highlighted the collaborations between machinistes and experimentalists, especially those of nucleon-nucleon, and showed how both sides benefitted from this. It wasn't always war!

The major user of polarised protons was the nucleon-nucleon group and often in the Comité des Expériences we exploited the fact that there was always a colleague who was willing to take beam time during any holiday period. It might be a coincidence but the moment that they finished their programme on proton-proton and proton-neutron elastic scattering, the Authorities closed the Laboratory down. The nucleon-nucleon studies represented perhaps the most fundamental of the research that was carried out at Saturne. This was done in a completely professional manner and in many cases the group had sufficient data in order to make direct reconstructions of the amplitudes without passing through theoretical models.

Where Saturne was really unique was in respect of its deuteron beams, which combined high intensity with large values of both vector and tensor polarisation. Without Saturne we could not have calibrated the AHEAD [19] or POLDER [20] deuteron tensor polarimeters which, as Garçon described in his talk, played vital roles in separating the deuteron form factors through measurements of the polarisation of the recoil deuteron in elastic electron-deuteron scattering. The Comité des Expériences and Management of Saturne always took the broad view that the advancement of Physics was the central theme of the Laboratory and, if this involved calibrating equipment for good experiments to be done elsewhere, then this was a price worth paying. It certainly paid off in the case of the deuteron form factor. Michel Garçon has shown us the first preliminary results from CEBAF on the value of T_{20} in elastic $ed \rightarrow e'\vec{d}'$. The collaboration could work to values of q^2 twice as big as earlier experiments with a much bigger efficiency than all competing techniques.

The POLDER polarimeter is based upon the deuteron charge exchange reaction $\vec{d}p \rightarrow \{pp\}n$, which was predicted to show a strong analysing power signal if the final proton-proton pair had small excitation energy and was produced with a small momentum transfer relative to the incident deuteron [21]. This reaction was investigated in the few hundred MeV range [22] but also in the GeV range through the remarkable technique of detecting both protons in the focal plane of the SPESIV spectrometer [23]. This was one of a whole series of charge-exchange experiments with different probes undertaken over years by the Copenhagen-Orsay group, and it is a matter of profound regret that the group leader, Carl Gaarde, was taken from us a few weeks before this meeting. Michelle Roy-Stéphan has reminded us of the enormous quantity of results, especially in the (^{3}He,t) reaction, achieved over 14 years and the successes in their interpretation has come through active collaborations with theorists. One of these, Madeleine Soyeur, showed how such results could be exciting for theorists and I believe that she communicated well this enthusiasm to the machinistes in the hall, even if they couldn't appreciate some of the finer points.

The method of detecting two protons in SPESIV could be extended to looking also at deep inelastic scattering, where a Δ-isobar had been produced on hydrogen or nuclear targets [24]. The $(\vec{d}, 2p)$ reaction is an interesting probe for $\Delta T = \Delta S = 1$ transitions but, inspired by Marcel Morlet, the Orsay group showed that measurements of vector spin transfer in $(\vec{d}, \vec{d}')$ can provide useful signals for $\Delta T = 0$, $\Delta S = 1$ states in the residual nucleus [25]. This needed a lot of intuition because the symmetry used here is only an approximate one and a complete separation would have required the combination of a tensor polarised beam and tensor polarimeter.

Of equally lasting importance was the whole series of measurements designed to investigate the few body problem in intermediate energy nuclear physics. The measurement of triple spin parameters in $\vec{d}\vec{p} \to \vec{p}d$ at 1.6 GeV [26] represented a real tour de force which is never likely to be repeated. More appealing are however the survey experiments which measure a single observable, such as the deuteron tensor analysing power T_{20}, as a function of beam energy, as was done for $\vec{d}p \to pd$ in the backward direction [27] or pion production through $\vec{d}p \to {}^3\mathrm{He}\,\pi^0$ [28,29]. Due to its D-state component, the deuteron has a geometric deformation looking in momentum space a bit like a pancake. This could be investigated by measuring the T_{20} dependence of the Fermi momentum in deuteron break up $\vec{d}p \to ppn$ [30] but the wealth of polarimeters at Saturne also permitted a polarisation measurement of a final proton [31].

The D-states in ^{6}Li could be studied in analogous break-up measurements [32] with beams of polarised ions of several GeV/c (which must be high compared to the Fermi momentum), and this seems to have been the only published experiment using such a beam. The MIMAS synchrotron was designed to furnish high quality polarised beams and also beams of ions up to ^{129}Xe (30^+) but the demand for such heavy ions has represented only about 10% of the total requests over many years, and even less of the publications. In part this might be that the competion from other facilities in the heavy ion field, such as Ganil or SIS, has proved more intense than for the light ions where Saturne was supreme. However, apart from Diogène, there were no specific heavy ion detectors at Saturne and groups tended to bring small equipment with them for multifragmentation or other studies [33]. Saturne was the great success that it was for light ions because it had a whole collection of outstanding spectrometers which were tailored for the purpose.

I have already stressed that physics must be exciting and no survey of the work at Saturne would be complete without mentioning "Le pari de Pascal" and the human desire to find something really revolutionary, even if the probability of success were infinitesimally small. If there were a good bridge player in the room, then he would tell you that you must think defensively, such that the international competitors should never be allowed to make too good a score. Thus Saturne could not allow other laboratories to make the earth-shattering discoveries which might have been made at Saclay. In other fields this maladie took the form of trying to repeat the discoveries of Pons and Fleischmann on cold fusion, but at Saturne it was the hunt for narrow dibaryon resonances [34], pion-nucleon bound states [35], abnormal production of pions [36], the anomalons [37] *etc.* It was really like something out of "Les vacances de M. Hulot". Nothing convincing was ever found which was not at the limits of systematic/statistical error bars. Colleagues will be relieved to know that this approach is being carried on in the successor laboratories and that I was equally unsuccessful in stopping the search for the d' at COSY and CELSIUS PAC's as I was with the narrow dibaryons at the Saturne Comité des Expériences.

It is true that just one such discovery might have provided some defence for Saturne and it is often these searches which bring out the best in technical ingenuity. One example of this is the famous wheel of Beurtey and Saudinos[38]. In order to make simultaneous measurements of A_y in proton-proton elastic scattering at many different energies, they constructed an energy degrader with many steps, looking something like a circular Escher staircase. By letting this rotate fast in the beam, high statistics data could be obtained at 16 different energies simultaneously. A short run was sufficient to kill off one dibaryon (but not French).

Over the years Saturne has been invaluable for the calibration of much other equipment in addition to polarimeters. It is true that Saturne will not be much remembered for such mundane work in the publications of physics carried out some time later at other laboratories. However people will remember Saturne when they find that they can no longer use it to test counters for LHC for example. Often the applied research at Saturne gave rise to interesting physics in related fields, and I have in mind here the some of the work described by Rolf Michel on pseudo-meteorites.

You will have noticed that I have not spoken at all about the three big programmes which have dominated the last one or two years of the running of Saturne-2, *viz* DISTO, SPESIVπ, and Nuclear Transmutation; it is much too early to judge their significance. We all of course hope that the information provided by the transmutation experiments will prove valuable in a search for a sensible method for dealing with nuclear waste. That really would be a worthwhile monument for Saturne to compete with the statue of the Ptolemy which was unveiled today outside the Grand Palais.

Pour terminer, je dois remercier Alain, Françoise, Simone et Bernard qui ont assuré le succès de cette réunion. Il est bien évident que le français n'est pas la langue maternelle, ni de M. Blair ni de moi. Néanmoins il va rester ma langue fraternelle et cela est d'une grande part dû aux collaborations fructueuses avec les gens autour de Saturne, beaucoup trôp nombreux à mentionner.

Saturne, nous allons tous nous souvenir de vous. Saturne, adieu! Merci!

References

1. R. Beurtey and A.-M. Gauriot, *De la Technique à la Physique des Energies Intermédiaires*, (Laboratoire National Saturne, 1987).
2. J. Thirion, in *High Energy Physics and Nuclear Structure*, ed. G. Tibell (North Holland, 1973).
3. J. Berger *et al.*, *Phys. Rev. Lett.* **61**, 919 (1988).
4. P. Berthet *et al.*, *Nucl. Phys.* A **443**, 589 (1985).
5. B. Mayer *et al.*, *Phys. Rev.* C **53**, 2068 (1996).
6. R. Frascaria *et al.*, *Phys. Rev.* C **50**, R537 (1994).
7. N. Willis *et al.*, *Phys. Lett.* B **406**, 14 (1997).
8. L. Goldzahl *et al.*, *Nucl. Phys.* A **533**, 675 (1991).
9. F. Plouin *et al.*, *Phys. Lett.* B **276**, 526 (1992).
10. R. Abegg *et al.*, *Phys. Rev.* D **50**, 92 (1994).
11. F. Plouin in *Production and Decay of Light Mesons*. ed. P. Fleury (World Scientific, Singapore, 1989).
12. R. Wurzinger *et al.*, *Phys. Rev.* C **51**, R443 (1993).
13. R. Wurzinger *et al.*, *Phys. Lett.* B **374**, 283 (1996).
14. A.M. Bergdolt *et al.*, *Phys. Rev.* D **48**, R2969 (1993).
15. E. Chiavassa *et al.*, *Phys. Lett.* B **337**, 192 (1994)).
16. F. Hibou *et al.*, (in preparation).
17. F. Hibou *et al.*, *Phys. Lett.* B (in press).
18. R. Bertini, spokesman for the DISTO collaboration (private communication), and W. Kühn (this meeting).
19. J.M. Cameron *et al.*, *Nucl. Instrum. Methods* A **305**, 257 (1991).
20. S. Kox *et al.*, *Nucl. Instrum. Methods* A **346**, 527 (1994).
21. D.V. Bugg and C. Wilkin, *Phys. Lett.* B **152**, 37 (1985).
22. S. Kox *et al.*, *Nucl. Phys.* A **556**, 621 (1993).
23. C. Ellegaard *et al.*, *Phys. Rev. Lett.* **59**, 974 (1987).
24. T. Sams *et al.*, *Phys. Rev.* C **51**, 1945 (1995).
25. B.N. Johnson *et al.*, *Phys. Rev.* C **51**, 1726 (1995).
26. V. Ghazikhanian *et al.*, *Phys. Rev.* C **43**, 1532 (1991).
27. J. Arvieux *et al.*, *Nucl. Phys.* A **431**, 613 (1984).
28. C. Kerboul *et al.*, *Phys. Lett.* B **181**, 28 (1986).
29. V.N. Nikulin *et al.*, *Phys. Rev.* C **54**, 1732 (1996).
30. V. Punjabi *et al.*, *Phys. Rev.* C **39**, 608 (1989).
31. S.L. Belostotski *et al.*, *Phys. Rev.* C **56**, 50 (1997).
32. V. Punjabi *et al.*, *Phys. Rev.* C **46**, 984 (1992).
33. C. Williams *et al.*, *Phys. Rev.* C **55**, R2132 (1997).
34. B. Tatischeff *et al.*, *Phys. Rev.* C **36**, 1995 (1987).

35. B. Tatischeff *et al.*, *Phys. Rev. Lett.* **79**, 601 (1997).
36. J. Julien *et al.*, *Z. Phys.* A **347**, 181 (1994).
37. M. Bedjidian *et al.*, *Z. Phys.* A **327**, 337 (1987).
38. R. Beurtey *et al.*, *Phys. Lett.* B **293**, 27 (1992).

EXPERIMENTS PERFORMED AT SATURNE - 2

1978 - 1997

All performed experiments, except numerous tests and short experiments not examined by the « Comité des Expériences »(PAC), are listed below according to their chronological number. The title and spokespersons' names are given, followed by the last or more significant related publications.

This list is not an exhaustive publication list, which would be much longer. It is intended to provide a track to most experimental work from SATURNE II. It was prepared by Michel Garçon and Simone Peresse from the material available to the « Comité des Expériences » and with the help of many users. We apologize for any missing information.

N° 1 Tests sur SPES I
G. BRUGE, A. BOUDARD
J. THIRION et al., Note CEA-N-1248 (1970)
R. BEURTEY et al., Nucl. Instr. Meth. 153 (1978) 257

N° 2 Diffusion de protons de 1.04 GeV sur SPES I
A. CHAUMEAUX, J.L. ESCUDIE
G. BRUGE, Journal de physique 40(1979) 635

N° 3 bis Mesure de T_{20} (180°) dans la diffusion d + p
J. ARVIEUX, A. BOUDARD, L. GOLDZAHL
J. ARVIEUX et al., Nucl. Phys. A431(1984) 613

N° 7 Etude de la réaction $^{10}B(p, \pi^+)^{11}B$ en fonction de l'énergie incidente
P. COUVERT
M. DILLIG et al. Nucl. Phys., A333 (1980) 477

N° 9 Production cohérente de résonances sur des noyaux légers
J.L. BOYARD, J.R. PIZZI
T. HENNINO et al., Phys. Rev. Lett. 48 (82) 997

N° 10 Etude de la production cohérente de π^+ et π^- par des ions légers d'énergie comprise entre 150 et 300 MeV/nucléon
F. HIBOU, N. WILLIS
E. ASLANIDES et al., Phys. Lett. 108B (1982) 91

N° 13 et 13bis Interaction alpha-alpha
J. DUFLO, L. GOLDZAHL
J. DUFLO, Phys. Rev. C36 (1987) 1425

N° 17 Jets croisés
M. GARÇON
M. GARÇON et al., Nucl. Phys. A445 (1985) 669
R. BURGEI et al., Nucl. Instr. Meth. 204 (1982) 53

N° 18 Radiographie par diffusion nucléaire
G. CHARPAK, G. SALAMON, J. SAUDINOS
J.C. DUCHAZEAUBENEIX et al., IEEE Trans. Nucl. Sci. NS-30 (1983) 601

N° 19 Etude d'optimisation du détecteur JANUS
J.J. ENGELMANN, A. RAVIART

N° 24 Etudes dosimétriques des faisceaux utilisés en radiographie
N. PARMENTIER, M. CHEMTOB

N° 35 Production de gaz rares par des protons de 1,2 à 3 GeV
H. SAUVAGEON
S. REGNIER et al., Phys. Rev. C26 (1982) 931

N° 37 Measurement of pd $\rightarrow$ ^{3}Heγ to test detailed balance
B.M.K. NEFKENS
W.J. BRISCOE et al., Phys. Rev. C32 (1985) 1956

N° 38bis Measurement of 2nd and 3rd order spin observables using a 1600 MeV tensor polarized deuteron beam, a polarized hydrogen target and a recoil proton polarimeter
G. IGO, M. BLESZYNSKI
V. GHAZIKHANIAN et al., Phys. Rev. C43 (1991) 1532

N° 39 Diffusion élastique et inélastique d'ions oxygène et de particules α par le plomb et le nickel aux énergies intermédiaires (40 à 120 MeV/A)
B. BONIN, M. BUENERD
B. BONIN et al., Nucl. Phys. A430 (1984) 349

N° 41 Etude de la fusion induite par ions lourds entre 50 et 150 MeV par nucléon et des processus en compétition
S. HARAR, C. VOLANT
F. SAINT-LAURENT et al., Nucl. Phys. A422 (1984) 307

N° 43 Mesure de multiplicités de pions et de protons dans les collisions d'ions lourds relativistes
J.P. ALARD, R. BABINET, P. GORODETZKY, J. GOSSET
G. MONTAROU et al., Phys. Rev. C44 (1991) 365
J.P. ALARD et al., Nucl. Instr. Meth. A261 (1987) 379

N° 44 Echange de charge et recherche de noyaux lourds autour de 0°
M. ROY-STEPHAN
D. BACHELIER et al., Phys. Lett. 172B (86) 23

N° 45 Mesure de la section efficace absolue de la diffusion de protons sur ^{12}C à 450 MeV et 1040 MeV avec une précision d'environ 1,5%
Th. BAUER
Non publiée

N° 47 Study of the reaction d + d $\rightarrow$ ^{4}He + π°
B.M. PREEDOM, J.P. EGGER
Non publiée

N° 49 Diffusion de protons de 1,04 GeV sur la série des isotopes du Samarium
J.P. TABET
J.P. TABET et al., Nucl. Phys. A386 (1982) 552

N° 50 Mesure de la composante (Δ^{++} nn) de ^{3}He par la réaction ^{3}He(p,t)Δ^{++} à l'avant

B. TATISCHEFF
B. TATISCHEFF et al., Phys. Lett. 77B (1978) 254

N° 50bis Etude de la composante baryonique (Δ^{++}, 2n) et éventuellement dibaryonique (T = 1) dans ^{3}He à l'aide des réactions de transfert ^{3}He(p,t) et ^{3}He(p, d)
B. TATISCHEFF
B. TATISCHEFF et al., Phys. Rev. C36 (1987) 36

N° 51 Diffusion élastique de protons sur le deutérium et l'hélium 3 entre 140° et 180° C.M. (700 MeV ≤ Ep ≤ 1,2 - 1,4 GeV

R. FRASCARIA

P. BERTHET et al., Phys. Lett. 106B n°6 (1981) 465

P. BERTHET et al., J. Phys. G, Nucl.Phys.8 (1982) L111

N° 52 Etude de la diffusion N-N auprès de Saturne 2 (N-N 1ère partie : p p élastique)

F. LEHAR

C.D. LAC et al., J. Phys. France 51 (1990) 2689

R. BERNARD et al., Nucl. Instr. Meth. A249 (1986) 176

N° 52 (2) Mesure de la section efficace différentielle et des paramètres de spin dans la région de l'interférence coulomb-nucléaire

F. LEHAR, A. MICHALOWICZ

J. BYSTRICKY et al., Lett. al Nuovo Cimento 40, 15 (1984) 466

S. DELLA TORRE-COLAUTTI et al., Nucl. Phys. A505 (1989) 561

J. BYSTRICKY et al., Nucl. Instr. Meth. A239 (1985) 131

N° 54 Tests du spectromètre 4 GeV/c

J.L. LACLARE, M. BOIVIN

E. GRORUD et al., Nucl. Instr. Meth. 188 (1981) 549

N° 57 Recherche d'états du baryonium près du seuil NN par détection de noyaux de recul avec SPES IV

M. ROY-STEPHAN, P. RADVANYI

Non publiée

N° 60 Recherche de résonances dibaryoniques

R. BERTINI, B. MAYER

B. MAYER et al., Nucl. Phys. A437 (1985) 630

R. BERTINI et al., Phys. Lett. B203 (1988) 18

N° 64 Recherche d'effet précurseur de la condensation des pions (opalescence critique) dans la diffusion inélastique de protons

C. GLASHAUSSER

J.L. ESCUDIE et al., Phys. Rev. C24 (1981) 792

N° 66 Une mesure de la réaction dd → ^{4}Heγ pour tester le principe de la réversibilité microscopique

B.M.K. NEFKENS

B.H. SILVERMAN et al., Phys. Rev. C29 (1984) 35

N° 67 The fission of ^{232}Th and ^{238}U induced by inelastic scattering of 1 GeV alpha particles

P. DAVID

P. DAVID et al., Z. Phys. A - Atomic Nuclei 326 (1987) 367

N° 68 Une étude des réactions p d → ^{3}He π°, pd → ^{3}He γet p d → t π$^+$ autour de Tp = 450 MeV

B.M.K. NEFKENS, P. COUVERT

B.H. SILVERMAN et al,. Nucl Phys. A444 (1985) 621

W.J. BRISCOE et al., Phys. Rev. C32 (1985) 1956

N° 70 Recherche de résonances dibaryoniques dans la diffusion élastique p-p (entre 600 et 1000 MeV)

D. LEGRAND

M. GARÇON et al., Phys. Lett. B183 (1987) 273

N° 75 Fonction de réponse des noyaux à l'excitation des particules α d'énergie intermédiaire

B. BONIN, B. BERTHIER

B. BONIN et al., Nucl. Phys. A430 (1984) 349

N° 76 Tests du faisceau polarisé

J. ARVIEUX

J. ARVIEUX et al., Nucl. Instr. Meth. A273 (1988) 48

N° 77 Tentative de mise en évidence d'ondes de choc dans une collision alpha-noyau

J.Y. GROSSIORD

K. CHAMCHAM et al., Int. Conf. on Selected Aspects of Heavy Ions Reactions, (Saclay, 1982)

N° 78 Diffusion élastique n-p aux petits angles

G. KOROLEV, Y. TERRIEN

B.H. SILVERMAN et al., Nucl. Phys. A499 (1989) 763

A.A. VOROBYOV et al., Nucl. Inst. Meth. A270 (1988) 419

N° 79 Premiers tests de la ligne d'analyse de SPES III et mesures préliminaires sur SPES III

J.L. LACLARE, Y. LE BORNEC

M.P. COMETS et al., Rapport IPNO DRE 88-41

N° 80 Etude des résonances dibaryoniques isoscalaires

R. FRASCARIA, B. TATISCHEFF, N. WILLIS

M.P. COMBES et al,. Nucl. Phys. A431 (1984) 703

N° 81 Tests de détecteurs

M. CONJEAUD, R. LEGRAIN

Y. BLUMENFELD et al., Proc. XXI Int. Winter Meeting on Nucl. Phys.(Bormio, 1983)

N° 82 Mesures d'étalonnage et expériences tests de SPES III

F. HIBOU, Y. LE BORNEC

Non publiée

N° 84 Mesure de la diffusion élastique de deutons polarisés à 200, 400 et 700 MeV

J. ARVIEUX, G. BRUGE, NGUYENVAN SEN

NGUYEN VAN SEN et al., Nucl. Phys. A464 ((1987) 717

N° 85 Réaction (^{3}He,t) aux énergies intermédiaires

C. GAARDE

A. BROCKSTEDT et al., Nucl. Phys. A530 (1991) 571

N° 87 Mesure de $\Delta\sigma_L$ dans la diffusion p-p pour des énergies comprises entre 400 et 2400 MeV

F. LEHAR

J. BYSTRICKY et al., Phys. Lett. 142B (1984) 130

M. ARIGNON et al., Nucl. Instr. Meth. A325(1985) 523

N° 88 Measurement of Aookk for p-p scattering from 725 to 1400 MeV

C. NEWSOM, J. YONNET

J. BYSTRICKY et al., Nucl. Phys. B258 (1985) 483

N° 89 Experimental confirmation of the phase-shift analysis predictions in the dibaryon region

J. DOMINGO, P. CATILLON

J. BYSTRICKY et al., Nucl. Phys. B258 (1985) 483

N° 90 Etude du fonctionnement machine en protons polarisés dans la gamme 900 à 1000 MeV

J.M. DURAND, J.L. LACLARE

P.A. CHAMOUARD, Rapport de synthèse en cours (1998)

N° 91 Study of the energy-loss of protons in Si-detectors for incoming momenta between 500 and 1500 MeV/c

P.G. RANCOITA

N° 92 Variation en fonction de l'énergie incidente de la production cohérente de pions dans l'expérience ^{3}He(^{3}He, π^+) ^{6}Li

L. BIMBOT, F. HIBOU, Y. LE BORNEC

Y. LE BORNEC et al., Phys. Lett. 133B (1983) 149

N° 95 Production cohérente de η_o très à l'arrière dans les systèmes p + d et d + d

P. BERTHET, R. FRASCARIA, L. GOLDZAHL

P. BERTHET et al., Nucl. Phys. A443 (1985) 589

J. BANAIGS et al., Phys. Rev. C32 (1985) 1448

N° 97 Mesure de la réaction pp $\rightarrow$ dπ^+ de 1 à 1.8 GeV

R. BERTINI, J.Y. GROSSIORD

R. BERTINI et al., Phys. Lett. B203 (1988) 18

N° 99 Mesure de la distribution angulaire de la section efficace différentielle et de la polarisation Ayo de la réaction $pd \rightarrow t\pi^+$ entre 500 et 1000 MeV
B. MAYER
B. MAYER et al., Phys. Lett. B181 (1986) 25

N° 100 Recherche sur l'anomalon
J.Y. GROSSIORD, A. GUICHARD
M. BEDJIDIAN et al., Zeit F. Phys. A327 (1987) 337

N° 101 Technique nouvelle pour la mesure de la polarisation des faisceaux de haute énergie (E > 1 GeV)
A. NAKACH, F. LEHAR, L. VAN ROSSUM
J. BYSTRICKY et al., Nucl. Instr. Meth. A234 (1985) 412

N° 102 Fission induite par des alphas sur des noyaux de masse moyenne - observation des impulsions transférées élevées - transition vers une fission asymétrique
H. OESCHLER, C. VOLANT
G. KLOTZ-ENGMANN et al., Nucl. Phys. A 499 (1989) 392

N° 103 Continued radiochemical research for anomalons
R. BRANDT
R. BRANDT et al., Phys. Rev. C45 (1992) 1194

N° 104 Mesure des paramètres de Wolfenstein dans la diffusion p-p entre 600 MeV et 3 Ge
J.M. FONTAINE, F. LEHAR
C.D. LAC et al., Nucl. Phys. B321 (1989) 269
M. ARIGNON et al., Nucl. Instr. Meth. A262(1987) 207

N° 105 Test de la symétrie de charge par la réaction $dd \rightarrow {}^4He + \pi^\circ$
F. PLOUIN
L. GOLDZAHL et al., Nucl. Phys. A533 (1991) 675

N° 106 Mesure simultanée des asymétries ε pp et εnp
F. LEHAR
A. de LESQUEN et al., Nucl. Phys. B304 (1988) 673

N° 107 Production anormale de pions dans l'interaction proton-noyau pour des protons d'énergie intermédiaire (~ 350 MeV
J. JULIEN, A.B. KUREPIN
J. JULIEN et al., Phys. Lett. B142 (1984) 340

N° 108 Mesure des pouvoirs d'analyse vectoriel et tensoriel dans la diffusion élastique deuton-proton (et deuton-noyau)
J. ARVIEUX, A. BOUDARD, J. CAMERON, M. GARÇON
M. GARÇON et al. Nucl. Phys. A458 (1986) 287

N° 109 Energy dependence of the ${}^{13}C$(p, d2p) reaction
J. McGILL, A. BOUDARD, C. GLASHAUSSER, C. MORRIS
Non publiée (?)

N° 110 Evolution des mécanismes de réactions induites par des ions lourds dans la région de transition
S. HARAR, C. VOLANT
M.S. NGUYEN, Thèse Orsay (1988)

N° 111 Mesure de sections efficaces de réaction ions lourds - noyaux en fonction de l'énergie
C. PERRIN
S. KOX et al., Phys. Rev. C35 (1987) 1678

N° 113 Recherche expérimentale de résonances multi-baryoniques par l'étude de spectres en masse manquante pour les réactions $p+p \rightarrow \pi^- X$ et $p+d \rightarrow \pi^- X$
N. WILLIS, P. FASSNACHT
M.P. COMBES-COMETS et al., Phys. Rev. C43 (1991) 973

N° 114 Diffusion élastique et inélastique d'ions lourds entre 100 et 200 MeV par nucléon

B. BONIN, M. BUENERD
J.Y. HOSTACHY et al., Phys. Lett. B184 (1987) 139
B. BONIN, J. Physique 48 (1987) 1479

N° 115 Réaction (d, ^{2}He)
C. GAARDE
C. ELLEGAARD et al., Phys. Rev. Lett. 59 (1987) 974

N° 116 Etude avec le détecteur Diogène des collisions à grande multiplicité Ne-noyau
J.P. ALARD, R. BABINET, P. GORODETZKY, J. GOSSET
M. DEMOULINS et al., Phys. Lett. B241 (1990) 476
Idem proposition n° 43

N° 117 Mesure de T_{20} à 0° et 180° C.M. et des sections efficaces correspondantes pour la réaction $p+d \rightarrow t+\pi^+$ ($He^3 + \pi^\circ$) de 700 MeV à 2200 MeV
J. BERGER, A. BOUDARD
A. BOUDARD et al., Phys. Lett. B214 (1988) 6

N° 118 Recherche de résonance dibaryonique autour de 350 MeV
J. SAUDINOS
Non publiée

N° 119 Détermination des densités de neutrons de ^{16}O et ^{18}O par diffusion élastique de protons polarisés de 1,04 GeV
G. BRUGE, A. CHAUMEAUX

N° 121 Recherche et étude de dibaryons d'étrangeté S = 1, du seuil ΛN au seuil ΣN
J.P. DIDELEZ, R. FRASCARIA, B. PREEDOM
R. FRASCARIA et al., Nuovo Cimento Vol. 102A, n° 2 (1989) 561

N° 122 Etude expérimentale des caractéristiques du SPES IV
M. BOIVIN, J.M. DURAND
Non publié

N° 123 Etude des structures étroites dans les masses invariantes de deux baryons
B. TATISCHEFF, Y. TERRIEN
B. TATISCHEFF et al., Phys. Rev. C36 (1987) 1995

N° 124 Recherche de résonances dibaryoniques étroites par la réaction $pp \rightarrow d\pi^+$ à 90° C.M.
B. MAYER, F. LEHAR
Non publiée

N° 125 Production cohérente et incohérente de π° et η° sur ^6Li
G. DELLACASA, A. MUSSO, R. BERTINI
E. SCOMPARIN et al., Journ. Phys. G19 (1993) L51

N° 126 Etude des structures étroites par réaction de transfert (p,d) dans les masses invariantes de deux baryons
B. TATISCHEFF
B. TATISCHEFF et al., Z. Phys. A, Atom. Nucl. 328 (1987) 147 ; Europhy. Lett. 4 (1987) 671

N° 128 Polarimètre de deutons pour SPES I
L. ANTONUK, D. HUTCHEON
Non publiée

N° 129 Test de faisabilité et de résolution de la détection de π° et η à Saturne
J.P. DIDELEZ, J.P. EGGER, G. PIGNAULT, B. SAGHAI
G. PIGNAULT, thèse, IPNO-T-84-07

N° 132 ($pp \rightarrow np\pi^+$) ($pp \rightarrow pp\pi^+\pi^-$) de 0.8 à 2.5 GeV
J. GOSSET, B. SAGHAI
C. COMPTOUR et al., Nucl. Phys. A579 (1994) 369

N° 133 Dependance en A de ($p+A \rightarrow \pi+X$) (recherche effets analogues à EMC)
F. HIBOU, Y. LE BORNEC
Non publiée

N° 134 Cassure du deuton d+A $\rightarrow$ p+X à 3.7 GeV/c
C. PERDRISAT, J. YONNET
V. PUNJABI et al., Phys. Rev. C39 (1989) 608

N° 135 Etude de la réponse isoscalaire de spin dans les noyaux par diffusion de deutons polarisés
C. GLASHAUSSER, M.MORLET,E.TOMASI-GUSTAFSSON
B.N. JOHNSON et al., Phys. Rev. C (1995) 1726
F.T. BAKER et al., Phys. Rep.289 (1997) 235

N° 137 Calibration du polarimètre "AHEAD"
J. CAMERON, L. ANTONUK
H. WITALA et al., Few-Body Systems 15 (1993) 67
J.M. CAMERON et al., Nucl. Instr. Meth. A305 (1991) 257

N° 138 Symétrie de charge par comparaison de T_{20} dans les réactions (d+p $\rightarrow$ ^{3}He π°) ($\rightarrow$ ^{3}H π^+)
L. GOLDZAHL, J. BERGER
Nouvelles de Saturne 13 (1989)

N° 140 Section efficace et pouvoir d'analyse des réactions np $\rightarrow$ (ppπ^-) et np $\rightarrow$ (d$\pi^+\pi^-$)
Y. TERRIEN
Y. TERRIEN et al., Phys. Lett. B294 (1992) 40

N° 141 Mécanisme de production de π° - Réactions noyau-noyau (^{20}Ne/A)
H. DABROWSKI, A. INSOLIA
L. BIMBOT et al., Z. Physik A333 (1989)393
R. FONTE et al., Nucl. Instr. Meth. A297 (1990) 410

N° 142 Magnetic Emulsion Chamber (MEC)
H. ITOH
T. HAYASHINO et al., Nucl. Instr. Meth. A271 (1988) 518

N° 144 Nucléon-Nucléon : 2ème partie (np)
F. PERROT, R. HESS, F. LEHAR
J. BALL et al., Z. Phys. C61 (1994) 579
J. BALL et al., Nucl. Instr. Meth. A327 (1993) 308

N° 145 Measurement of polarization parameters for the reaction dp $\rightarrow$ ppn in complete kinematics
S.L. BELOSTOTSKY, A. BOUDARD
S.L. BELOSTOTSKI et al., Phys. Rev. C56 (1997) 50

N° 147 Séparation et étude de noyaux exotiques produits par fragmentation d'ions lourds relativistes
J.P. DUFOUR
B. BLANK et al., Zeit. Phys. A343 (1992) 375
B. BLANK et al., Nucl. Instr. Meth. A286 (1990) 160

N° 148 Etude de la décroissance et de l'absorption de la résonance Δ dans le noyau
T. HENNINO, C. GAARDE
T. HENNINO et al., Phys. Lett. B303 (1993) 236

N° 152 Emission of particle unstable fragments and the space-time extent of the emitting system in high energy nucleus-nucleus collisions
J. POCHODZALLA
G.J. KUNDE et al., Phys. Rev. Lett. 70 (1993) 2545

N° 153 Etude du mécanisme de multifragmentation induit par ions lourds
B. BERTHIER

N° 155 Etude de la production anormale de pions neutres de faible énergie dans les réactions A(p, π°)X pour 300 < T< 420 MeV
J. JULIEN
J. JULIEN et al., Z. Physik A347 (1994) 181

N° 156 Etude avec le détecteur Diogène des collisions à grande multiplicité argon-noyau
J.P. ALARD, M.C. LEMAIRE, J. POITOU
J. POITOU et al., Nucl. Phys. A536 (1992) 767
Idem proposition n° 43

N° 157 Mesure de la masse du η - Calibrations de l'énergie de Saturne
P. FLEURY
F. PLOUIN et al., Phys. Lett. B276 (1992) 526

N° 158 Response of CsI((Tl) - crystals to charged particles and neutrons
A. GOBBI

N° 159 Etude de résonances dibaryoniques au SPES III
B. TATISCHEFF
B. TATISCHEFF et al., Phys. Rev. C45 (1992) 2005

N° 160 Etudes des collisions centrales induites par des ions lourds
C. VOLANT
Non publiée

N° 161 Transition from binary to multifragmentation regime in the intermediate energy region
K. KWIATOWSKI, V.E. VIOLA, E. HOURANI, E. POLLACO
S.J. YENNELLO et al., Phys. Rev. C48 (1993) 1092

N° 162 Cooperative particle production in heavy ion collisions
E. GROSSE
M. DEBOWSKI et al., Z. Phys. A356 (1996) 313
M. DEBOWSKI et al., Phys. Lett. B413 (1997) 8

N° 163 Recherche de la résolution optimale de SPES I
B. BONIN
Non publiée

N° 166 Réaction H(d, 2p) avec des deutons polarisés à 200 MeV
NGUYEN VAN SEN
S. KOX et al., Nucl. Phys. A556 (1993) 621

N° 167 Proton induced spallation reactions at 1200 MeV
R. MICHEL
R. MICHEL et al., Nucl. Instr. Meth. Phys. Res. B103 (1995) 183

N° 168 A study of multifragmentation in nuclear emulsions
B. JAKOBSSON
F. SCHUSSLER et al., Nucl. Phys. A584 (1995) 704

N° 169 Test of ribbon targets in internal target mode in the Saturne ring
O. SCHULT
H.R. KOCH et al., Nucl. Instr. Meth. A271 (1988) 375

N° 170 Spallation cross-sections for proton induced reactions with heavy element targets
K. SUEMMERER
K. SUEMMERER et al., Phys. Rev. C42 (1990) 2546
B. SZWERYN et al., Radiochemica Acta 47 (1989) 33

N° 171 Etude des interactions proton-noyau $\Delta S = 1$ $\Delta T = 0$ et $\Delta S = 1$ $\Delta T = 1$ par diffusion inélastique de protons polarisés entre 200 et 800 MeV
N. MARTY, M. MORLET
A. WILLIS et al., Phys. Rev. C43 (1991) 2177

N° 172 Simulation of the interaction of galactic protons with meteorites
R. MICHEL
R. MICHEL et al., Planet Space Sci. 43 (1995) 557

N° 173 Mesure continue, en fonction de l'énergie, de l'asymétrie de la diffusion pp entre 130 et 260 MeV
R. BEURTEY
R. BEURTEY et al., Phys. Lett. B293 (1992) 27

N° 174 Etude de la production du méson η dans les collisions p + p et p + ^{12}C
O. BING
A.M. BERGDOLT et al., Phys. Rev. D48 (1993) R296

F. HIBOU et al., Preprint IReS 98-05 (Strasbourg 1998)

N° 175 Proton induced radiation damage in solid-state detectors
J. BRUCKNER
M. KOENEN et al., IEEE Trans. Nucl. Sci. 38 (1991) 209

N° 177 Polarisation vectorielle du deuton et coefficients de transfert de polarisation dans la réaction pp $\rightarrow d\pi^{+}$
B. MAYER, D. HUTCHEON
J. YONNET et al., Nucl. Phys. A562 (1993) 352

N° 180 Test de modules pour le projet de détecteur 4π de SIS à Darmstadt
J.P. ALARD, J.P. COFFIN, A. GOBBI, A. OLMI

N° 182 Expérience exploratoire de la production de $_{\Lambda}{}^{3}H$ par réaction P + D $\rightarrow K^{+}{}_{\Lambda}{}^{3}H$
R. FRASCARIA, R. SIEBERT
R. SIEBERT et al., Nucl. Phys. A567 (1994) 819

N° 186 Etude de la production de mésons lourds dans les réactions p + d $\rightarrow He^{3}$ + X et d + d $\rightarrow He^{4}$ + X
Y. LE BORNEC
En cours de publication

N° 188 Réaction d'échange de charge en ions lourds
M. ROY-STEPHAN
M. ROY-STEPHAN et al., Nucl. Phys. A488 (1988) 187c
C. ELLEGAARD et al., Phys. Lett. B231 (1989) 365
T. HENNINO, 5ème JES (1989) 95

N° 189 Etude du bremsstrahlung neutron-proton à 200 MeV
H. NIFENECKER, J.A. PINSTON
F. MALEK et al., Phys. Lett. B266 (1991) 255

N° 190 Spin structure of the Δ excitation
C. GAARDE, P. ZUPRANSKI
C. ELLEGAARD et al., Phys. Lett. B231 (1989) 365

N° 191 Etude de la transition entre la fusion incomplète et la multifragmentation pour le système $^{12}C+^{197}Au$ entre 100 et 400 MeV/u
C. NGO
U. MILKAU et al., Phys. Rev. C44 (1991) R1242

N° 192 Etude de l'interaction proton-noyau à 800 MeV et 2 GeV
P. GORODETZKY, M.C. LEMAIRE, G. MONTAROU
M.C. LEMAIRE et al., Phys. Rev. C43 (1991) 2711

N° 193 Test of the detector system for fast neutrons
H. EMLING

N° 194 Etude des états de trous profonds dans les noyaux moyens et lourds par réaction (d, ^{3}He) ou (d, t) en faisceaux polarisés
J. VAN DE WIELE, H. LANGEVIN-JOLIOT, J. GUILLOT
H. LANGEVIN-JOLIOT et al., Phys. Rev. C58 (1998) 2192

N° 195 Essais et étalonnage du polarimètre POMME
B. BONIN
B. BONIN et al., Nucl. Instr. Meth. A288 (1990) 379 et 389

N° 197 Etude des réactions p + d $\rightarrow {}^{3}He$ + X aux seuils pour X = ω, η' et pour mX = 1 à 1,5 GeV
F. PLOUIN
F. PLOUIN, Production and decay of light mesons, (3-4 Mars 1988) Ed. P. Fleury, World Scientific

N° 198 Mesure de quelques désintégrations fondamentales du η - Tests préliminaires et $\eta \rightarrow \mu^{+}\mu^{-}$
B. MAYER, B. NEFKENS
R. ABEGG et al., Phys. Rev. D50 (1994) 92
B. MAYER et al., Phys. Rev. C53 (1996) 2068

N° 199 Détermination des paramètres g_0 de l'interaction résiduelle isoscalaire du spin
A. WILLIS, M. MORLET
Non publiée (tests non concluants)

N° 201 Charge and mass identification of complex fragments
A. OLMI, A. GOBBI

N° 202 Polarisation du proton dans A(d, p)X à 0° et 2,1 et 1,25 GeV
C.F. PERDRISAT, J. YONNET
N.E. CHEUNG et al., Phys. Lett. B284 (1992) 210
N.E. CHEUNG et al., Nucl. Instr. Meth. A363 (1995) 563

N° 203 Réactions périphériques avec des ions Kr de 200 MeV/A
C. STEPHAN
C. STEPHAN et al., Phys. Lett. B262 (1991) 6

N° 204 The (^{6}Li, ^{6}He) reaction
J.L. BOYARD, C. GAARDE
Non publiée

N° 207 Etude de la dépolarisation du faisceau de p
P.A. CHAMOUARD
Saturne 2 : résultats et bilan huit ans après la mise en service de Mimas - JES7 (Ramatuelle - Janvier 1996)

N° 208 Accélération de Kr^{84}
P.A. CHAMOUARD, J. FAURE
Non publiée

N° 209 Total cross section of the reaction $pp \rightarrow pp\pi^0$
J.P. DIDELEZ, E. HOURANI, G. BLANPIED
J.P. DIDELEZ et al,. Nucl. Phys. A535 (1991) 445
G. RAPPENECKER et al., Nucl. Phys. A590 (1995) 763
C. LIPPERT et al., Nucl. Instr. Meth. A333 (1993) 413

N° 210 Test d'une couronne du mur de plastique à zéro degré équipant le détecteur 4π de SIS à Darmstadt
N. BASTID

N° 211 Préparation d'états de Rydberg sphériques doubles
J.P. BRIAND

N° 212 Etude des réactions pp → ΔN et ΔΔ à 1500 MeV et 1800 MeV
M.P. COMETS, N. WILLIS
B. TATISCHEFF et al.,Phys. Rev. Lett. 79 (1997) 601

N° 213 DISTO
R. BERTINI
F. BALESTRA et al., Phys. Rev. Lett. 81 (1998) 4533
F. BALESTRA et al., Nucl. Instr. Meth. A (1999)

N° 214 Studies of deeply bound π^- states by (d, ^{2}He) pion transfer reactions
T. YAMAZAKI, P. RADVANYI
Non publiée

N° 215 Etude des effets dynamiques des ions légers sur les composants électroniques
J. BUISSON
B. DOUCIN et al., IEEE Trans. Nucl. Sci. 41 (1994) 593

N° 216 Measurement of 1.6 GeV asymmetries with a K-type polarized hydrogen target and a tensor and vector polarized deuteron beam
G. IGO
V. GHAZIKHANIAN et al., Phys. Rev. C43 (1991) 1532

N° 218 Response of Tl sheet detector to ion particles
T. WADA, O. SAAVEDRA

N° 220 Recherche de l'excitation de la résonance Roper (1440) dans la diffusion inélastique de particules α
H.P. MORSCH, R. FRASCARIA

H.P. MORSCH, Z. Phys. A350 (1994) 167

N° 221 Absolute calibration of neutron detectors efficiency

J. ARENDS, J.P. DIDELEZ, E. HOURANI

N. HARPES, Diplomarbeit, Bonn-IR-91-50

N° 222 Meson production near threshold from the ϕ(1020) to the f1 (1285)

R. JAHN, R. SIEBERT

R. WURZINGER et al., Phys. Lett. B374 (1996) 283

N° 223 The (^{6}Li, ^{6}He) reaction

J.L. BOYARD, C. GAARDE

Non publiée

N° 224 Etude de la multifragmentation des noyaux induite par des Argons et des Kryptons de 200 MeV/u dans des émulsions

F. SCHUSSLER

N° 225 Determination of p-n scattering amplitude in the energy region from 1.7 to 2.7 GeV and the search for a structure around T_{Kin} ~ 2.1 GeV

J. BALL, J.M. FONTAINE, M. FINGER, R. HESS, H. SPINKA

J. BALL et al., Phys. Lett. B320 (1994) 206

D. ADAMS et al., Acta Polytechnica (Prague) 36 (1996) 11

J. BALL et al., Nucl. Instr. Meth. A381 (1996) 4

N° 226 Recherche de la limite d'intensité de Mimas en protons et deutons avec Amalthée

P.A. CHAMOUARD

Evolution des accélérateurs du LNS depuis 1984 - JES6 (Le Mont Ste Odile)

N° 228 Multifragmentation emission in proton induced reactions

E. POLLACO

K.B. MORLEY et al., Phys. Rev. C54 (1996) 737

E. RENSHAW-FOXFORD et al., Phys. Rev. C54 (1996) 749

N° 229 Study of multi-fragment nuclear decays with 4π coverage

C.G. GELBKE

C. WILLIAMS et al., Phys. Rev. C55 (1997) R2132

G.F. PEASLEE et al., Phys. Rev. C49 (1994) R2271

N° 230 Accélération de I_{127} dans Mimas

P.A. CHAMOUARD

Non publiée

N° 231 Etude de la collection de charges dans les semi-conducteurs

Y. PATIN

Y. PATIN et al., IEEE Trans. Nucl. Sci. 41 (1994) 517

N° 232 Production of residual nuclei in the transition from free preequilibrium to spallation

R. MICHEL

R. MICHEL et al., Nucl. Instr. Meth. Phys. Res. B129 (1997) 153

R. MICHEL et al., Nucl. Instr. Meth. Phys. Res. B114 (1996) 91

N° 233 Cassure en vol de ^{6}Li polarisés

C.F. PERDRISAT, J. YONNET

V. PUNJABI et al., Phys. Rev. C46 (1992) 984

A. COURTOIS et al., Nucl. Instr. Meth. A311 (1992) 10

N° 234 Etude de l'irradiation avec des protons d'un détecteur pour l'astronomie des rayons X

B. PARLIER, E. COSTA

N° 235 Calibration du polarimètre POLDER pour la mesure de G_c du deuton à CEBAF

S. KOX, C. FURGET

S. KOX et al., Nucl. Instr. Meth. A346 (1994) 527

N° 236 Simulation experiments from planetary γ-rays spectroscopy by means of thick target proton irradiations

J. BRUCKNER

N° 237 Etude des réactions pp $\rightarrow$ ppη et (p, η) sur les noyaux à Tp > 1.26 GeV
N. DE MARCO, E. VERCELLIN
E. CHIAVASSA et al., Phys. Lett. B337 (1994) 192

N° 238 Mesure absolue de l'intensité extraite
P.A. CHAMOUARD
P. AUSSET et al., LNS/SM PAC/RC 92/29
P.A. CHAMOUARD, PAC/RC 93/15

N° 239 Dépôt d'énergie d'excitation dans les réactions périphériques avec des ions Kr de 200 MeV/A
C. STEPHAN
L. TASSAN-GOT et al., Nucl. Phys. A583 (1995) 453

N° 240 Test de comptage près de la cible pour la détection des reculs dans la diffusion inélastique de particules α
H.P. MORSCH, J.L. BOYARD
Non publiée

N° 242 Polarimètre SD2
J. YONNET
Non publiée

N° 243 I : Dynamics and decline of the pression process at high energies and low spins
II : Study of peripherical collisions between two heavy nuclei : bridging the gap between two energy regions
J. GALIN, D. HILSCHER
L. PIENKOWSKI et al., Phys. Lett. B336 (1994) 147

N° 244 Etude de la réaction p(p, π^- π^-)X
B. TATISCHEFF
Non publiée

N° 245 Etude de Mimas et Saturne à q/A = 0.25
P.A. CHAMOUARD
Non publiée

N° 246 Production de $\pi°$ par la réaction dp $\rightarrow$ ^{3}He $\pi°$ près du seuil
B. MAYER, B. NEFKENS
V.N. NIKULIN et al., Phys. Rev. C54 (1996) 1732

N° 247 Etude des mécanismes de formation des atomes creux
J.P. BRIAND

N° 248 Isoscalar spin response in the continuum
R.J. PETERSON
M.D. HOLCOMB et al., Phys. Rev. C57 (1998) 1778

N° 249 Polarization transfer in elastic backward pd scattering dp $\rightarrow$ pd, a joint proposal to study the spin structure of the reaction from 200 to 2300 MeV at Saturne in Saclay and from 2000 to 7200 MeV at the synchrophasotron in Dubna
C.F. PERDRISAT, V. PUNJABI, I. SITNIK
V. PUNJABI et al., Phys. Lett. B350 (1995) 178
V. LADYGIN et al., soumis à Nucl. Instr. Meth.

N° 250 Etude de la réponse isoscalaire de spin dans le domaine des résonances baryoniques
C. DJALALI, M. MORLET
En cours de publication.

N° 251 Recherche de l'excitation de la résonance Roper (1440) dans les noyaux
H.P. MORSCH, J.L. BOYARD
H.P. MORSCH et al., Zeit Phys. A350 (1994) 167

N° 253 Mesure de la polarisation tensorielle et de la probabilité de spin dans la réaction ^{12}C(d, d')^{12}C à 400 MeV avec POLDER
E. TOMASI-GUSTAFSSON, M. MORLET, S. KOX
C. FURGET et al., Phys. Rev. C51 (1995) 1562

N° 254 Test sur fond continu
T. YAMAZAKI, H. LANGEVIN-JOLIOT
Non publiée

N° 257 Mesure de sections efficaces pour le rayonnement cosmique
A. SOUTOUL
Y. CASSAGNOU et al., Proc. 24th Int. Cosmic Ray Conf., (Rome 1995) p. 176

N° 258 Mesure directe du rapport d'embranchement pour la désintégration du méson eta en deux photons
M. GARÇON, M. CLAJUS, L. LYTKIN
R. ABEGG et al., Phys. Rev. D53 (1996) 11

N° 259 Tests du nouveau post-accélérateur RFQ2 en ions lourds
P.A. CHAMOUARD
P.A. CHAMOUARD et al., 17th Int. LINAC Conference, (Tsukuba, Japan - 21-26 Août 1994) et rapport interne 94/21

N° 260 Tests du prototype des nouvelles chambres de SPES I
J. GUILLOT
Rapport interne IPN Orsay IPNO-DRE 92-10

N° 261 Induced radioactivity by cosmic protons on CsI and I_2Hg detectors
J.L. FERRERO
J.L. FERRERO et al., Astron. & Astrop. Suppl. Series 120 (1996) 1
E. PORRAS et al., Nucl. Instr. Meth. B111 (1996) 315

N° 262 CAPRICE
M. SUFFERT
P. CARLSON et al., Nucl. Instr. Meth. A349 (1994) 577

N° 263 Etude des effets nucléaires sur les composants électroniques en silicium en vue de leur emploi dans les détecteurs LHC
T. MOUTHUY
L. BLANQUART et al., 3rd Int. Work. Semiconductor Pixel Detectors forParticles and X-Rays, (Bari, 24-27 Mars 1996)

N° 266 Faut-il une cible dans le vide pour NA50 ?
J.Y. GROSSIORD
P. BELLAICHE et al., Nucl. Instr. Meth. A398 (1997) 180

N° 267 Tests de détecteur SPES III
N. WILLIS
Non publiée

N° 269 Recherche des pions cohérents dans les réactions d'échange de charge
T. HENNINO, M. KAGARLIS
Non publiée

N° 271 Recherche et développement en ultra-violet
P. BESSON, R. ALEKSAN
R. ALEKSAN et al., Nucl. Instr. Meth. A385 (1997) 438

N° 275 Residual nuclide production relevant for accelerator-based transmutation
R. MICHEL
R. MICHEL et al., Nucl. Instr. Meth. Phys. Res. B103 (1995) 183

N° 276 Calibration of the neutron polarimeter for G^n
R. MADEY
R. MADEY et al., Bull. Am. Phys. Soc. 41 (1996) 1260

N° 278 AI Coherent pion production in charge-exchange reactions with SPES 4π : (^{3}He, t)
J.L. BOYARD, R. KUNNE
L. FARHI, Thèse Univ. Paris VII (1997)

N° 278 AII Coherent pion production in charge exchange reactions with SPES 4π : (^{12}C, ^{12}N)
J.L. BOYARD, R. KUNNE
Expérience en cours d'analyse

N° 278 AIII The ρ-meson in the medium
C. GAARDE, J.L. BOYARD
Expérience en cours d'analyse

N° 278 B Exclusive study of the Roper resonance N*(1440 MeV) in alpha-proton scattering
J.L. BOYARD, R. KUNNE
Expérience en cours d'analyse

N° 278 C Exclusive study of the Roper resonance N*(1440 MeV) in deuteron-proton scattering
R. KUNNE, E. STROKOWSKI
Expérience en cours d'analyse

N° 279 Time of flight cold silicon counter for antinucleon detection
A. CODINO

N° 280 Etude de la réaction dd $\rightarrow \alpha\eta$ au voisinage du seuil
N. WILLIS, Y. LE BORNEC, A. ZGHICHE
N. WILLIS et al.,Phys. Lett. B406 (1997) 14

N° 282 Etude des neutrons de spallation
S. LERAY, Y. PATIN
X. LEDOUX et al., Preprint DAPNIA/SPhN-98-77
F. BORNE et al., Nucl. Instr. Meth. A385 (1997) 339
E. MARTINEZ et al., Nucl. Instr. Meth. A385 (1997) 345

N° 285 η production near threshold in dd $\rightarrow \alpha\eta$
R. FRASCARIA, F. PLOUIN
R. FRASCARIA et al., Phys. Rev. C50 (1994) R537

N° 286 Aerogel detector prototype test measurements
F. GARIBALDI, G. LOLOS

N° 287 Test of extended detector elements through quasi-elastic p scattering
E. KUHLMANN
E. KUHLMANN et al., KFA-Jülich Annual Report 1993,
Juel-2879 (1994) 17

N° 290 Mesures des observables tensorielles liées à la polarisation du deuton de recul dans la réaction pp $\rightarrow d\pi^+$
C. FURGET, S. KOX
En cours

N° 291 Calibration of HYPOM at high energy
E. TOMASI-GUSTAFSSON, L. GOLOVANOV
E. TOMASI-GUSTAFSSON et al., Nucl. Instr. Meth. A402 (1998) 361
L. GOLOVANOV et al., Nucl. Instr. Meth. A381 (1996) 15

N° 294 Simulation de l'environnement spatial dans le cadre du programme INTEGRAL
B. CORDIER

N° 296 Etude des noyaux résiduels de spallation
J. FREHAUT
Expérience en cours d'analyse.

N° 297 Integral neutron measurements for APT
N.S.P. KING, J. FREHAUT
J. FREHAUT et al., Rapport CEA-R-5809 (1998)

N° 299 Tests of the new SPES 4 beam line
J.L. BOYARD
Non publiée

N° 300 Etude de la formation d'un halo lors du transport d'un faisceau de protons
P.A. CHAMOUARD

N° 301 Etude des neutrons de spallation sur cibles épaisses

J. FREHAUT, S. LERAY, J.P. SCHAPIRA
Expérience en cours d'analyse
mêmes que n° 282

N° 304 Mesures de production de noyaux radioactifs et de neutrons rapides à 100 MeV
F. CLAPIER, A. VILLARI
N. PAUWELS et al., Rapport IPNO 98-04

N° 305 Measurement of the deuteron tensor polarization in the ^{3}He → d + X reaction at zero angle
I. SITNIK, C.F. PERDRISAT
Expérience en cours d'analyse

PUBLICATIONS « GRAND PUBLIC »

- **R. BEURTEY, J. SAUDINOS**
  « Des forces nucléaires classiques à l'interaction forte élémentaire »
  Clefs CEA 18 (1990) 19
- **A. BOUDARD, M. GARCON, S. PLATCHKOV**
  « Quelle est la forme du deuton ? »
  La Recherche 235 (1991) 1094
- **J.L. BOYARD, P. RADVANYI, M. SOYEUR**
  « SATURNE, a versatile hadron facility »
  Nuclear Physics News 2/4 (1992) 15
- **Collaboration ETA**
  « Source intense de mésons êta à Saturne »
  Courrier du CERN 32 (Mai 1992) 10
- **B. DESPLANQUES, H.P. MORSCH, J. ARVIEUX**
  « Peut-on comprimer un nucléon ? »
  Images de la Physique CNRS (1993) 18
- **A. MAGGIORA**
  « Spin and strangeness at Saturne »
  Nuclear Physics News 5/4 (1995) 23
- **M. MORLET, J. VAN DE WIELE**
  « Observables de spin et structure des noyaux »
  Images de la Physique CNRS (1996) 48
- **E. TOMASI-GUSTAFSSON, M. GARCON, J. MARTIN**
  « Deutons polarisés »
  Scintillations-DAPNIA/CEA 30 (Mars 1997)

Some Pictures of SATURNE

Participants to the colloquium “Les 20 ans de Saturne-2” in the main hall of the “Palais de la Découverte”.

A 3D sketch of the accelerators and experimental areas.

The EBIS (Electron Beam Ion Source) heavy ion source DIONE.

A basement view of the polarized particle source HYPERION. The massive pillars support and isolate the room above containing the source which can be at a potential up to ~380 kV.

The light ion source AMALTHEE (here open) is of the Cockroft type and can reach ~750 kV.

The inner of the RFQ (Radio Frequency Quadrupole) used after the heavy ion source DIONE to reach an energy of ~200 keV by mass unit.

The main control room (PCP) of the sources, accelerators and lines.

The synchrotron SATURNE-2.

The preinjector MIMAS.

The magnetic spectrometer SPES 1 (one quadrupole and one dipole).

The magnetic spectrometer SPES 2 (one quadrupole and two dipoles).

The magnetic spectrometer SPES 3: It was designed from the magnet of the old 30 MeV Saclay cyclotron.

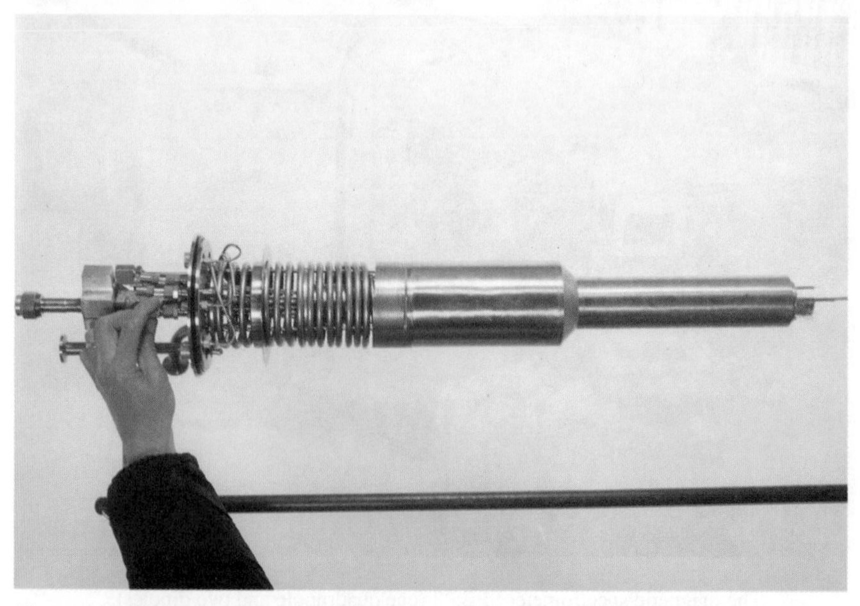

The cryogenic frozen spin target of the Nucleon–Nucleon experiment. The temperature of the target could be lowered down to 30 mK.

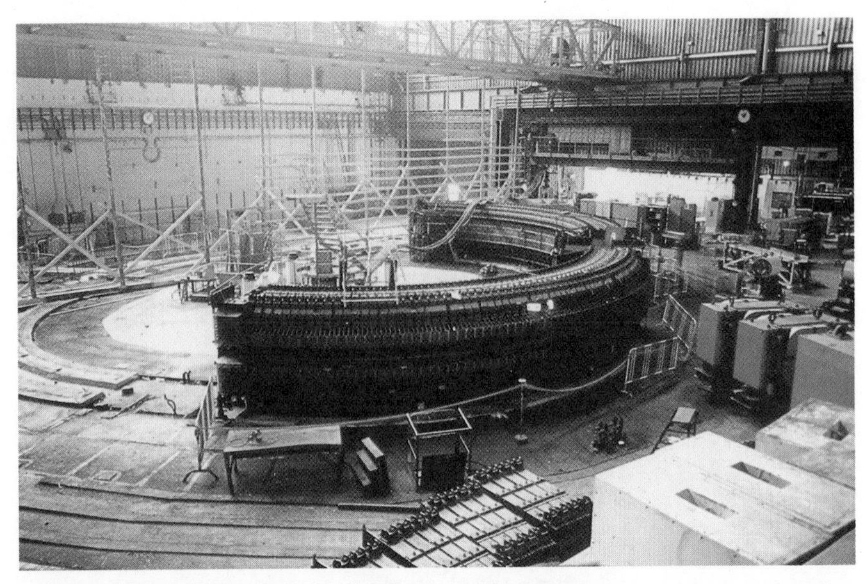

Part of the original weak focusing synchrotron SATURNE (1) from which the 4 dipoles of the magnetic spectrometer SPES 4 were built.

The DIOGENE detector used to study the high multiplicity of charged particles produced in heavy ion collisions.

Figure in tangle-board display of the gate SAF 82-0010 from top view... and... ... acoustic transmission SNL...

The PTT... and to provide shade the image of physical... and... above the visible...

Figure 1-15 (top view, left) Copyright, J.H. Temperature...
Figure 1-15 (top view, right, right view...) Atmosphere transmitted (475-850...

LIST OF PARTICIPANTS

ABGRALL Yvon
CEN-*Bordeaux*

ARVIEUX Jacques
IPN-Orsay

AUSSET Patrick
IPN-Orsay

BALL Jacques
DAPNIA, Saclay

BEAUVAIS Pierre-Yves
DAPNIA, Saclay

BERTINI Raimondo
SATURNE, Saclay

BIMBOT Louis
IPN Orsay

BOUDARD Alain
DAPNIA, Saclay

BOULET Armand
Fontenay-aux-Roses

BOYARD Jean-Louis
IPN Orsay

BROCHEN Jean-Claude
Saclay

CACAUT Daniel
DAPNIA, Saclay

CATZ Henri
DSE/SEE, Saclay

CESARSKY Catherine
DSM, Saclay

CHAMOUARD Paul-André
SATURNE, Saclay

COUVERT Pierre
DSM/LMCE, Saclay

CRESPIN Sylvain
DAM, Bruyères-le-Châtel

DAHL Rasmus
NBI Copenhague

DEGUEURCE Louis
Saclay

DELLACASA Giuseppe
INFN Turin

DETRAZ Claude
IN2P3

DUCHAZEAUBENEIX Jean-Claude
Saclay

EDELHEIT Geneviève
IN2P3

FABBRO Bernard
DAPNIA, Saclay

FAIVRE Jean-Claude
DAPNIA, Saclay

FAURE Jean
SATURNE, Saclay

FELTESSE Joël
DAPNIA, Saclay

FONTAINE Gérard
IN2P3

FOURNIER Guy
CEA-DSE, Paris

GAILLARD Gérald
EUROPSCAN, S.A.

GARCON Michel
DAPNIA, Saclay

GASTINEAU Bernard
DAPNIA, Saclay

GENDREAU Gabriel
GANIL

GLASHAUSSER Charles
Rutgers University, Piscataway

GORODETZKY Philippe
LPC, Collège de France

GOSSET Jean
DAPNIA, Saclay

HAMEL Jean-Louis
DAPNIA, Saclay

HARAR Samuel
DSM, Saclay

HASEGAWA Takeo
Miyazaki University

HEINZ Sophie
SATURNE, Saclay

HENNINO Thierry
IPN Orsay

HENRIOT Claude
Saclay

HIBOU François
IRS Strasbourg

JOURDAIN Jean-Claude
IPN Orsay

KIENLE Paul
Université de Munich, Garching

KILIAN Kurt
KFA, Jülich

KUEHN Wolfgang
KFA, Jülich

LACLARE Jean-Louis
Projet SOLEIL

LAGET Jean-Marc
DAPNIA, Saclay

LAGNIEL Jean-Michel
DAPNIA, Saclay

LARSEN Jens Syrak
NBI Copenhague

LECOLLEY Jean-François
LPC Caen

LEGRAIN Robert
DAPNIA, Saclay

LEHAR François
DAPNIA, Saclay

LELEUX Gérard
Saclay

LELUC Catherine
Université de Genève

LEMAIRE M.Claude
DAPNIA, Saclay

LERAY Sylvie
DAPNIA, Saclay

LOMBARD Roselyne
DAPNIA, Saclay

LUGOL J.Claude
DAPNIA, Saclay

MARTINO Jacques
DAPNIA, Saclay

MAYER Benjamin
DAPNIA, Saclay

METSCH Bernard
Université de Bonn

MICHEL Rolf
Université de Hanovre

MORLET Marcel
IPN Orsay

MORSCH Hans-Peter
KFA, Jülich

NAHAMA Fred
Saclay

NETTER Francis
Saclay

NGHIEM Phi
Projet SOLEIL

NGO Christian
CEA-DSE, Paris

OSET Eulogio
Université de Valencia (Burjassot)

PISKUNOV Nikolaï
JINR Dubna

POITOU Jean
DSM/LMCE, Saclay

RADVANYI Pierre
IPN Orsay

RAMSTEIN Béatrice
IPN Orsay

ROCHE Guy
LPC Clermont-Ferrand (Aubière)

ROY-STEPHAN Michèle
IPN Orsay

SAUDINOS Jean
Saclay

SCHMITT Hans
Université de Freiburg

SOYEUR Madeleine
DAPNIA, Saclay

STEPHAN Claude
IPN Orsay

TERRIEN Yves
DAPNIA, Saclay

THIRION Jacques
Saclay

TIBELL Gunnar
Université d' Uppsala

TKATCHENKO André
IPN Orsay

TKATCHENKO Malgorzata
Projet SOLEIL

VOLANT Claude
DAPNIA, Saclay

VORORYOV Alexei
PNPI Gatchina

WILKIN Colin
University College London